Methods in Molecular Biology

Series Editor
John M. Walker
School of Life Sciences
University of Hertfordshire
Hatfield, Hertfordshire, AL10 9AB, UK

For further volumes:
http://www.springer.com/series/7651

Stable Isotope Labeling by Amino Acids in Cell Culture (SILAC)

Methods and Protocols

Edited by

Bettina Warscheid

Institute of Biology II, University of Freiburg, Freiburg, Germany

Humana Press

Editor
Bettina Warscheid
Institute of Biology II
University of Freiburg
Freiburg, Germany

ISSN 1064-3745 ISSN 1940-6029 (electronic)
ISBN 978-1-4939-1141-7 ISBN 978-1-4939-1142-4 (eBook)
DOI 10.1007/978-1-4939-1142-4
Springer New York Heidelberg Dordrecht London

Library of Congress Control Number: 2014944734

Printed on acid-free paper

Humana Press is a brand of Springer
Springer is part of Springer Science+Business Media (www.springer.com)

Preface

Since its introduction about a decade ago, SILAC—stable isotope labeling by amino acids in cell culture—combined with mass spectrometry has not merely been integrated into the spectrum of quantitative methods applied in proteomics and protein research laboratories. In fact, it has become a prime tool of functional proteomics research, which allows us to address important questions in the fields of biology, biotechnology, medicine, and beyond.

SILAC could obtain this status by virtue of its versatility. This book tries to provide a synopsis of the large array of different SILAC methods by presenting a set of protocols that have been established by renowned scientists and their working groups. These protocols describe basic applications such as the labeling of various model organisms but also highly advanced strategies relying on SILAC, e.g., for the analysis of protein interactions, the mapping of posttranslational modifications, or the characterization of subcellular proteomes.

The book aims at applicability, and so all chapters entail step-by-step instructions that are easy to follow.

Chapter 1 provides a historical overview in which Matthias Mann outlines, from a very personal perspective, important steps in the development and implementation of SILAC as well as further significant innovations in the use of this unique method over the last 12 years. The main feature of the SILAC technology is the potential to label the entire proteome with defined combinations of stable isotopes during an organism's growth. At first applied to mammalian cell lines, its applicability to various other organisms has since been demonstrated as well. Chapter 2 is about a protocol for the effective labeling of both Gram positive and negative bacteria using only lysine in its "light" and "heavy" version. Chapter 3 continues with the description of an experimental procedure for complete SILAC labeling of the yeast *Saccharomyces cerevisiae*, a widely used model system for higher eukaryotes, including details on the generation of auxotrophic strains and the SILAC-based analysis of membrane protein complexes. SILAC applied to the protozoan *Trypanosoma brucei*, a further unicellular eukaryotic organism, enables new studies on the parasite's unique biology. A straightforward protocol for the metabolic labeling of both the procyclic and the bloodstream form of *T. brucei* with SILAC amino acids in a cell culture system can be found in Chapter 4.

Since the complete incorporation of SILAC amino acids into proteomes requires approximately five cell doublings, issues of partial labeling arise when working with nondividing cells. Chapter 5 addresses this problem by introducing an elegant multiplex SILAC labeling approach. It allows for the quantification of partially labeled proteins from nondividing cell types as exemplified here by primary neurons.

The following four chapters deal with the applicability of SILAC to the metabolic labeling of multicellular organisms including the higher plant model system *Arabidopsis thaliana* as well as *Drosophila melanogaster*, *Mus musculus*, and *Caenorhabditis elegans*. In Chapter 6, limitations of the SILAC technology in plant cells are discussed and an alternative protocol for labeling whole *A. thaliana* plants using ^{15}N salts is provided. SILAC labeling of *D. melanogaster*, as described in Chapter 7, relies on feeding "heavy" lysine-labeled yeast to flies, which can easily be implemented for quantitative analyses. Chapter 8 informs about the details how to generate SILAC mice and how to utilize them as spike-in standard for quantitative proteomics studies of organs. And Chapter 9 presents an innovative protocol for SILAC-based quantitative phosphoproteome analyses in *C. elegans* along with RNAi-mediated gene knockdown.

The high potential SILAC offers for the global study of signaling networks is highlighted in Chapter 10 providing the reader with state-of-the-art knowledge and practical information about how to conduct large-scale quantitative and time-resolved phosphoproteomics studies using SILAC. Detailed experimental procedures for the global analysis of dynamic changes in protein ubiquitination and methylation are described in Chapters 11 and 12, respectively.

A further important field of application for SILAC is the study of protein interactomes, in which SILAC-based protein quantification provides an effective measure to reliably distinguish between specific interaction partners and co-purified background binders in affinity-based protein purification or coimmunoprecipitation experiments. Complementary to Chapter 3, the potential and versatility of SILAC approaches for the study of protein–protein interactions are discussed in Chapters 13–16, i.e., the comparative analysis of human protein complexes (Chapter 13), the identification of stable and dynamic interaction partners exemplified by the human 26S proteasome complex (Chapter 14), the characterization of nuclear protein–protein interactions in mammalian cells (Chapter 15), and the delineation of dynamic processes involved in the assembly of the human splicosome (Chapter 16).

Chapter 17 turns to a protocol for protein interaction studies in autotrophic organisms, in which $^{14}N/^{15}N$ labeling combined with coimmunoprecipitation and antigen competition is used.

At first introduced as a label-free approach, the original protein correlation profiling method has been refined using SILAC. Dynamic aspects of protein interactomes and organellar proteomes can be analyzed by such spike-in SILAC standard-enhanced variations, for which protocols are presented in Chapters 18 and 19, respectively.

An important innovation of the SILAC technology has been its implementation in the study of cancer tissues, which is based on the generation of SILAC-labeled reference proteomes used as spike-in standards for relative protein quantification. In Chapter 20, the reader finds detailed information about how to properly design and successfully conduct such super-SILAC experiments for quantitative proteome analyses of tissue samples. Chapter 21 provides an optimized protocol for SILAC labeling of *D. melanogaster*, both in cell culture and in tissues, which allows tackling various questions in genetics and developmental biology. As outlined in Chapter 22, SILAC was also applied to the global study of secreted proteins, a new promising approach to identify protein biomarker candidates for human diseases. Furthermore, employed in pulse experiments, SILAC is a powerful technology for the large-scale analysis of protein turnover rates. Chapter 23 is about a protocol for pulsed SILAC that enables the identification of microRNA-mediated changes in the synthesis rates of proteins. The protocol is complemented by the description of computational and experimental approaches for the analysis of potential microRNA targets identified in a pulsed SILAC study.

The SILAC technology is intimately linked to high-resolution mass spectrometry facilitating the generation of a wealth of data about both the identity and the abundance of proteins in biological samples. Efficient processing and computational analysis of large SILAC-encoded mass spectrometry datasets is therefore a key step in quantitative proteomics studies. Chapter 24 presents an easy-to-follow protocol for the quantitative analysis of SILAC-based proteomics data using the freely available software MaxQuant.

In sum, the methods and experimental strategies described in this volume hopefully give an impression of the amazing diversity of SILAC applications. Smartly combined with different molecular and cell biology or biochemical techniques, SILAC is and will be a most valuable and multifunctional tool for many research studies aiming at a better understanding of cellular organization, protein interaction, and signaling networks as well as protein translation, turnover, and expression changes in health and disease.

The chapters also provide important information about practical aspects of sample fractionation, enrichment, and/or further processing steps as well as—to different extent—details on how to conduct LC/MS and downstream data analyses. In doing so, the book hopefully will serve students and experienced scientists alike as a valuable reference of how to make use of the SILAC technology for their own research.

I would like to thank the editors of the book series for their initial suggestion and their ongoing support in all editorial matters and also members of my group, especially Silke Oeljeklaus and Ida Suppanz, for providing much needed and welcomed assistance.

Finally and most importantly, I would like to cordially thank all contributors for sharing their knowledge!

Freiburg, Germany ***Bettina Warscheid***

Contents

Contributors

H. Irem Baymaz • *Department of Molecular Cancer Research, University Medical Center Utrecht, Utrecht, The Netherlands*

Andrea C. Becker • *Department of Dermatology, Medical Center - University of Freiburg, Freiburg, Germany; Freiburg Institute for Advanced Studies (FRIAS), University of Freiburg, Freiburg, Germany; ZBSA Center for Biological Systems Analysis, University of Freiburg, Freiburg, Germany*

Blagoy Blagoev • *Department of Biochemistry and Molecular Biology, Center for Experimental BioInformatics, University of Southern Denmark, Odense M, Denmark*

Karsten Boldt • *Division of Experimental Ophthalmology and Medical Proteome Center, Center of Ophthalmology, Institute for Ophthalmic Research, University of Tübingen, Tübingen, Germany*

Tiziana Bonaldi • *Department of Experimental Oncology, European Institute of Oncology (IEO), Milan, Italy*

Jia-Xuan Chen • *Max Delbrück Center for Molecular Medicine, Berlin, Germany*

Olivera Cirovic • *Institute of Cell Biology, University of Bern, Bern, Switzerland; The Graduate School for Cellular and Biomedical Sciences, University of Bern, Bern, Switzerland*

Jürgen Cox • *Department for Proteomics and Signal Transduction, Max Planck Institute of Biochemistry, Martinsried, Germany*

Alessandro Cuomo • *Department of Molecular Oncology, European Institute of Oncology, Milan, Italy*

Katrin Deinhardt • *Institute for Life Sciences and Centre for Biological Sciences, University of Southampton, Southampton, UK*

Jörn Dengjel • *Department of Dermatology, Medical Center - University of Freiburg, Freiburg, Germany; Freiburg Institute for Advanced Studies (FRIAS), University of Freiburg, Freiburg, Germany; ZBSA Center for Biological Systems Analysis, University of Freiburg, Freiburg, Germany; BIOSS Centre for Biological Signalling Studies, University of Freiburg, Freiburg, Germany*

Kasper Engholm-Keller • *Department of Biochemistry and Molecular Biology, University of Southern Denmark, Odense M, Denmark; Children's Medical Research Institute, Westmead, NSW, Australia*

Nils J. Færgeman • *Department of Biochemistry and Molecular Biology, University of Southern Denmark, Odense M, Denmark*

Leonard J. Foster • *Department of Biochemistry & Molecular Biology and Centre for High-Throughput Biology, University of British Columbia, BC, Vancouver, Canada*

Chiara Francavilla • *Faculty of Health Science, Department of Proteomics, Novo Nordisk Foundation Center for Protein Research, University of Copenhagen, Copenhagen, Denmark*

Julius Fredens • *Department of Biochemistry and Molecular Biology, University of Southern Denmark, Odense M, Denmark*

TAMAR GEIGER • *The Department of Human Molecular Genetics and Biochemistry, Sackler Faculty of Medicine, Tel Aviv University, Tel Aviv, Israel*
CHRISTIAN J. GLOECKNER • *Division of Experimental Ophthalmology and Medical Proteome Center, Center of Ophthalmology, Institute for Ophthalmic Research, University of Tübingen, Tübingen, Germany; Department of Protein Science, Helmholtz Zentrum München, German Research Center for Environmental Health, Neuherberg, Germany*
OMID HEKMAT • *Faculty of Health Science, Department of Proteomics, Novo Nordisk Foundation Center for Protein Research, University of Copenhagen, Copenhagen, Denmark*
DOROTHEA HEMME • *Molecular Biotechnology and Systems Biology, TU Kaiserslautern, Kaiserslautern, Germany; Max Planck Institute of Molecular Plant Physiology, Potsdam-Golm, Germany*
JEANETTE HENNINGSEN • *Department of Biochemistry and Molecular Biology, Center for Experimental BioInformatics (CEBI), University of Southern Denmark, Odense, Denmark*
HEIKO HERMEKING • *Experimental and Molecular Pathology, Institute of Pathology, Ludwig-Maximilians-Universität München, Munich, Germany*
SORAYA HÖLPER • *Max Planck Institute for Heart and Lung Resarch, Bad Nauheim, Germany*
LAN HUANG • *Department of Physiology & Biophysics, University of California, Irvine, CA, USA*
MARKUS KALLER • *Experimental and Molecular Pathology, Institute of Pathology, Ludwig-Maximilians-Universität München, Munich, Germany*
KARIN KÖHL • *Max-Planck-Institut für Molekulare Pflanzenphysiologie, Golm, Germany*
IRINA KRATCHMAROVA • *Department of Biochemistry and Molecular Biology, Center for Experimental BioInformatics (CEBI), University of Southern Denmark, Odense, Denmark*
ANDERS R. KRISTENSEN • *Department of Biochemistry & Molecular Biology and Centre for High-Throughput Biology, University of British Columbia, BC, Vancouver, Canada*
MARCUS KRÜGER • *Max Planck Institute for Heart and Lung Resarch, Bad Nauheim, Germany*
MARTIN RØSSEL LARSEN • *Department of Biochemistry and Molecular Biology, University of Southern Denmark, Odense M, Denmark*
HO-TAK LAU • *Department of Pharmacology, University of Washington, Seattle, WA, USA*
KAREN A. LEWIS • *Department of Pharmacology, University of Washington, Seattle, WA, USA*
REINHARD LÜHRMANN • *Department of Cellular Biochemistry, Max Planck Institute for Biophysical Chemistry, Göttingen, Germany*
BORIS MACEK • *Proteome Center Tübingen, University of Tübingen, Tübingen, Germany*
MATTHIAS MANN • *Department for Proteomics and Signal Transduction, Max Planck Institute of Biochemistry, Martinsried, Germany*
ANNEMARIE MATTHES • *Max-Planck-Institut für Molekulare Pflanzenphysiologie, Golm, Germany*
JAKOB MØLLER-JENSEN • *Department of Biochemistry and Molecular Biology, University of Southern Denmark, Odense M, Denmark*
TIMO MÜHLHAUS • *Molecular Biotechnology and Systems Biology, TU Kaiserslautern, Kaiserslautern, Germany; Max Planck Institute of Molecular Plant Physiology, Potsdam-Golm, Germany*

CHAN HYUN NA • *Department of Structural Biology, St. Jude Proteomics Facility, St. Jude Children's Research Hospital, Memphis, TN, USA; Department of Developmental Neurobiology, St. Jude Proteomics Facility, St. Jude Children's Research Hospital, Memphis, TN, USA*

THOMAS A. NEUBERT • *Department of Biochemistry and Molecular Pharmacology and Kimmel Center for Molecular Medicine at the Skirball Institute, New York University School of Medicine, New York, NY, USA*

TORSTEN OCHSENREITER • *Institute of Cell Biology, University of Bern, Bern, Switzerland*

SILKE OELJEKLAUS • *Department of Biochemistry and Functional Proteomics, Faculty of Biology and BIOSS Centre for Biological Signalling Studies, University of Freiburg, Freiburg, Germany*

JESPER V. OLSEN • *Faculty of Health Science, Department of Proteomics, Novo Nordisk Foundation Center for Protein Research, University of Copenhagen, Copenhagen, Denmark*

SHAO-EN ONG • *Department of Pharmacology, University of Washington, Seattle, WA, USA*

JUNMIN PENG • *Department of Structural Biology, St. Jude Proteomics Facility, St. Jude Children's Research Hospital, Memphis, TN, USA; Department of Developmental Neurobiology, St. Jude Proteomics Facility, St. Jude Children's Research Hospital, Memphis, TN, USA*

YAIR POZNIAK • *The Department of Human Molecular Genetics and Biochemistry, Sackler Faculty of Medicine, Tel Aviv University, Tel Aviv, Israel*

MONIKA RAABE • *Bioanalytical Mass Spectrometry Group, Max Planck Institute for Biophysical Chemistry, Göttingen, Germany*

AARON RUHS • *Max Planck Institute for Heart and Lung Resarch, Bad Nauheim, Germany*

ROBERTA SANFILIPPO • *Telethon Dulbecco Institute c/o STEBICEF Department, Palermo, Italy*

CARLA SCHMIDT • *Department of Chemistry, University of Oxford, Oxford, UK*

MICHAEL SCHRODA • *Molecular Biotechnology and Systems Biology, TU Kaiserslautern, Kaiserslautern, Germany; Max Planck Institute of Molecular Plant Physiology, Potsdam-Golm, Germany*

WALTRAUD X. SCHULZE • *Department of Plant Systems Biology, University of Hohenheim, Stuttgart, Germany*

ANDREAS SCHUMMER • *Department of Biochemistry and Functional Proteomics, Faculty of Biology, and BIOSS Centre for Biological Signalling Studies, University of Freiburg, Freiburg, Germany*

MATTHIAS SELBACH • *Max Delbrück Center for Molecular Medicine, Berlin, Germany*

FREDERIK SOMMER • *Molecular Biotechnology and Systems Biology, TU Kaiserslautern, Kaiserslautern, Germany; Max Planck Institute of Molecular Plant Physiology, Potsdam-Golm, Germany*

BOUMEDIENE SOUFI • *Proteome Center Tübingen, University of Tübingen, Tübingen, Germany*

CORNELIA G. SPRUIJT • *Department of Molecular Cancer Research, University Medical Center Utrecht, Utrecht, The Netherlands*

IDA SUPPANZ • *Department of Biochemistry and Functional Proteomics, Faculty of Biology, and BIOSS Centre for Biological Signalling Studies, University of Freiburg, Freiburg, Germany*

MATTHIAS D. SURY • *Max Delbrück Center for Molecular Medicine, Berlin, Germany*

HAIYAN TAN • *Department of Structural Biology, St. Jude Proteomics Facility, St. Jude Children's Research Hospital, Memphis, TN, USA; Department of Developmental Neurobiology, St. Jude Proteomics Facility, St. Jude Children's Research Hospital, Memphis, TN, USA*

YVES TEXIER • *Department of Molecular Epigenetics, Helmholtz Zentrum München, German Research Center for Environmental Health (GmbH), München, Germany*

STEFKA TYANOVA • *Department for Proteomics and Signal Transduction, Max Planck Institute of Biochemistry, Martinsried, Germany*

MARIUS UEFFING • *Division of Experimental Ophthalmology and Medical Proteome Center, Center of Ophthalmology, Institute for Ophthalmic Research, University of Tübingen, Tübingen, Germany; Department of Protein Science, Helmholtz Zentrum München, German Research Center for Environmental Health, Neuherberg, Germany*

HENNING URLAUB • *Bioanalytical Mass Spectrometry Group, Max Planck Institute for Biophysical Chemistry, Göttingen, Germany; Bioanalytics, Department of Clinical Chemistry, University Medical Center, Göttingen, Germany*

THOMAS VACCARI • *IFOM, Istituto FIRC di Oncologia Molecolare, Milan, Italy*

MICHIEL VERMEULEN • *Department of Molecular Cancer Research, University Medical Center Utrecht, Utrecht, The Netherlands*

DANIEL VEYEL • *Molecular Biotechnology and Systems Biology, TU Kaiserslautern, Kaiserslautern, Germany; Max Planck Institute of Molecular Plant Physiology, Potsdam-Golm, Germany*

XIAORONG WANG • *Department of Physiology & Biophysics, University of California, Irvine, CA, USA*

BETTINA WARSCHEID • *Department of Biochemistry and Functional Proteomics, Faculty of Biology, and BIOSS Centre for Biological Signalling Studies, University of Freiburg, Freiburg, Germany*

ZHIPING WU • *Department of Structural Biology, St. Jude Proteomics Facility, St. Jude Children's Research Hospital, Memphis, TN, USA; Department of Developmental Neurobiology, St. Jude Proteomics Facility, St. Jude Children's Research Hospital, Memphis, TN, USA*

GUOAN ZHANG • *Department of Biochemistry and Molecular Pharmacology and Kimmel Center for Molecular Medicine at the Skirball Institute, New York University School of Medicine, New York, NY, USA*

FELIX VON ZWEYDORF • *Division of Experimental Ophthalmology and Medical Proteome Center, Center of Ophthalmology, Institute for Ophthalmic Research University of Tübingen, Tübingen, Germany*

Chapter 1

Fifteen Years of Stable Isotope Labeling by Amino Acids in Cell Culture (SILAC)

Matthias Mann

Abstract

Here I describe the history of the Stable Isotope Labeling by Amino Acids in Cell culture (SILAC) technology. Although published in 2002, it had already been developed and used in my laboratory for a number of years. From the beginning, it was applied to challenging problems in cell signaling that were considered out of reach for proteomics at the time. It was also used to pioneer proteomic interactomics, time series and dynamic posttranslational modification studies. While initially developed for metabolically accessible systems, such as cell lines, it was subsequently extended to whole animal labeling as well as to clinical applications—in the form or spike-in or super-SILAC. New formats and applications for SILAC labeling continue to be developed, for instance for protein-turnover studies.

Key words Stable isotope labeling, Proteomics, Quantitation, Posttranslational modifications, Mass spectrometry, Electrospray, Protein analysis

The original paper describing stable isotope labeling by amino acids in cell culture (SILAC) was published about 12 years ago [1], but the technology had been developed and applied in my laboratory in Odense, Denmark, for a number of years before that. At the time we were doing research on tyrosine phosphorylation, an interest brought into the group by Akhilesh Pandey [2, 3]. Akhilesh was a post-doc, who had come from Harvey Lodish's laboratory at the Whitehead Institute at MIT, where such questions were studied. Together with Blagoy Blagoev and Hanno Steen, Akhilesh used anti-tyrosine antibodies to precipitate proteins phosphorylated upon growth factor stimulation. The idea arose that it would be nice to precisely quantify tyrosine phosphorylated peptides by using metabolic labeling of a cell line with heavy or light forms of tyrosine. If done on a control population and a population that was stimulated, we would encode the phospho-state in the isotopic forms of the peptides and we would be able to quantify by the ratios in a peptide pair. In the event, we actually did not go for tyrosine labeling at first, but we found that we could easily and

Bettina Warscheid (ed.), *Stable Isotope Labeling by Amino Acids in Cell Culture (SILAC): Methods and Protocols*, Methods in Molecular Biology, vol. 1188, DOI 10.1007/978-1-4939-1142-4_1, © Springer Science+Business Media New York 2014

cheaply order leucine or deuterated leucine and put that in the media. At the time it was not clear to us or to others if this would actually work. One worry was that even though we were using essential amino acids, somehow the label would distribute itself throughout the proteome via catabolic processes. Fortunately, this and other concerns about the SILAC approach did not materialize. Later on, we switched to arginine and lysine labeling, which was a marriage made in heaven with the trypsin enzyme that is the mainstay of bottom-up proteomics [4]. In the case of arginine, which is not strictly an essential amino acid, there is indeed some scrambling from arginine to proline, but this does not affect the accuracy of quantification. In any case it can be addressed by titrating the amount of arginine (unfortunately not possible in commercial kits) or in other ways [5–7].

At this time—between 2000 and 2002—a number of isotopic techniques, which are anyway very common in small molecule work, had already been applied to proteins [8]. Single proteins had long been isotope-labeled for structural studies and isotopic encoding with amino acids had also been used to count the number of leucines in proteins around the same time [9]. Fortunately for us, we were the first to publish amino acid labeling for protein quantification. The main competitors in this area were the ICAT technology from the Aebersold group [10], in which cysteines were chemically reacted with an isotope-labeled linker and a biotin affinity handle. Compared to ICAT one great advantage of SILAC was that no chemical reactions were involved, so there could be no side reactions and worries about over- or under-labeling the sample. More importantly, SILAC could isotopically label the entire proteome and peptidome and this happened already at the level of the living cell. Therefore, if cell populations were mixed together directly after perturbing one of them, any subsequent manipulations could not influence the SILAC ratios of the peptide pairs, which were frozen at the time of the original experiments. This last point was also the decisive advantage compared to label-free quantification, which at the time was already heavily used in many laboratories including ours [11, 12]. By its nature, label-free quantification was much less accurate and reliable, especially on the low-resolution ion traps that were common in proteomics at that time.

While the advantages of SILAC were quite clear after the initial proof of concept, the technology was also from the start seen as somewhat restricted. The very name suggested that the method was restricted to cell lines only and not applicable to tissues or body fluids, which were of keen interest to many proteomics laboratories. Although difficult to imagine now, mass spectrometry laboratories also viewed cell culture as an exotic technology. Being chemists, they were much more comfortable with a derivatization reaction. Nevertheless, the intrinsic advantages of SILAC as a quantification method meant that the results of a SILAC experiment were far more robust, reliable, and in depth as those of other approaches. The success of SILAC experiments was difficult to argue with, and most of the groundbreaking inroads of proteomics into cell biology were in fact performed with this technology [13–15], including the pioneering interaction studies and studies of large-scale quantitative phosphoproteomics by Blagoy Blagoev and Jesper Olsen [16, 17]. A precondition for SILAC to have the influence that it did was the development of software to extract the quantitative values. My laboratory did this as a matter of necessity at first, because obviously nobody had developed SILAC quantification software at that point. Peter Mortensen soon joined this effort and the open source software that we developed has been used by almost everybody in the beginning years of SILAC [18]. Upon the move of my laboratory from Denmark to Munich, Jürgen Cox began developing the MaxQuant suite of programs, which propelled quantitative proteomics into entirely new dimensions [19] (*see* also Chapter 24).

One of the achievements that I am most proud of is the quantification of the proteome of an entire model organism (budding yeast) [20], something which was considered by many to be impossible even in principle. Admittedly, this was quite a tour de force, which leveraged our sample preparation technology, the SILAC technology, the Orbitrap instruments, and MaxQuant to the utmost. However, recently, my group has shown that basically the entire yeast genome can be analyzed in a few hours in a single LC MS/MS run [21], a speedup in analysis time in some ways reminiscent to that involved in going from the first human reference genome to current routine sequencing of individual genomes. Indeed, we have argued that the entire human cell line proteome will be analyzable in a similar fashion in the near future [22]. The single-shot or single-run analysis of the yeast proteome quantified control against heat shock. The experiment was performed with a measurement time of only about 1 day and did not even require SILAC labeling of the yeast to be analyzed. Instead it made use of spike-in SILAC [23], in which a proteome is metabolically labeled offline in larger batches, and aliquotted out to serve as an internal reference for each experiment. Interestingly, in this format, SILAC is also extremely cheap, involving only cents in reagent costs per run. Thus, setting aside the initial purchase of the mass spectrometer, an experiment that would normally require a specific antibody and a western blotting setup (actually 6,000 of them), can now be performed with less material, less time and much more accurately by MS-based proteomics.

The spike-in format also answers one of the main criticisms of SILAC, namely that it could not be used in the clinic. It turns out that in cancer tissue analysis, for instance, a reference proteome can be constructed by mixing multiple SILAC-labeled cell lines.

When picked from different stages and grades of the cancer in question, it covers a large proportion of the proteome [24]. Even the ones that are not covered can be quantified in a label-free approach, which is somewhat less accurate but allows for quantification of extremely large ratios [25], which is difficult for current mass spectrometers in the case of peptide pairs generated by metabolic labeling. In our laboratory, we have used this super-SILAC concept (because the SILAC standard is generated by a super-set of cancer cell lines) on more than 200 tissues from cancer patients. We also investigated if super-SILAC could distinguish two types of a blood cancer by their protein expression patterns. The particular case involves large B-cell lymphoma, in which faulty rearrangements of the B-cell genome during antibody maturation leads to uncontrolled proliferation. It is very difficult to accurately distinguish the so-called GBC type from the ABC-type by standard histological assessment, but transcriptome measurements had already shown success. By constructing a B-cell super-SILAC mix, Sally Deeb was able to clearly separate the ABC from the GBC subtypes even using the relatively straightforward methods of principal component analysis (PCA) [26]. To my knowledge, these experiments established for the first time that proteomics can reliable differentiate between subtle subtypes of cancers. Also interesting from an analytical and clinical perspective, the samples could be distinguished already on the basis of the single, in-depth LC MS/MS runs mentioned above. This suggests that super-SILAC, combined with modern high-resolution shotgun proteomics can in principle be used in the clinic as a reliable and rapid way to diagnose cancer subtypes.

A similar approach was successful in retrieving secreted proteins from breast cancer cells from human plasma. This involved enriching glycosylated peptides by the N-glyco FASP method [27], and it works accurately because of the inherent ability of SILAC to mix samples at a very early stage of sample preparation—in this case, proteins secreted from breast cancer cell lines with plasma.

As mentioned above, SILAC has from the beginning competed with label-free quantification, and in our laboratory, we have used both approaches side by side for more than 10 years. The advantages of label-free quantification are of course that it does not require any sample manipulation whatsoever. Furthermore, since one proteome is measured at a time, a greater depth of analysis can be achieved in very complex proteomes (conversely, SILAC offers some multiplexing capabilities such as twofold or threefold multiplexing [28]). During the last few years, several developments have greatly boosted the label-free quantification approaches. The first is that high resolution instruments are almost exclusively employed in quantitative proteomics, today, in stark contrast to the situation only a few years ago. Thus separation of precursor peptide ions, a prerequisite of accurate quantification, is much more universally available than before. Secondly and just as importantly, algorithms have become much more sophisticated in analyzing and normalizing label-free data. For this reason the difference between label-free and SILAC approaches in terms of quantitative accuracy is shrinking, especially in single-run formats or other formats where there is little up-front sample preparation or fractionation. Nevertheless, SILAC is still the gold standard in quantitative proteomics in terms of quantitative reliability and robustness. In our laboratory it continues to be the method of choice if very accurate ratios are needed, if the ratios are small (say less than a factor of two), or when complex and variable steps of sample preparation are involved.

To make this tangible, consider the case of protein interactions. For a standard interaction experiment, the ratios between background and specific binders would typically be a factor four or more. Therefore, this task is routinely performed in a label-free format in our laboratory [29]. However, in the case when stretches of DNA or RNA or modified peptides are the bait, the ratios of specific binders can be quite low, and in this case we use SILAC labeling [30–32].

Furthermore, there are entire areas of biological inquiry that are predestined for metabolic labeling in the SILAC format. Among these are protein turnover studies. In this case, the SILAC labels are switched dynamically [33, 34] and only one cell population is analyzed. The SILAC ratios then contain information about protein synthesis and degradation. This experiment can also be performed as a triple encoding SILAC experiment, where the light population serves as control, the medium as reference for normal turnover and the heavy state encodes the perturbation, such as microRNA ablation [35]. Apart from these applications in proteome dynamics, another unique application is in the encoding of two different but spatially connected cell types. In this case, one can identify from which cell type a given protein came, such as in cross-signaling between two cells that communicate with each other [36] or in other ways exchange protein material with each other.

On a more general note, the balance between application of label-free and SILAC approaches will clearly also hinge on further developments in the sequencing speed and dynamic range of future mass spectrometers. Already instrumental capabilities are very close to allowing for the analysis of entire human cell line proteomes [37, 38]. Therefore, it is not unreasonable to imagine that sequencing speed will not be a limiting factor in years to come, or at least not in the current form. Likewise, higher dynamic range instruments and instruments that can rapidly inject different parts of the mass range for combined analysis [39], could lead to much higher and even more accurate quantification of SILAC ratios. In this connection the recent reports of SILAC using nuclear encoding are also interesting [40]. Here SILAC is used with labels of the same nominal weight—the precursor peaks are collapsed in MS scans of normal resolution and in MS/MS scans, but they are resolved and differentially quantifiable in ultrahigh resolution MS scans. This last example shows how the basic SILAC principle can even now—after more than 10 years—be extended into new formats with unique properties.

References

1. Ong SE, Blagoev B, Kratchmarova I et al (2002) Stable isotope labeling by amino acids in cell culture, SILAC, as a simple and accurate approach to expression proteomics. Mol Cell Proteomics 1:376–386
2. Mann M, Hendrickson RC, Pandey A (2001) Analysis of proteins and proteomes by mass spectrometry. Annu Rev Biochem 70: 437–473
3. Steen H et al (2002) Tyrosine phosphorylation mapping of the epidermal growth factor receptor signaling pathway. J Biol Chem 277(2): 1031–1039
4. Ong SE, Kratchmarova I, Mann M (2003) Properties of 13C-substituted arginine in stable isotope labeling by amino acids in cell culture (SILAC). J Proteome Res 2(2):173–181
5. Bendall SC et al (2008) Prevention of amino acid conversion in SILAC experiments with embryonic stem cells. Mol Cell Proteomics 7(9):1587–1597
6. Park SK et al (2009) A computational approach to correct arginine-to-proline conversion in quantitative proteomics. Nat Methods 6(3): 184–185
7. Bicho CC et al (2010) A genetic engineering solution to the "arginine conversion problem" in stable isotope labeling by amino acids in cell culture (SILAC). Mol Cell Proteomics 9(7): 1567–1577
8. Julka S, Regnier F (2004) Quantification in proteomics through stable isotope coding: a review. J Proteome Res 3(3):350–363
9. Martinovic S et al (2002) Selective incorporation of isotopically labeled amino acids for identification of intact proteins on a proteome-wide level. J Mass Spectrom 37(1):99–107
10. Gygi SP et al (1999) Quantitative analysis of complex protein mixtures using isotope-coded affinity tags. Nat Biotechnol 17(10):994–999
11. Lasonder E et al (2002) Analysis of the Plasmodium falciparum proteome by high-accuracy mass spectrometry. Nature 419(6906):537–542
12. Andersen JS et al (2003) Proteomic characterization of the human centrosome by protein correlation profiling. Nature 426(6966): 570–574
13. Mann M (2006) Functional and quantitative proteomics using SILAC. Nat Rev Mol Cell Biol 7(12):952–958
14. Choudhary C et al (2009) Lysine acetylation targets protein complexes and co-regulates major cellular functions. Science 325(5942):834–840
15. Walther TC, Mann M (2010) Mass spectrometry-based proteomics in cell biology. J Cell Biol 190(4):491–500
16. Blagoev B et al (2003) A proteomics strategy to elucidate functional protein–protein interactions applied to EGF signaling. Nat Biotechnol 21(3):315–318
17. Olsen JV et al (2006) Global, in vivo, and site-specific phosphorylation dynamics in signaling networks. Cell 127(3):635–648
18. Mortensen P et al (2010) MSQuant, an open source platform for mass spectrometry-based quantitative proteomics. J Proteome Res 9(1): 393–403
19. Cox J, Mann M (2008) MaxQuant enables high peptide identification rates, individualized p.p.b.-range mass accuracies and proteome-wide protein quantification. Nat Biotechnol 26(12):1367–1372
20. de Godoy LM et al (2008) Comprehensive mass-spectrometry-based proteome quantification of haploid versus diploid yeast. Nature 455(7217):1251–1254
21. Nagaraj N et al (2012) System-wide perturbation analysis with nearly complete coverage of the yeast proteome by single-shot ultra HPLC runs on a bench-top Orbitrap. Mol Cell Proteomics 11(3):M111.013722
22. Mann M et al (2013) The coming age of complete, accurate, and ubiquitous proteomes. Mol Cell 49(4):583–590
23. Geiger T et al (2011) Use of stable isotope labeling by amino acids in cell culture as a spike-in standard in quantitative proteomics. Nat Protoc 6(2):147–157
24. Geiger T et al (2010) Super-SILAC mix for quantitative proteomics of human tumor tissue. Nat Methods 7(5):383–385

25. Meissner F et al (2013) Direct proteomic quantification of the secretome of activated immune cells. Science 340(6131):475–478
26. Deeb SJ et al (2012) Super-SILAC allows classification of diffuse large B-cell lymphoma subtypes by their protein expression profiles. Mol Cell Proteomics 11(5):77–89
27. Zielinska DF et al (2010) Precision mapping of an in vivo N-glycoproteome reveals rigid topological and sequence constraints. Cell 141(5): 897–907
28. Blagoev B et al (2004) Temporal analysis of phosphotyrosine-dependent signaling networks by quantitative proteomics. Nat Biotechnol 22(9):1139–1145
29. Hubner NC et al (2010) Quantitative proteomics combined with BAC TransgeneOmics reveals in vivo protein interactions. J Cell Biol 189(4):739–754
30. Mittler G, Butter F, Mann M (2009) A SILAC-based DNA protein interaction screen that identifies candidate binding proteins to functional DNA elements. Genome Res 19(2):284–293
31. Vermeulen M et al (2010) Quantitative interaction proteomics and genome-wide profiling of epigenetic histone marks and their readers. Cell 142(6):967–980
32. Butter F et al (2012) Proteome-wide analysis of disease-associated SNPs that show allele-specific transcription factor binding. PLoS Genet 8(9): e1002982
33. Doherty MK et al (2009) Turnover of the human proteome: determination of protein intracellular stability by dynamic SILAC. J Proteome Res 8(1):104–112
34. Schwanhausser B et al (2009) Global analysis of cellular protein translation by pulsed SILAC. Proteomics 9(1):205–209
35. Selbach M et al (2008) Widespread changes in protein synthesis induced by microRNAs. Nature 455(7209):58–63
36. Jorgensen C et al (2009) Cell-specific information processing in segregating populations of Eph receptor ephrin-expressing cells. Science 326(5959):1502–1509
37. Nagaraj N et al (2011) Deep proteome and transcriptome mapping of a human cancer cell line. Mol Syst Biol 7:548
38. Beck M et al (2011) The quantitative proteome of a human cell line. Mol Syst Biol 7:549
39. Michalski A et al (2011) Mass spectrometry-based proteomics using Q Exactive, a high-performance benchtop quadrupole Orbitrap mass spectrometer. Mol Cell Proteomics 10(9): M111.011015
40. Rose CM et al (2013) Neutron encoded labeling for peptide identification. Anal Chem 85(10): 5129–5137

Chapter 2

Stable Isotope Labeling by Amino Acids Applied to Bacterial Cell Culture

Boumediene Soufi and Boris Macek

Abstract

Stable isotope labeling by amino acids in cell culture (SILAC) is a widely used approach in quantitative proteomics; however, due to limitations such as required auxotrophy for the amino acids employed for labeling, it was thus far rarely employed in bacteria. Although limitations of SILAC in microbiological applications are significant and restrict its use exclusively to cells cultured in minimal media, we and others have successfully used it to fully label proteomes of model bacteria and measure their relative expression dynamics under different experimental conditions. Here we provide a brief overview of applications of SILAC in bacteria and describe a detailed protocol for SILAC labeling of *Escherichia coli* and *Bacillus subtilis* cells in culture, which in many cases can be applied to other members of both gram-positive and gram-negative bacterial species.

Key words SILAC, Quantitative proteomics, Mass spectrometry, Bacteria, Prokaryotes

1 Introduction

1.1 Quantitative Proteomics in Bacteria

Since its onset in the early 1990s, mass spectrometry-based proteomics has been revolutionizing biomedical research while also having a significant impact in the microbiology field [1]. Bacteria are especially suitable for global quantitative gene expression analyses, due to their relatively simple and small genomes and proteomes; for example, the theoretical proteome of *E. coli* consists of approximately 4,100 protein-coding genes, of which about 2,700 are estimated to be expressed under standard laboratory growth conditions [2].

Traditional proteomics approaches relied on 2D gel electrophoresis for protein separation and quantification, whereas more modern quantitative proteomic approaches are mostly gel-free and use stable (nonradioactive) isotope labeling to quantify directly from mass spectra [3]. The stable isotope labeling approaches in quantitative proteomics have been extensively reviewed elsewhere [4–6]. Briefly, two major strategies for introducing a stable isotope

Bettina Warscheid (ed.), *Stable Isotope Labeling by Amino Acids in Cell Culture (SILAC): Methods and Protocols*, Methods in Molecular Biology, vol. 1188, DOI 10.1007/978-1-4939-1142-4_2, © Springer Science+Business Media New York 2014

label into the proteome exist: chemical labeling, where a labeled compound is attached by a chemical reaction (in vitro) to specific reactive groups on protein/peptide termini or side chains of amino acids (e.g., N-terminal protein/peptide amino groups or thiol group on cysteines), and metabolic labeling, where the cell, tissue, or organism is grown in a defined medium with a single stable isotope-labeled source, and thus metabolically incorporates the label. In both approaches, protein samples from two states that are to be compared are labeled with either the unlabeled ("light") or labeled ("heavy") version of a compound. The differentially labeled samples are mixed and digested prior to MS measurement, in which intensities and fragmentation patterns of peptides are measured and used for subsequent identification and quantitation. If applicable, metabolic labeling, which includes ^{15}N labeling and SILAC, is preferred due to the high efficiency of labeling and straightforward sample preparation which minimizes quantitation errors due to sample handling. Application of ^{15}N labeling to bacteria is relatively straightforward and involves culturing of the cells on a sole source of nitrogen (usually ammonium salt) in a minimal medium and has previously been applied to bacteria [7, 8]. It does not require the creation of auxotrophic strains, but data analysis and interpretation are rather difficult due to the variable mass difference between labeled peptide pairs (difference depends on the number of N atoms in each peptide) and different isotope distributions between two members of a pair. Conversely, SILAC-labeled samples are easier to analyze and quantify than ^{15}N-labeled samples, as SILAC introduces a label of equal size (mass) in all peptides and thus keeps the isotope patterns of heavy and light signals almost identical. Every peptide is detected by a mass spectrometer system with high resolution capabilities in the form of two MS peaks (light and heavy) and relative quantitation is achieved by calculating the ratio between the measured signals (Fig. 1a). Protein samples can also be extended to three states which are labeled with either unlabeled ("light") or labeled ("medium-heavy" and "heavy") versions of a compound (Fig. 1b).

However, SILAC labeling also has notable limitations. In order to fully incorporate the labeled amino acid, the cell must not be able to synthesize it endogenously. Due to a robust bacterial metabolism, this effectively means that most of the bacterial species must be rendered auxotrophic for a specific amino acid prior to the SILAC experiment, e.g., by knocking out one or more genes involved in its biosynthesis pathway (*E. coli* is a notable exemption, see below). SILAC is almost completely inapplicable to photosynthetic bacteria, although some approaches applied to SILAC labeling of plant cells cultured in the dark [9] may be also applicable to bacteria. To ensure that there is only one source of the labeled amino acid in a cell's environment, minimal media must be employed in all bacterial SILAC cell cultures. This obviously prevents

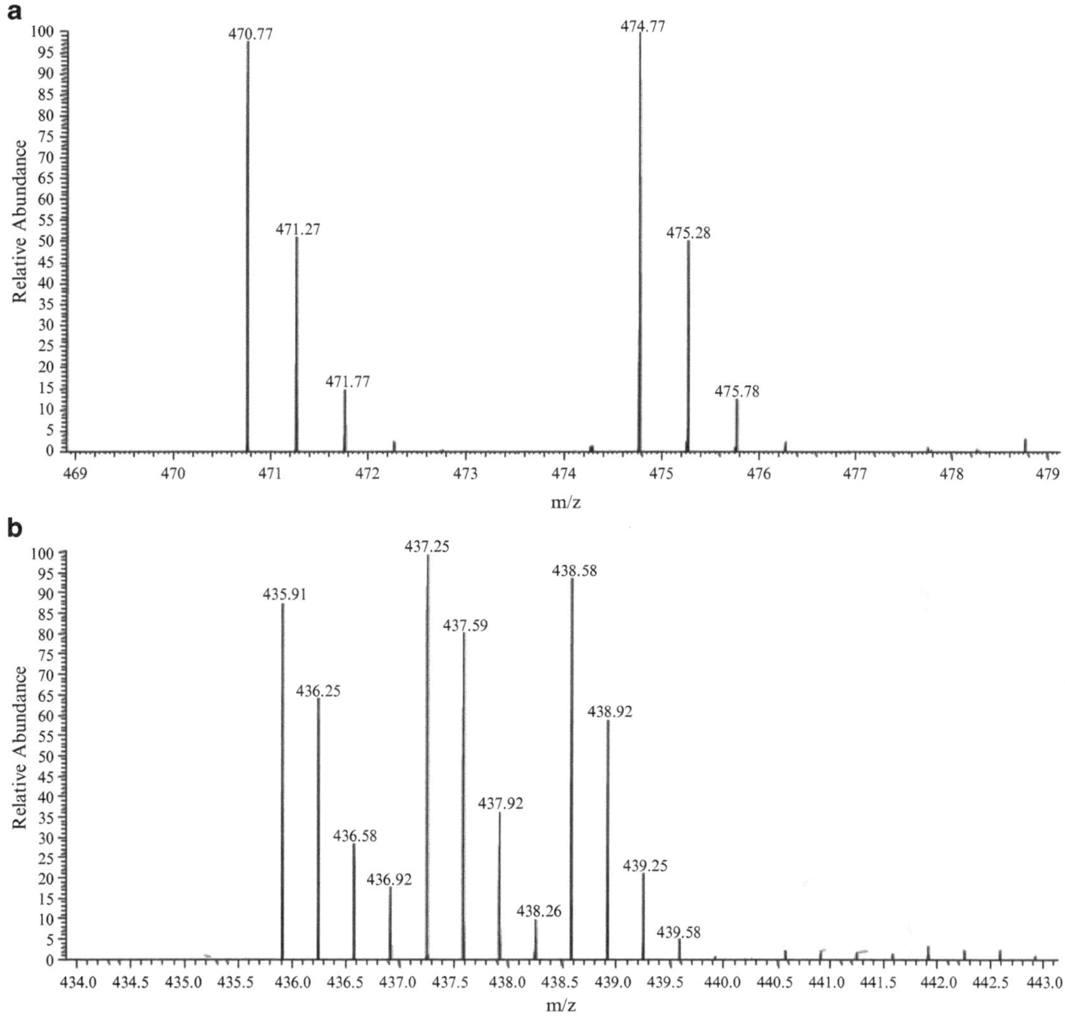

Fig. 1 Typical MS spectra of SILAC signals. (**a**) MS full scan corresponding to a double SILAC labeling experiment using Lys0 and Lys8 amino acids resulting in a mass shift of 8 Da between signal peaks. (**b**) MS full scan corresponding to a triple SILAC labeling experiment using Lys0, Lys4, and Lys8 amino acids resulting in a mass shift of 4 Da between signal peaks

the use of SILAC in quantitative proteomic analysis of environmental samples and clinical isolates (note that this limitation also applies to ^{15}N labeling). Despite of these obvious limitations, we and others have shown that a full SILAC labeling of a bacterium is possible and can be used to address fundamental molecular physiology processes, such as starvation or exchange of carbon sources [7, 10]. Once applicable to study a bacterial organism and process of interest, SILAC is a simple and cost-effective way to perform quantitation of a bacterial proteome to an unprecedented depth. It can be performed in several ways, such as super-SILAC [11], pulsed SILAC [12], and dynamic SILAC [13] and, together with appropriate data processing software, enables global, comprehensive quantitation of the complete bacterial proteome or sub-proteome (e.g., phosphoproteome, acetylome, membrane proteome).

1.2 Overview of the SILAC Labeling Protocol

Here we describe SILAC labeling of two commonly used bacterial model organisms, *E. coli* and *B. subtilis*. In order to avoid metabolic loss of label and quantitation bias due to conversion of arginine (Arg) to proline (Pro), in our laboratory we perform SILAC labeling of bacterial cultures through lysine (Lys) only; however, we note that several approaches to avoid Arg-Pro conversion in animal cells have been described (e.g., addition of unlabeled proline to cell cultures or knockout (KO) of Arg and Pro biosynthesis) [14, 15]. The labeling process itself is quite straightforward and requires replacing the unlabeled amino acid with the same amount of the labeled amino acid in the minimal medium and letting the cells divide in such a medium at least 5–6 times. In case of *B. subtilis*, we use a LysA KO mutant strain that is auxotroph for lysine. Creation of such auxotrophic strains is a prerequisite for SILAC experiments in most bacterial species; however, in case of *E. coli* we and others have observed that the full SILAC labeling is possible even in the wild-type (WT) strain [16]. To ensure full incorporation of the labeled amino acid, we first inoculate a small SILAC pre-culture, which we run into mid-logarithmic phase and use to inoculate the main SILAC culture used in the experiment. In every SILAC experiment, we perform two critical steps of quality control: the incorporation level of the SILAC label (to ensure complete proteome labeling) and mixing check (to minimize possible quantitation error due to erroneous mixing). Finally, we use the MaxQuant software suite [17–19] to perform identification and quantification of the data.

2 Materials

2.1 Bacterial Strains

In order to ensure the highest level of incorporation of the amino acid isotope, the bacterial strain employed should be an auxotrophic mutant of the corresponding amino acid biosynthetic pathway. However, in some cases, for example in the lysine biosynthetic pathway in some strains of *E. coli*, efficient incorporation is achieved also in the WT prototrophic strain [16]. Preliminary incorporation tests should be performed in order to determine if an auxotrophic mutant is required or not.

2.2 SILAC Amino Acids

In principle, every amino acid can be used for SILAC labeling, but Arg and Lys are preferred due to compatibility with trypsin and LysC digestion protocols. If a bacterial strain is an auxotrophic mutant for lysine, then typically Lys4 and Lys8 are used.

Moreover, the conventional SILAC format can easily be extended to a triple labeling format (e.g., Lys0, Lys4, and Lys8). Common suppliers of arginine and lysine amino acids as well as their stable isotopes can be purchased from many companies (*see* **Note 1**).

2.3 Bacterial Growth Media

Several different types of chemically defined minimal media exist for the growth of bacteria. The Belitsky Minimal Medium (BMM) [20] with minor modifications and the M9 medium are routinely used in our laboratories for *B. subtilis* and *E. coli*, respectively. The minimal medium chosen will depend on the specific type of bacteria being studied, and therefore its respective optimal minimal medium should be used.

2.3.1 BMM Synthetic Growth Medium for B. subtilis

The medium consists of 8 mM $MgSO_4$, 27 mM KCl, 7 mM sodium citrate, 50 mM Tris–HCl pH 7.5, 2 mM $CaCl_2$, 0.6 mM KH_2PO_4, 15 mM $(NH_4)_2SO_4$, 10 μM $MnSO_4$, 1 μM $FeSO_4$, 0.5 % glucose, 250 μM thymine (optional: proline, *see* **Note 2**), 0.025 % light amino acid for light medium, and 0.025 % heavy amino acid for labeled medium.

2.3.2 M9 Minimal Medium for Growth of E. coli

The medium consists of M9 Minimal Salts (5×, Sigma), 1 % thiamine, 1 mM $MgSO_4$, 10 μM $CaCl_2$, 0.5 % glucose (optional: proline, *see* **Note 2**), 0.025 % light amino acid for light medium, and 0.025 % heavy amino acid for labeled medium.

2.3.3 LB Plates

The LB plates were prepared from LB media (Roth) and 15 g/l of Agar (Roth).

2.4 Bacterial Lysis

1. Y-PER™ Yeast Protein Extraction Reagent (Thermo Scientific).
2. Lysozyme (chicken egg white isolate).
3. Sonifier.

2.5 Chloroform–Methanol Protein Precipitation

1. Milli-Q water.
2. Chloroform (>99.8 % GC grade).
3. Methanol (>99.8 % GC grade).

2.6 Determination of Protein Concentration

1. Bradford Protein Assay Dye Reagent Concentrate (Bio-Rad).
2. Folded filter paper.
3. BSA or lysozyme can be used as a protein standard.

2.7 In-solution Protein Digestion

1. Denaturation buffer: 6 M urea, 2 M thiourea, 1 % (w/v) *n*-octylglucoside, 10 mM HEPES, pH 8.0 (*see* **Note 3**).
2. Reduction buffer: 1 M dithiothreitol (DTT) in 50 mM ammonium bicarbonate (ABC) (*see* **Note 4**).
3. Alkylation buffer: 550 mM iodoacetamide (IAA) in 50 mM ABC.

4. Lysyl Endopeptidase LysC.
5. Trypsin, sequencing grade, modified.
6. 20 mM ABC.

2.8 StageTips

1. Methanol (>99.8 % GC grade).
2. Solvent A*: 2 % acetonitrile (ACN)/1 % trifluoroacetic acid (TFA).
3. Solvent A: 0.5 % acetic acid.
4. Solvent B: 80 % ACN in 0.5 % acetic acid.
5. Empore™ 47 mm C18 Disk (3 M, Cat. No. 2215).

2.9 Liquid Chromatography Mass Spectrometry

1. HPLC solvent "A": 0.5 % acetic acid.
2. HPLC solvent "B": 80 % ACN in 0.5 % acetic acid.
3. HPLC loading solvent: 2 % ACN/1 % TFA.
4. 15 cm fused silica column emitters with inner diameter of 75 μM (New Objective).
5. Reversed phase material for nano-HPLC: Reprosil-Pur C18-AQ, 3 μM resin (Dr. Maisch GmbH).
6. EASY-nLC II system (Thermo).
7. High resolution MS instrument. For example, LTQ Orbitrap Elite MS instrument (Thermo).

2.10 Data Analysis

1. MaxQuant Software suite [17–19].

3 Methods

3.1 Bacterial Growth in SILAC Media

1. Streak out bacteria for single colonies on a standard LB plate (*see* **Note 5**).
2. Generate bacterial pre-cultures as follows: inoculate the minimal medium with a single colony containing 0.025 % of the appropriate SILAC amino acid ("heavy") (*see* **Note 6**), and in a separate flask, inoculate the minimal medium with a single colony containing 0.025 % of the non-labeled form of the amino acid ("light").
3. Grow bacterial pre-cultures under appropriate growth conditions to an OD600 of 0.3–0.5 (corresponding to a growth phase between early to mid-exponential growth).
4. Use pre-cultures in early to mid-exponential growth phase (*see* **Note 7**) to inoculate minimal media containing 0.025 % of either heavy or light amino acids to a starting OD600 between 0.03 and 0.05 and grow to desired OD600.

Table 1
Volume of Y-PER reagent (Thermo Scientific) to add per milligram of cell pellet

Wet cell pellet weight (mg)	Y-PER reagent volume (μl)
50	125–250
100	250–500
250	625–1,250
500	1,250–2,500

3.2 Bacterial Lysis

This protocol is slightly modified and optimized for efficient protein extraction of *B. subtilis* and *E. coli* cells based on the Y-PER™ Yeast Protein Extraction Reagent protocol. However, it is also effective for both gram-positive and gram-negative bacteria in general. There are various other types of lysis buffers available, but caution must be taken with regard to the amount of common enzymes (lysozyme, DNAse) present in large amounts in many of these buffers. They can lead to a false estimation of the amount of bacterial proteins in the sample.

1. Pellet cells by centrifugation at approximately 3,000 × *g* for 5 min at 4 °C (*see* **Note 8**).
2. Resuspend the cells in an appropriate amount of Y-PER Reagent as indicated in Table 1. Pipette up and down until the mixture is homogeneous.
3. Prior to the addition of the Y-PER™ reagent to the cell pellet, add lysozyme to a final concentration of 50 μg/ml. This can be added from a lysozyme stock solution (e.g., a 5 mg/ml lysozyme stock solution) (*see* **Notes 9** and **10**).
4. Agitate the mixture at 37 °C for 20 min.
5. Sonicate the mixture for 30 s at an amplitude of 40 %.
6. Pellet the cell debris by centrifuging at 14,000 × *g* for 30 min and remove supernatant immediately after end of centrifugation to avoid contamination with cellular debris.

3.3 Chloroform–Methanol Protein Precipitation (See Note 11)

1. Mix the sample with 4 volumes of methanol and vortex.
2. Add 1 volume of chloroform and vortex.
3. Add 3 volumes of water and vortex.
4. Spin 1 min at maximum speed (*see* **Note 12**).
5. Discard aqueous phase (*see* **Note 13**).
6. Add 4 volumes of methanol and vortex.
7. Spin for 2 min at maximum speed.

8. Remove methanol and let the precipitate dry.
9. Dissolve the protein sample in the lowest volume of denaturation buffer possible (optimally, this should be around 1–2 μg of protein per μl of denaturation buffer).

3.4 Determination of Protein Concentration

1. Dilute Bradford Protein Assay-stock solution 1:5 with water.
2. Filter through a folded filter (*see* **Note 14**).
3. Prepare standard calibration curve and proceed with standard Bradford procedures (*see* **Notes 15** and **16**).

3.5 In-solution Digestion

This protocol is just one of many possible methods of digestion. Other possibilities include in-gel digestion and further in-solution fractionation through isoelectric focusing of the bacterial peptides.

1. Add reduction buffer to the sample to a final concentration of 1 mM DTT; incubate 1 h at room temperature (*see* **Note 17**).
2. Add alkylation buffer to the sample to a final concentration of 5.5 mM IAA; incubate for 1 h at room temperature in the dark.
3. Check the pH (should be 8.0); adjust with NaOH or HCL if necessary.
4. Add 1 μg of LysC per 100 μg of protein and incubate for 3 h at room temperature (*see* **Note 18**).
5. Dilute sample with 4 volumes of 20 mM ABC.
6. Check the pH (should be 8.0); adjust if necessary (*see* **Note 19**).
7. Add additional 1 μg of LysC per 100 μg of sample protein and incubate overnight at room temperature (*see* **Note 20**).

3.6 Quality Control: Incorporation of the SILAC Amino Acid and Mixing Check

1. Determine the efficiency of incorporation of the amino acid isotope by performing MS analysis (*see* Subheading 3.8). Calculate the median value of the heavy/light ratios by using the formula: $\text{Incorporation} = [H/L]_{median}/(1 + [H/L]_{median})$. Multiply by 100 to get median incorporation in percentage (Fig. 2). To achieve optimal conditions for SILAC experiments, the labeling efficiency should be more than 95 %.
2. For mixing check, mix heavy and light samples in a 1:1 ratio (based on Bradford measurements) followed by MS analysis and data processing to verify accurate 1:1 mixing of heavy and light samples (Fig. 3).

3.7 Peptide Cleanup via StageTips

1. Activate StageTips with 200 μl of methanol.
2. Wash/equilibrate with 200 μl of solvent A*.
3. Add sample about 10 % of solvent A* can be added to the sample (*see* **Notes 21** and **22**).
4. Wash with 200 μl of solvent A (*see* **Note 23**).

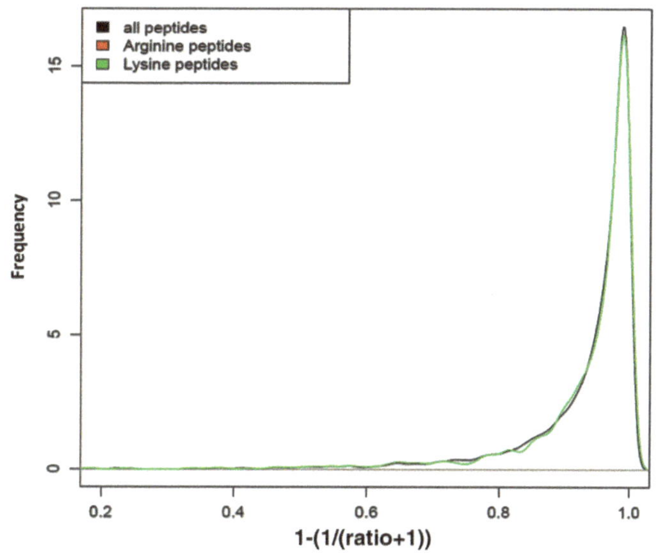

Fig. 2 Efficiency of Incorporation. The incorporation efficiency of the amino acid stable isotopes can be determined on all identified SILAC peptides (pairs) using the following formula: 1-(1/(ratio+1)). As seen above, if the level of incorporation is near to 100 %, the median incorporation level will be close to 1. This calculation should be done separately for arginine- and lysine-containing peptides because the arginine and lysine stable isotopes may have different labeling efficiencies. In the example above, there are no ratios for arginine containing peptides due to the fact that only Lys8 was used

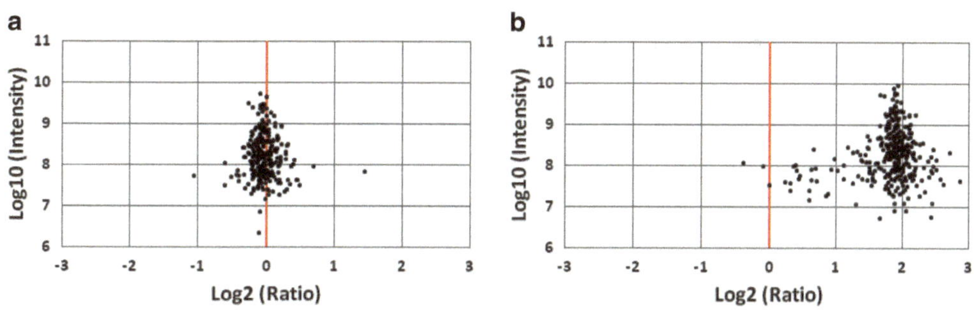

Fig. 3 Importance of accurate mixing of differentially labeled SILAC samples. (**a**) Scatterplot of peptide ratios versus intensities illustrates an example of accurate 1:1 mixing of heavy to light amino acids. (**b**) Scatterplot of peptide ratios versus intensities illustrates a poor example of heavy to light amino acid mixing with an approximately fourfold higher level of heavy compared to light amino acid; while relatively small mixing errors (<30 %) are routinely dealt with by processing software, extreme mixing errors will result in wrong quantitation values

5. Elute peptides with 50 μl of solvent B directly into an Eppendorf tube.
6. Dry peptides in a SpeedVac for approx. 10–12 min to approx. 5 μl of sample (*see* **Note 24**).
7. Add 1 μl of solvent A* to the sample.

3.8 Mass Spectrometry Analysis

Peptide samples are analyzed using online nanoflow liquid chromatography tandem mass spectrometry (LC-MS/MS). In our laboratory, nanoLC-MS/MS experiments are performed on an EASY-nLC II system connected to an LTQ-Orbitrap Elite through a nanoelectrospray ion source; however, almost any other LC-MS setup (including those using MALDI sources) can be used for analysis.

1. Load peptides directly onto a 15 cm long 75 μm-inner diameter analytical column packed with reversed-phase C18 Reprosil AQUA-Pur 3 μm particles using the "Intelliflow" setting (variable flow rate up to a maximum pressure of 280 bar).
2. Reduce flow rate to 200 nl/min after loading, and then separate peptides with a segmented linear gradient of ACN from 5 to 33 to 50 % in 0.5 % acetic acid for either 60, 100, 150, or 240 min (*see* **Note 25**).
3. Use the following MS settings on the LTQ-Orbitrap Elite: target values 1E6 charges (Orbitrap analyzer) and 5E3 charges (linear ion trap). Operate the instrument in the positive ion mode with the following acquisition cycle: a full scan recorded in the Orbitrap mass analyzer at a resolution R of 120,000 followed by MS/MS (CID Rapid Scan Rate) of the 20 most intense peptide ions in the LTQ mass analyzer.

3.9 Data Processing and Analysis

1. Download the latest version of the MaxQuant software (*see* **Note 26**).
2. Load the appropriate RAW files (*see* **Note 27**).
3. Open the Experimental Design tab. This will write a folder located in the same location as where the RAW files are and will be named "Combined". This folder will contain an experimentaldesigntemplate.txt file (*see* **Note 28**).
4. Set number of threads located at the bottom left hand corner of the MaxQuant software (*see* **Note 29**).
5. Select the Group-Specific parameters tab and specify the following: variable modifications of interest (if any), the multiplicity (corresponding to how many amino acid isotopes were used), and the specific protease used for the experiment. All other parameters in this tab should initially be left to their default settings unless further specific changes are required.
6. Open the MS/MS & Sequences tab. Specify the appropriate FASTA file corresponding to the organism being studied by selecting the Add File tab (*see* **Note 30**). The MS/MS and all other parameters should be left at their default settings unless desired otherwise.
7. Open the Identification & Quantification tab. Open "Select file" under the experimental design tab and select the experimental design file that was created earlier. All other parameters

should be left at their default settings unless desired otherwise (*see* **Note 31**).

8. Open the performance tab and then click the start tab located at the bottom of the MaxQuant screen. If the setup was successful, the analysis shall begin. The entire progress of the analysis can be monitored under the performance tab. When the analysis is complete, MaxQuant will indicate this with a "Done" button. All data is located in the combined folder inside the txt folder. All of the .txt files can be opened with Microsoft Excel™.

4 Notes

1. SILAC amino acids can be purchased from several companies such as EURISO-TOP, Sigma-Aldrich, Silantes, and Cambridge Isotope Laboratories.
2. As previously documented in eukaryotes, there is also the possibility of arginine to proline conversion in bacteria [14, 15]. This can cause difficulties in the quantitation due to the abundance of non-quantifiable satellite peaks caused from the proline signal. In order to circumvent this, proline may be added to the minimal media (amount must be optimized according to the bacterial strain used) in order to discourage the bacteria to convert arginine to proline since proline would already exist at a saturated level in the minimal media.
3. All solvents used in this procedure should be prepared with Milli-Q water. pH of Buffers can be adjusted with NaOH or HCl until the correct pH is obtained.
4. The denaturation, reduction, and alkylation buffers may be stored in aliquots at −20 °C and remain stable for several months.
5. LB Plates are prepared using standard 25 g/l of LB media and 15 g/l of agar. Mixture is then autoclaved, and once cooled (to approximately 55 °C), the mixture is poured into petri dishes, allowed to harden, and stored at 4 °C in the dark until use. Incubation temperatures vary according to the type of bacteria used. Furthermore, antibiotics may also be used if necessary (e.g., if bacteria employed contains antibiotic resistance markers).
6. It was established in *B. subtilis* and *E. coli* that a final amino acid concentration of 0.025 % does not compromise bacterial growth.
7. Pre-cultures need to be in early to mid-exponential growth to ensure efficient and accurate reproducibility between experiments; an "overgrown" stationary overnight bacterial culture contains a mixed bacterial cell population composed of dormant, living,

and dead bacterial cells, whereas a culture in the exponential growth phase contains a higher proportion of healthy, synchronized bacterial cells, and therefore improving reproducibility between biological replicates of experiments.

8. Cells may be processed immediately after centrifugation or the cell pellet may be frozen at −20 °C or −80 °C.
9. For lysozyme stock solutions, make a 5 mg/ml lysozyme stock solution with 50 mM Tris–HCl, pH 7.5, and split into 100 μl aliquots. Store these aliquots at −20 °C until ready for use.
10. At this step, protease inhibitors may be added.
11. Alternative methods for precipitation/purification may be used.
12. Spin at a minimum of 4,000 × *g* in order to ensure good separation of the protein mixture.
13. Proteins will remain between the hydrophobic and hydrophilic phases.
14. Keep diluted and filtered Bradford reagent in a glass bottle (at room temperature, the reagent can be kept for approximately 2 weeks).
15. Standard calibration curves may be generated with known amounts of a standard protein such as bovine serum albumin.
16. Other methods of protein determination may be employed such as the BCA kit, NanoDrop technologies etc.
17. Avoid heating up the samples during protein solubilization and digestion due to the fact that the high concentration of urea will lead to carbamylation of the free amino groups present.
18. Other proteases may also be used depending on the desired protein cleavage. Note that a combination of heavy lysine SILAC labeling and LysC protease digestion results in one labeled amino acid per fully cleaved peptide (i.e., all peptides except for the protein C-termini are quantifiable).
19. The pH may be adjusted accordingly with a very low volume of 1 M Tris–HCl.
20. If cells are labeled with lysine only (like presented in this protocol), LysC is used as the protease for the overnight digestion step. LysC cleaves at the C terminal end of lysine residues, allowing for quantitation of all resulting peptides. However, if arginine and lysine SILAC amino acids are used for labeling, trypsin should be utilized as the protease for the overnight digestion step.
21. Solvent A* contains TFA that is a good ion pairing reagent and increases the binding efficiency of the peptides to the C18 material in the stage tip.

22. Take extra caution to ensure that the StageTip does not become dry (critical until the sample is loaded onto the tip).
23. StageTips can be stored at 4 °C until measurement.
24. Ensure that the sample is not dried completely.
25. The total acquisition time can be varied typically in a gradient range within 100 or 240 min depending on the desired level of peptide coverage/identifications. For example, during an incorporation check, a shorter MS method shall suffice (60 min) due to the fact that detection of approximately 1,000 peptides will be enough for the accurate determination of the level of the SILAC amino acid incorporation.
26. MaxQuant software can be downloaded at www.maxquant.org. The following publications provide both relevant and useful information that pertains to the theory and usage of this software [17–19, 21, 22]. The protocol instructions described for the MaxQuant software suite is based on version 1.3.0.5. It is important to note that there are other programs available that could be used as an alternative for downstream data analysis.
27. The RAW files should be copied to a local folder where the processing is to occur.
28. The experimental designtemplate.txt file can be opened with the Microsoft Excel™ program and modified according to which experiment(s) correspond to which RAW file(s). Slices (fractions) may also be designated if further fractionation methodologies were performed (examples include off gel fractionation or in gel fractionation).
29. The number of threads may be set according to the number of RAW files. However, the total number of threads should not exceed the total capacity of the operating computer (number of cores). For each thread used, approximately 1.5 GB of RAM is required. For optimal processing performance, at least twice the size of the total size of all RAW files being analyzed should be available on the hard disk space. Further information may be found at: http://www.maxquant.org/requirements.htm.
30. The FASTA files of the organism which correspond to all known annotated genes of the organism must be added and therefore the microorganism being studied must be fully sequenced.
31. The Misc tab contains advanced features of the MaxQuant software that will not be discussed in this chapter with one exception: When performing incorporation and mixing checks for quality control purposes, the Re-quantify option should be unchecked as it could skew the true value of the incorporation rate if left checked.

Acknowledgements

The authors wish to thank Alejandro Carpy and the other PCT members for useful comments on the manuscript. Our work is financed by the Juniorprofessoren-Programm of the Landesstiftung BW, the SFB766 of the Deutsche Forschungsgemeinschaft, and PRIME-XS consortium.

References

1. Seraphin B, Hettich R (2012) Microbial proteomics: the quiet revolution. Curr Opin Microbiol 15(3):348–350
2. Iwasaki M, Miwa S, Ikegami T et al (2010) One-dimensional capillary liquid chromatographic separation coupled with tandem mass spectrometry unveils the Escherichia coli proteome on a microarray scale. Anal Chem 82(7):2616–2620
3. Otto A, Bernhardt J, Hecker M et al (2012) Global relative and absolute quantitation in microbial proteomics. Curr Opin Microbiol 15(3):364–372
4. Ong SE, Mann M (2005) Mass spectrometry-based proteomics turns quantitative. Nat Chem Biol 1(5):252–262
5. Bantscheff M, Schirle M, Sweetman G et al (2007) Quantitative mass spectrometry in proteomics: a critical review. Anal Bioanal Chem 389(4):1017–1031
6. Macek B, Mann M, Olsen JV (2009) Global and site-specific quantitative phosphoproteomics: principles and applications. Annu Rev Pharmacol Toxicol 49:199–221
7. Dreisbach A, Otto A, Becher D et al (2008) Monitoring of changes in the membrane proteome during stationary phase adaptation of Bacillus subtilis using in vivo labeling techniques. Proteomics 8(10):2062–2076
8. Otto A, Bernhardt J, Meyer H et al (2010) Systems-wide temporal proteomic profiling in glucose-starved Bacillus subtilis. Nat Commun 1:137
9. Schutz W, Hausmann N, Krug K et al (2011) Extending SILAC to proteomics of plant cell lines. Plant Cell 23(5):1701–1705
10. Soufi B, Kumar C, Gnad F et al (2010) Stable isotope labeling by amino acids in cell culture (SILAC) applied to quantitative proteomics of Bacillus subtilis. J Proteome Res 9(7):3638–3646
11. Geiger T, Wisniewski JR, Cox J et al (2011) Use of stable isotope labeling by amino acids in cell culture as a spike-in standard in quantitative proteomics. Nat Protoc 6(2):147–157
12. Schwanhausser B, Gossen M, Dittmar G et al (2009) Global analysis of cellular protein translation by pulsed SILAC. Proteomics 9(1): 205–209
13. Doherty MK, Hammond DE, Clague MJ et al (2009) Turnover of the human proteome: determination of protein intracellular stability by dynamic SILAC. J Proteome Res 8(1):104–112
14. Bendall SC, Hughes C, Stewart MH et al (2008) Prevention of amino acid conversion in SILAC experiments with embryonic stem cells. Mol Cell Proteomics 7(9):1587–1597
15. Lossner C, Warnken U, Pscherer A et al (2011) Preventing arginine-to-proline conversion in a cell-line-independent manner during cell cultivation under stable isotope labeling by amino acids in cell culture (SILAC) conditions. Anal Biochem 412(1):123–125
16. Frohlich F, Christiano R, Walther TC (2013) Native SILAC: metabolic labeling of proteins in prototroph microorganisms based on lysine synthesis regulation. Mol Cell Proteomics 12(7):1995–2005
17. Cox J, Mann M (2008) MaxQuant enables high peptide identification rates, individualized p.p.b.-range mass accuracies and proteome-wide protein quantification. Nat Biotechnol 26(12):1367–1372
18. Cox J, Matic I, Hilger M et al (2009) A practical guide to the MaxQuant computational platform for SILAC-based quantitative proteomics. Nat Protoc 4(5):698–705
19. Cox J, Neuhauser N, Michalski A et al (2011) Andromeda: a peptide search engine integrated into the MaxQuant environment. J Proteome Res 10(4):1794–1805
20. Stulke J, Hanschke R, Hecker M (1993) Temporal activation of beta-glucanase synthesis in Bacillus subtilis is mediated by the GTP pool. J Gen Microbiol 139(9):2041–2045
21. Kirchner M, Selbach M (2012) In vivo quantitative proteome profiling: planning and evaluation of SILAC experiments. Methods Mol Biol 893:175–199
22. Tyanova S, Mann M, Cox J (2014) MaxQuant for In-Depth Analysis of Large SILAC Datasets. In:Stable Isotope Labeling by Amino Acids in Cell Culture (SILAC), Warscheid B (ed) 1188: 351–364

Chapter 3

SILAC Labeling of Yeast for the Study of Membrane Protein Complexes

Silke Oeljeklaus, Andreas Schummer, Ida Suppanz, and Bettina Warscheid

Abstract

Despite their simplicity compared to multicellular organisms, single-celled yeasts such as the baker's yeast *Saccharomyces cerevisiae* are widely recognized as model organisms for the study of eukaryotic cell biology. To gain deeper insights into the molecular mechanisms underlying cellular processes, it is of utmost interest to establish the interactome of distinct proteins and to thoroughly analyze the composition of individual protein complexes and their dynamics. Combining affinity purification of epitope-tagged proteins with high-resolution mass spectrometry and quantitative proteomics strategies, in particular stable isotope labeling by amino acids in cell culture (SILAC), represents an unbiased and powerful approach for a most accurate characterization of protein complexes. In this chapter, we provide detailed protocols for the generation of yeast strains (*S. cerevisiae*) amenable to SILAC-labeling, for epitope tagging of a protein of interest for affinity purification, and for the SILAC-based characterization of membrane protein complexes including the identification of stable core components and transient interaction partners.

Key words SILAC, Yeast, Affinity purification–mass spectrometry, Quantitative proteomics, Membrane protein complex, Gene disruption, Epitope tagging

1 Introduction

In 1996, the baker's yeast *Saccharomyces cerevisiae* was the first eukaryotic organism having its genome completely sequenced [1], and it soon evolved into a prime model organism for the study of eukaryotic cell biology. Many fundamental biological structures and processes remained conserved throughout evolution, and for a significant number of disease-associated human genes, homologs in yeast exist. Knowledge gained in yeast studies may therefore often be transferred to higher eukaryotic systems [2, 3]. From the experimental point of view, the appeal of working with *S. cerevisiae* lies in short generation times, its ability to grow in defined media under controlled conditions, and the ease of genetic manipulation. The last feature has been exploited in a number of large-scale

Bettina Warscheid (ed.), *Stable Isotope Labeling by Amino Acids in Cell Culture (SILAC): Methods and Protocols*, Methods in Molecular Biology, vol. 1188, DOI 10.1007/978-1-4939-1142-4_3, © Springer Science+Business Media New York 2014

ventures aiming at elucidating the function and localization of each of the roughly 6,000 predicted yeast proteins by deleting each open reading frame (ORF) [4] as well as by tagging the ORFs with the sequence coding for the green fluorescent protein (GFP) [5] or the tandem affinity purification (TAP) tag [6, 7], for example. The clone collections generated in these studies are largely available to the scientific community.

About a decade ago, *S. cerevisiae* was the object of several large-scale protein interaction screens combining biochemical affinity purification of thousands of TAP- or Flag-tagged proteins with mass spectrometry for the identification of interaction partners [6, 8–10] each resulting in the identification of tremendous numbers of protein–protein interactions not known at the time. These pioneering studies certainly made a considerable contribution to a better understanding of cellular processes as well as the modular organization of the proteome into protein complexes consisting of core components and attachments. However, they are also compromised by a number of limitations [discussed in detail in ref. 11], in particular the unambiguous discrimination between specific interaction partners and co-purified contaminants following a "qualitative" affinity purification–mass spectrometry (AP-MS) approach. Furthermore, attempts to isolate protein complexes of higher purity by employing, for instance, two-step purification protocols and/or high stringent washing conditions, inevitably run the risk of losing specific low abundant as well as weakly associated and transient interaction partners.

These drawbacks can be overcome by integrating quantitative proteomics techniques, in particular stable isotope labeling, into the AP-MS workflow. Among various chemical and metabolic labeling techniques available, stable isotope labeling by amino acids in cell culture (SILAC) is arguably the most robust and accurate alternative. ^{2}H-, ^{13}C-, and/or ^{15}N-containing ("heavy") variants of selected amino acids are incorporated into the proteome of cells or entire organisms during protein biosynthesis introducing a predictable mass shift into proteins compared to those from cells grown in the presence of normal, unlabeled ("light") amino acids. Differentially SILAC-labeled cells or protein extracts to be compared are combined for subsequent sample preparation and liquid chromatography-tandem MS (LC-MS/MS) analysis. Relative peptide and protein quantification based on intensities or peak areas of SILAC-encoded proteolytic peptide pairs in mass spectra using suitable software tools then reveals differences in protein abundance between the samples. By allowing for mixing the samples at early time-points in the workflow and, thus, minimizing experimental variations resulting from separate sample handling, SILAC generally facilitates the generation of most accurate data.

In a generic SILAC experiment, the amino acids selected for labeling should be completely incorporated into the proteins to ease data analysis for relative protein quantification. Since *S. cerevisiae* is able to metabolically synthesize all amino acids, it is highly recommended to use a strain auxotrophic for the amino acid(s) chosen. The first studies employing full metabolic labeling of the yeast proteome made use of different variants of deuterated leucine [12–14]. However, the chromatographic behavior of peptides containing deuterated leucine and their unlabeled counterparts differs slightly resulting in a shift in retention time [15] which has to be taken into account for accurate protein quantification. Today, the use of heavy arginine and lysine—first employed for labeling of yeast cells to study the pheromone signaling pathway [16]—is typically preferred since protein digest with the standard protease trypsin results in at least one labeled amino acid per peptide (except for the C-terminus of a protein), thereby maximizing the number of SILAC peptide pairs available for protein quantification. However, heavy arginine is often metabolically converted into heavy proline [15, 17, 18]. This bidirectional metabolic pathway encompasses the enzymes arginase (Car1p), ornithine aminotransferase (Car2p), and delta-1-pyrroline-5-carboxylate reductase (Pro3p) sequentially catalyzing the synthesis of proline from arginine via ornithine, glutamate γ-semialdehyde and delta-1-pyrroline-5-carboxylate (http://pathway.yeastgenome.org/). Arginine-to-proline conversion is generally not taken into account for quantification and, depending on the extent of the conversion, may considerably compromise the accuracy of quantitative data. Strategies to reduce arginine-to-proline conversion range from titrating the amount of heavy arginine and/or adding unlabeled proline to the medium, omitting proline-containing peptides from the quantification [19], deleting a gene coding for an enzyme essential for proline biosynthesis [18] to experimentally correcting for arginine-to-proline conversion by using $^{15}N_4$-arginine combined with normal lysine for the light and $^{13}C_6{}^{15}N_4$-arginine/$^{13}C_6{}^{15}N_2$-lysine for the heavy condition [20]. In the latter approach, heavy proline is formed in both light and heavy condition ($^{15}N_1$- and $^{13}C_5{}^{15}N_1$-proline), thus providing an internal correction. Alternatively, proteomes can be labeled with heavy lysine only.

Using SILAC combined with AP-MS, a cell population expressing an epitope-tagged version of the protein of interest (POI) is grown in heavy medium while control cells expressing the endogenous POI are grown in light medium (or vice versa). Following mixing of cells, affinity purification, LC/MS, and relative quantification, proteins specifically enriched with a tagged POI from heavy labeled cells are identified by high heavy-to-light ratios while co-purified contaminants, which are derived from both light and heavy labeled cell populations, exhibit an abundance ratio of approx. 1. The potential to reliably discriminate between specific interaction partners and nonspecifically binding proteins allows for the purification of protein complexes following a simple one-step purification protocol and/or applying mild washing conditions (low salt and detergent concentrations) and, thus, supports the identification of specific weak or low abundance interactors. While this classical approach fully exploits the advantage of SILAC, i.e., mixing of differentially labeled samples at the earliest time-point possible, it also enables the exchange of labeled and unlabeled proteins only transiently associated with the POI during affinity purification. As a consequence, specific dynamic interaction partners may be misclassified as co-purified contaminants. Alternatively, affinity purification may be performed separately from POI-tagged as well as control cells and differentially labeled samples are combined just prior to LC/MS analysis preventing the exchange of transient binding partners. For a thorough characterization of protein complexes, SILAC data obtained from the two distinct purification tracks can be combined to facilitate the identification of both stable core components and proteins transiently attached to the complex [21–24].

SILAC-based AP-MS has proven to be extremely well suited for the in-depth characterization of yeast membrane protein complexes and interaction networks without the need of purifying complexes to homogeneity: it enabled us to discover ER-to-peroxisome contact sites involved in the regulation of peroxisome biogenesis [25], to decipher the interactome of Pex14p, a central component of the peroxisomal protein import machinery [24], to uncover new functions for known mitochondrial proteins [26] and novel components of mitochondrial membrane protein complexes [27, 28] as well as to identify an entire new mitochondrial membrane protein complex (MINOS; [29]) and a supercomplex formed by the import and export translocases TOM and SAM of the outer mitochondrial membrane promoting the formation of β-barrel proteins [30].

In this chapter, we provide a protocol for the generation of yeast strains auxotrophic for arginine and lysine as well as the replacement of a given gene in this *arg*Δ*lys*Δ mutant strain with a construct coding for the corresponding protein C-terminally fused to a cleavage site for the tobacco etch virus protease (TEV) and the Protein A (PA) affinity tag. This allows for the expression of the tagged POI under the native promoter avoiding potential artifacts caused by overexpression of proteins from plasmids. The manipulation of the yeast genome described here requires only a minimum of expertise in standard cloning techniques and can be performed in any laboratory equipped with the basic instruments for molecular biology. We further describe a generic workflow for SILAC AP-MS experiments allowing for the comprehensive analysis of protein complexes including the definition of the core complex and transient interaction partners. We focus on the characterization of membrane protein complexes, which initiate and mediate a large number of fundamental biological processes but, due to their physicochemical properties, are difficult to purify and, thus, are still underrepresented in quantitative protein interaction studies.

2 Materials

2.1 Generation of an argΔlysΔ Strain and Genomic Epitope Tagging

2.1.1 Genetic Manipulation of S. cerevisiae

1. A haploid yeast strain auxotrophic for histidine, leucine, and uracil which does not contain a kanMX marker (e.g., BY4741, available from Euroscarf; http://web.uni-frankfurt.de/fb15/mikro/euroscarf/) (*see* **Note 1**).
2. Plasmids with two different loxP-flanked marker cassettes such as pUG27 containing the *Schizosaccharomyces pombe* $his5^+$ marker and pUG73 containing the *Kluyveromyces lactis* LEU2 marker for gene disruption of *ARG4* and *LYS1*, respectively (Table 1A) (*see* **Note 2**).
3. Plasmid pYM8 with the kanMX6 marker for epitope tagging of a gene of interest with TEV-PA (Table 1B).
4. A Cre recombinase expression plasmid for marker rescue by Cre/loxP recombination, e.g., pSH47 carrying the URA3 marker (Table 1C) (*see* **Note 3**).
5. Oligonucleotide pairs Arg4_KO_fwd/rev and Lys1_KO_fwd/rev (Table 2A) for disruption of *ARG4* and *LYS1*, respectively (*see* **Note 4**).
6. Oligonucleotide pairs Arg4_A/D and Lys1_A/D (Table 2B) for confirmation of both gene disruption and marker rescue.
7. Oligonucleotides ORF_TPA_fwd and ORF_TPA_rev (Table 2C) as PCR primers for amplification of TEV-PA-kanMX6 from pYM8. Design the oligonucleotides in a way that the target gene and the tag are fused in frame resulting in expression of the POI-TEV-PA fusion protein.

Table 1
Plasmids for gene disruption (A), TEV-PA epitope tagging (B), and marker rescue by Cre/loxP recombination (C)

	Plasmid	Features	Selection marker	Euroscarf accession #	Ref.
A	pUG27	loxP-P_{AgTEF1}-Sphis5-T_{AgTEF1}-loxP, bla	Histidine	P30115	[39]
	pUG73	loxP-P_{KlLEU2}-KlLEU2-T_{KlLEU2}-loxP, bla	Leucine	P30118	[39]
	pUG6	loxP-P_{AgTEF1}-kanMX-T_{AgTEF1}-loxP, bla	Geneticin	P30114	[40]
	pUG66	loxP-P_{AgTEF1}-ble-T_{AgTEF1}-loxP, bla	Phleomycin	P30116	[39]
	pUG72	loxP-P_{KlURA3}-KlURA3-T_{KlURA3}-loxP, bla	Uracil	P30117	[39]
B	pYM8	TEV-2xProteinA-ADH_{Term}, kanMX, bla	Geneticin	–	[35]
C	pSH47	CEN6, ARSH4, $GAL1_{Pro}$-Cre-CYC_{Term}, P_{ScURA3}-ScURA3-T_{ScURA3}, bla	Uracil	P30119	[39]
	pSH63	CEN6, ARSH4, $GAL1_{Pro}$-Cre-CYC_{Term}, P_{ScTRP1}-ScTRP1-T_{ScTRP1}, bla	Tryptophan	P30121	[39]
	pSH65	CEN6, ARSH4, $GAL1_{Pro}$-Cre-CYC_{Term}, P_{AgTEF1}-ble-T_{ScCYC1}, bla	Phleomycin	P30122	[39]

Table 2
Oligonucleotide sequences of primers used for gene disruption of *Arg4* and *Lys1* (A), confirmation of gene disruption and marker rescue (B) as well as epitope tagging (C)

	Name	Sequence (5′–3′)[a]
A	Arg4_KO_fwd	GAGCTCAAAAGCAGGTAACTATATAACAAGACTAAGGCAAAC ATGcagctgaagcttcgtacgc
	Arg4_KO_rev	TACCAGACCTGATGAAATTCTTGCGCATAACGTCGCCATCTG CTAgcataggccactagtggatctg
	Lys1_KO_fwd	AGATAACAACGAAAACGCTTTATTTTTCACACAACCGCAAAAA TGcagctgaagcttcgtacgc
	Lys1_KO_rev	TAAATGTCAGCGTAACGATAATGTATATACTTTAAATGTAAACT Agcataggccactagtggatctg
B	Arg4_A	TTTTCTTTACTCTTCCAAACCCTCT
	Arg4_D	GAGTGCGAATGGGTATAAACTAAGA
	Lys1_A	AGTACTTGAGCTATAATGACCCTGC
	Lys1_D	TACAAATTTGGCAACCAAGATAGTT
C	ORF_TPA_fwd	42–46 bp upstream of stop codon of ORF + cgtacgctgcaggtcgac
	ORF_TPA_rev	Reverse complement of a 42–46 bp sequence in the 3′ region of ORF (0–200 bp downstream of the stop codon) + atcgatgaattcgagctcg

[a]Sequences in lower case anneal to pUG27, pUG73, or pYM8, sequences in upper case are homologous to genomic DNA of *ARG4* or *LYS1*

8. Oligonucleotides ORF_C and ORF_D for confirmation of epitope tagging of the gene of interest: a forward primer binding within the coding region and a reverse primer binding within the 3′ region of the ORF, downstream of the site of homologous recombination (*see* **Note 5**).
9. A proofreading DNA polymerase, e.g., the Phusion® High Fidelity DNA polymerase.
10. Standard material and equipment for PCR and the analysis of PCR products.
11. Standard material for the generation of competent yeast cells and transformation of yeast (*see* **Note 6**).
12. Zymolyase 20T for direct PCR.

2.1.2 Media for Yeast Culture and Selection of Transformants

1. YPD medium: 1 % (v/w) yeast extract, 2 % (w/v) peptone, 2 % (w/v) glucose.
2. YPD medium containing geneticin (G418; 200 mg/l).
3. YPG medium: 1 % (w/v) yeast extract, 2 % (w/v) peptone, 3 % (w/v) glycerol.
4. PMGal medium: 0.17 % (w/v) yeast nitrogen base (YNB) without amino acids, 0.5 % (w/v) ammonium sulfate, 2 % (w/v) peptone, 2 % (w/v) glucose.

5. SC medium: 0.17 % (w/v) YNB without amino acids, 0.5 % (w/v) ammonium sulfate, 2 % (w/v) glucose, 2 % (w/v) amino acid drop-out (DO) mix.
6. Amino acids required for the DO mix ("Hopkins mix") [31]: adenine (hemisulfate salt), 0.5 g; L-alanine, L-arginine HCl, L-asparagine (monohydrate), L-aspartic acid, L-cysteine HCl, L-glutamine, L-glutamic acid (monosodium salt), glycine (sodium salt), L-histidine HCl, myo-inositol, L-isoleucine, L-lysine, L-methionine, para-aminobenzoic acid, L-phenylalanine, L-proline, L-serine, L-threonine, L-tryptophan, L-tyrosine, uracil, L-valine, 2.0 g each; L-leucine, 4.0 g. Omit the selected amino acid(s) to prepare SC-his, SC-leu, SC-his/-leu, SC-arg, SC-lys, or SC-ura and mix components well with mortar and pestle (*see* **Note 7**).
7. Agar; add 2 % (w/v) to liquid media for agar plates.

2.2 Characterization of Membrane Protein Complexes by SILAC AP-MS

2.2.1 Yeast Strains and SILAC Media

1. Isogenic *S. cerevisiae* strains generated as described in Subheading 3.1 with lysine and arginine auxotrophies expressing (a) the TEV-PA-tagged POI (referred to as "POI-tagged strain" or "POI-tagged cells") and (b) the endogenous version of the POI as negative control, respectively (*see* **Note 8**).
2. Amino acids for stable isotope labeling: $^{12}C_6{}^{14}N_4$-L-arginine (Arg0), $^{13}C_6{}^{15}N_4$-L-arginine (Arg10), $^{12}C_6{}^{14}N_2$-L-lysine (Lys0), and $^{13}C_6{}^{15}N_2$-L-lysine (Lys8) (*see* **Note 9**).
3. Light and heavy SC medium (*see* **Note 10**): 0.17 % (w/v) YNB without amino acids, 0.5 % (w/v) ammonium sulfate, 2 % (w/v) glucose, 2 % (w/v) DO-arg/-lys (*see* Subheading 2.1.2), adjusted to pH 6.0 with KOH; light medium is supplemented with Arg0 and Lys0, heavy medium with Arg10 and Lys8 (25 mg/ml each) (*see* **Note 11**).

2.2.2 Cell Lysis and Affinity Purification of Membrane Protein Complexes

1. Lysis buffer: 20 mM Tris–HCl, 80 mM sodium chloride (pH 7.5) supplemented with dithiothreitol (DTT; 1 mM) and the following protease and phosphatase inhibitors: antipain (5 μg/ml), aprotinin (2 μg/ml), benzamidine (0.16 mg/ml), bestatin (0.35 μg/ml), chymostatin (6 μg/ml), leupeptin (2.5 μg/ml), pepstatin A (1 μg/ml), phenylmethylsulfonyl fluoride (PMSF; 174 μg/ml), sodium fluoride (0.4 μg/ml); prepare freshly before use (*see* **Note 12**).
2. Glass beads (0.4–0.6 mm in diameter).
3. Reagents and bovine serum albumin (BSA) protein standard for protein determination, e.g., using the Bradford assay [32].
4. Glycerol.
5. Digitonin (Merck) for the solubilization of membrane proteins; 10 % (v/v) stock solution (*see* **Note 13**).

6. Washing buffer: lysis buffer containing 10 % (v/v) glycerol and 0.1 % (w/v) digitonin.
7. Elution buffer: lysis buffer containing 10 % (v/v) glycerol and 0.1 % (w/v) digitonin *without* inhibitors (*see* **Note 14**).
8. MobiCol "classic" spin columns with 35 μm filters (MoBiTec, Göttingen, Germany).
9. AcTEV protease (Life Technologies).
10. Ni-NTA agarose.
11. IgG-coupled Sepharose beads (*see* **Note 15**).
12. Ultracentrifuge, adequate rotor, and ultracentrifuge tubes.
13. Spectrophotometer.
14. Rotating wheel.
15. Thermomixer.

2.2.3 Tryptic In-Solution Digestion and Sample Preparation for LC-MS Analysis

1. Acetone (100 %, ice-cold).
2. Urea buffer: 8 M urea dissolved in 50 mM NH_4HCO_3.
3. 50 mM Tris(2-carboxy-ethyl)phosphine (TCEP).
4. 50 mM iodoacetamide dissolved in 50 mM NH_4HCO_3.
5. Trypsin (modified sequencing grade), 15 ng/μl dissolved in 20 mM NH_4HCO_3.
6. 100 mM DTT in H_2O.
7. 0.1 % (v/v) trifluoroacetic acid (TFA) (*see* **Note 16**).
8. Acetonitrile (ACN).
9. Glass vials with cap, septa, and inserts for LC analysis (CS-Chromatographie Service GmbH, Langerwehe, Germany; article number 300101, 300305, and 300405).
10. Vacuum concentrator (SpeedVac).

2.2.4 LC-MS/MS and Data Analysis

1. 0.1 % (v/v) TFA.
2. Solvent A: 0.1 % (v/v) formic acid (FA).
3. Solvent B: 86 % (v/v) ACN in 0.1 % (v/v) FA.
4. Nano (U)HPLC system (e.g., the UltiMate 3000RSLCnano; Thermo Scientific, Idstein, Germany), preferentially equipped with a C18 μ-pre-column and a C18 reversed-phase (RP) nano LC column.
5. ESI-MS instrument (*see* **Note 17**).
6. Software for protein identification and quantification such as MaxQuant/Andromeda (http://www.maxquant.org) [33, 34] (*see* **Note 18**).
7. Protein sequence database for *S. cerevisiae* (*see* **Note 19**).

3 Methods

3.1 Generation of an argΔlysΔ Strain and Genomic Epitope Tagging

We describe here the generation of an *S. cerevisiae* strain suitable for SILAC AP-MS studies following a series of gene disruption and replacement steps as depicted in Fig. 1a. To render the strain auxotrophic for arginine and lysine (*arg*Δ*lys*Δ), the genes *ARG4* and *LYS1* coding for the enzymes argininosuccinate lyase and saccharopine dehydrogenase involved in the biosynthesis of arginine and lysine, respectively, are disrupted. A generic strategy for disruption of a selected target gene is shown in Fig. 1b.

The loxP-flanked marker cassette is amplified from a plasmid by PCR using long chimeric primers that introduce sequences of 44–45 bp homologous to the 5′ and 3′ region of the target gene at

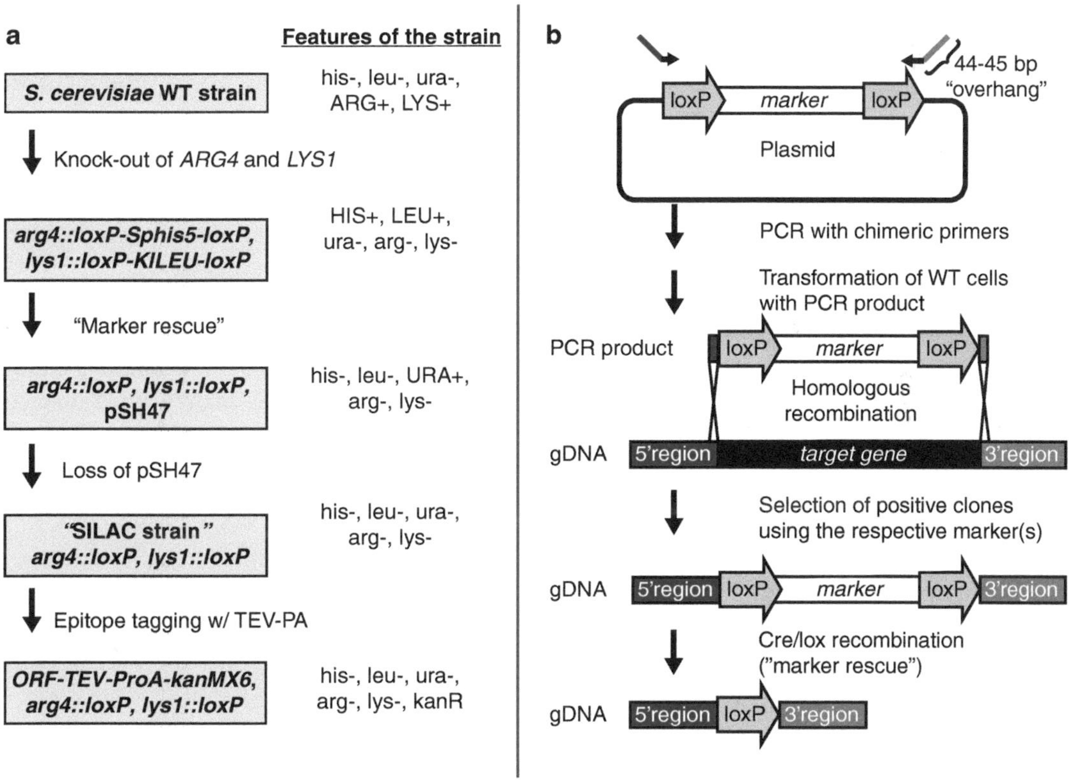

Fig. 1 Strategy for the genetic manipulation of *Saccharomyces cerevisiae*. (**a**) Overall workflow for the generation of an arginine and lysine auxotrophic yeast strain and genomic epitope tagging of the gene coding for a protein of interest comprising a series of gene disruption and replacement steps. For details, see Subheading 3.1. (**b**) Strategy for gene disruption. The loxP-flanked cassette of a selection marker is amplified from a plasmid by PCR using long chimeric primers that introduce a sequence "overhang" of 44–45 bp homologous to the 5′ and 3′ region of the target gene. This gene is replaced by the marker cassette via homologous recombination. Following selection of positive clones using the respective marker, the marker cassette is removed via Cre/lox recombination. As a result of this "marker rescue," a single loxP site remains integrated

the ends of the PCR product ("overhang"). To generate *arg*Δ *lys*Δ yeast strains, *Arg4* is replaced with a loxP-Sphis5-loxP and *Lys1* with a loxP-KlLEU2-loxP marker cassette via homologous recombination. Double transformants carry an *S. pombe* his5+ as well as a *K. lactis* LEU2 marker for nutritional selection. To allow for further genetic manipulations of the newly created *arg*Δ *lys*Δ strain using the same nutritional markers, the marker cassettes are removed ("marker rescue") by transforming the yeast cells with the plasmid pSH47 carrying the Cre recombinase gene. As a result of the Cre/loxP recombination, one of the loxP-sites flanking the his5+ and LEU2 markers remains integrated at the former *ARG4* and *LYS1* loci. Following marker rescue and selection for pSH47 loss, the ORF of interest is tagged via homologous recombination using a TEV-PA-kanMX6 insertion cassette amplified with appropriate primers using the plasmid pYM8 as template.

3.1.1 Yeast Cultivation for Genetic Manipulation: General Practice

1. Sterilize all media by autoclaving.
2. For the generation of competent yeast cells, inoculate 50 ml of YPD medium with cells from fresh agar plates and incubate overnight at 30 °C and with vigorous shaking.
3. Plate freshly transformed cells on appropriate agar plates and incubate for 2–3 days at 30 °C.
4. For selection of positive transformants, streak out single colonies on appropriate agar plates and incubate overnight at 30 °C.

3.1.2 Gene Disruption of ARG4 and LYS1

1. Amplify the loxP-flanked *S. pombe* his5+ marker cassette with the primer pair Arg4_KO_fwd/rev using the plasmid pUG27 as template as well as the loxP-flanked *K. lactis* LEU2 cassette using the plasmid pUG27 and the primer pair Leu1_KO_fwd/rev by PCR (*see* **Note 20**).
2. Co-transform competent yeast cells with both PCR products (*see* **Note 6**).
3. Select for double transformants on SC-his/-leu plates (*see* **Notes 21** and **22**).
4. Streak out single colonies on YPD, SC-his, SC-leu, SC-arg, SC-lys, and YPG plates.
5. Select clones that are able to grow on SC-his, SC-leu, and YPG plates (*see* **Note 23**), but unable to grow on SC-arg and SC-lys plates.
6. Confirm the correct integration of the his5+ marker cassette into the *ARG4* locus by direct PCR (*see* **Note 24**) of clones grown on the YPD plate (*see* **Note 25**) using the primer pair Arg4_A/D. Use the wild type as control (genomic DNA or a clone grown on YPD). Expected product sizes are 2.197 bp for *Δarg4::loxP-HisMX-loxP* and 2.132 bp for *ARG4* wild type (*see* **Note 26**).

7. Confirm the correct integration of the LEU2 marker cassette into the *LYS1* locus by direct PCR of clones grown on the YPD plate using the primer pair Lys1_A/D. Expected product sizes are 3.096 bp *for Δlys1::loxP-KlLEU2-loxP* and 1.786 bp for *LYS1* wild type.

3.1.3 Marker Rescue by Cre/loxP Recombination

1. Transform competent *Δarg4/Δlys1* cells with plasmid pSH47 expressing the Cre recombinase.
2. Select for transformants on SC-ura plates (*see* **Note 27**).
3. Culture cells of a positive colony in 2 ml of PMGal medium for 3 h at 30 °C with shaking to induce expression of the Cre recombinase.
4. Plate 200 μl of a 1:100 and a 1:10^3 dilution in PMGal medium on YPD plates.
5. Streak out single colonies on YPD, SC-his, and SC-leu plates.
6. Select clones that are no longer able to grow on SC-his and SC-leu.
7. Confirm successful Cre/loxP recombination by direct PCR of clones grown on the YPD plate using the primer pairs Arg4_A/D and Lys1_A/D, respectively. Expected product sizes are 848 bp for *Δarg4::loxP* and 823 bp for *Δlys1::loxP*.
8. Culture cells of a *Δarg4::loxP/Δlys1::loxP* positive colony in 2 ml of YPD medium for 3 h at 30 °C with shaking.
9. Plate 200 μl of a 1:100 and a 1:10^3 dilution in YPD medium on YPD plates.
10. Streak out single colonies on YPD and SC-ura plates to identify uracil auxotrophic colonies that have lost the vector pSH47.
11. Assess the incorporation of stable isotope-coded amino acids into the proteome of the newly generated yeast strain as well as the degree of arginine-to-proline conversion as described in **Note 28**.

3.1.4 Genomic Epitope Tagging of a Gene of Interest with TEV-PA

1. Amplify the TEV-PA-kanMX6 insertion cassette using the plasmid pYM8 as template and the primer pair ORF_TPA_fwd/rev by PCR.
2. Transform competent *Δarg4::loxP/Δlys1::loxP* yeast cells with the PCR product.
3. Select for transformants on YPD plates containing G418.
4. Streak out single G418-resistant colonies on YPD, YPD containing G418, and YPG plates.
5. Select clones that are resistant to G418 and able to grow on YPG.
6. Confirm the correct integration into the target locus by direct PCR of clones grown on the YPD plate using the primer pair ORF_C/D (*see* **Note 29**).

7. Sequence the PCR product using the primer ORF_C to ensure that no mutations in the sequence of the tag occurred and that the tag is fused in-frame with the target gene (*see* **Note 30**).
8. If appropriate assays are available, test for the functionality of the fusion protein (*see* **Note 31**).

3.2 Characterization of Membrane Protein Complexes by SILAC AP-MS

In this part of the chapter, we describe a generic protocol for a comprehensive and most accurate characterization of membrane protein complexes. It relies on affinity purification of proteins from crude membrane preparations of differentially labeled yeast populations and can be applied to virtually any membrane protein of choice. Protein complexes are purified (1) after mixing (AP-AM, affinity purification after mixing) as well as (2) prior to mixing (AP-PM) (Fig. 2a). This dual-track strategy allows for defining core components of the complex and transiently associated interaction partners (Fig. 2b)—information that may be of crucial significance for the functional role of the protein complex as well as individual proteins of the complex.

Before starting such an extensive SILAC AP-MS study, however, it is highly recommended to optimize the conditions for the affinity purification in label-free experiments.

3.2.1 Cultivation of Yeast for SILAC AP-MS

1. Prepare light and heavy SILAC medium; sterilize by autoclaving.
2. Plate the arginine and lysine auxotrophic POI-tagged and control strains on separate YPD plates and incubate them for 2–3 days at 30 °C.
3. Per liter of main culture, inoculate 10 ml of SC medium in 100-ml Erlenmeyer flasks with cells from fresh YPD plates and incubate cultures for approx. 8 h at 30 °C and 160 rpm (first starter culture) (*see* **Note 32**). Grow POI-tagged cells in heavy and control cells in light SC medium. Reverse label when performing replicates (*see* **Note 33**).
4. Transfer first starter cultures into separate 1-l Erlenmeyer flasks and dilute them with 190 ml of light or heavy SC medium (second starter culture). Incubate the cultures overnight at 30 °C and 160 rpm.
5. For the main culture, 1 l of light or heavy SC medium in 5-l Erlenmeyer is inoculated with cells of the second starter cultures; adjust the OD_{600} to 0.1. Cultures are incubated for approx. 8 h under the same conditions.
6. Change medium by collecting the cells (centrifugation for 8 min at 7,000 × *g*) and carefully resuspending them in fresh light or heavy medium (*see* **Note 34**). Incubate the cultures further overnight as described.

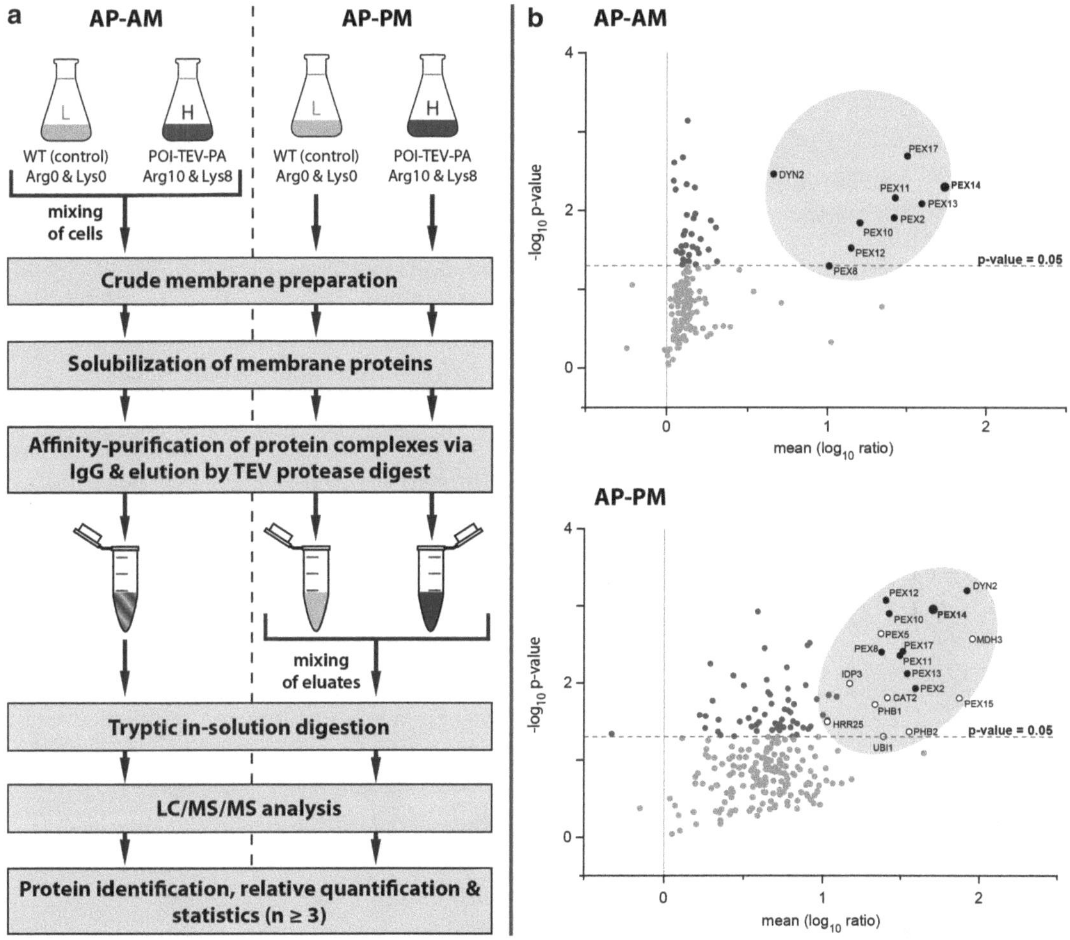

Fig. 2 In-depth characterization of membrane protein complexes from yeast following a dual-track SILAC approach. (**a**) Experimental design. Yeast cells auxotrophic for lysine and arginine expressing either the wild type or the TEV-Protein A-tagged version of the protein of interest (POI-TEV-PA) are grown in medium containing normal arginine and lysine (Arg0, Lys0) or the respective $^{13}C^{15}N$-labeled variants (Arg10, Lys8). Membrane protein complexes are affinity-purified from solubilized crude membrane fractions using IgG Sepharose either after mixing of differentially labeled cells (affinity-purification after mixing, AP-AM) or prior to mixing (AP-PM). Protein complexes are eluted from the matrix by TEV protease digestion, tryptically digested in solution and analyzed by nano HPLC/ESI-MS/MS for peptide and protein identification as well as SILAC-based relative protein quantification. For statistically sound data, at least 3 independent replicates should be performed. (**b**) Classification of stable core components of a protein complex and transient interaction partners. Complexes of Pex14p, a central component of the peroxisomal matrix protein import machinery, were purified and analyzed following the dual-track strategy described in A [24]. Interaction data are visualized by plotting the mean $\log_{10}$ ratios of proteins quantified in at least two out of three independent replicates against their *p*-values ($-\log_{10}$) determined in AP-AM and AP-PM experiments, respectively. Integration of both datasets allows for discriminating between the Pex14p core complex (*filled black circles*) and proteins transiently associated with the complex (*open circles*): while stable core components are identified as specific interaction partners in both AP-AM and -PM experiments, transient interaction partners show high abundance ratios in AP-PM experiments only. *Dark grey circles*, further proteins with a *p*-value <0.05; *light grey circles*, proteins with a *p*-value >0.05

7. Harvest the cells by centrifugation (7,000 × *g* for 8 min), resuspend them in deionized water and combine cells of equal strains and equal SILAC labeling.
8. Wash the cells twice with ionized water.
9. Determine the wet weight of the cells.

3.2.2 Cell Lysis and Affinity Purification of Membrane Protein Complexes

1. For *AP-AM* experiments, mix equal amounts of differentially SILAC-labeled POI-tagged and control cells (based on wet cell weight) immediately after harvesting. For *AP-PM* experiments, carry out cell lysis and purification steps described in the following separately for differentially labeled cells.
2. Resuspend cells in 2 ml of lysis buffer per g wet weight and transfer suspension to 50-ml tubes with a maximum of 20–25 ml (corresponding to approx. 7–8 g cells) per tube.
3. To each tube, add 3 ml of glass beads per g wet cell weight.
4. Vortex 12× 1 min with at least 1 min cooling on ice in between to mechanically disrupt the cells.
5. Pellet glass beads and unbroken cells by centrifugation (10 min at 2,000 × *g* and 4 °C).
6. Transfer supernatants (i.e., the homogenate) to ultracentrifugation tubes and obtain membrane fractions by ultracentrifugation (60 min at 100,000 × *g* and 4 °C) (*see* **Note 35**).
7. Remove the supernatants (i.e., the cytosolic fraction) and resuspend the pellets (i.e., the membrane fraction) on ice in lysis buffer (*see* **Note 36**).
8. Determine the protein concentration using, for example, the Bradford assay with BSA (0.25–2.0 mg/ml) as standard.
9. For the solubilization of membrane proteins, adjust the protein concentration of the membrane fractions to 3.35 mg/ml using lysis buffer, glycerol, and 10 % (w/v) digitonin to reach a final concentration of 10 % (v/v) glycerol and 1 % (w/v) digitonin (*see* **Note 37**). When performing *AP-PM* experiments, continue with equal total protein amounts obtained from POI-tagged and control cells.
10. Incubate the samples for 90 min at 4 °C with slight agitation, e.g., on a rotating wheel.
11. Separate solubilized membrane proteins from insolubilized material by ultracentrifugation (60 min at 100,000 × *g* and 4 °C).
12. Add 37.5 μl of IgG-coupled Sepharose per 50 mg of protein (as determined before solubilization of membrane proteins) and incubate overnight at 4 °C with slight agitation. Wash IgG-Sepharose beads twice with lysis buffer before use.

13. Spin down IgG-Sepharose beads and proteins bound to it by centrifugation (5 min at 100 × *g* and 4 °C) (*see* **Note 38**) and remove the supernatant (i.e., flow-through).
14. Transfer the beads to MobiCols and wash with a total of 75–100 bed volumes of washing buffer by repeated resuspension of the beads in small volumes of washing buffer and short centrifugation steps (30 s at 100 × *g* and 4 °C). Perform the last washing steps with *elution buffer* to remove the inhibitors.
15. To detach the protein complexes from the IgG-Sepharose, resuspend the beads in two bed volumes of elution buffer, add TEV protease (100 units per 100 μl of IgG-Sepharose) and incubate for 2 h at 1,100 rpm and 16 °C on a thermomixer.
16. To remove the TEV protease via its N-terminal polyhistidine tag, add Ni-NTA agarose (50 μl per 100 μl of IgG-Sepharose) equilibrated with elution buffer and incubate for an additional 30 min under the same conditions (*see* **Note 39**).
17. Collect the eluates by centrifugation (2 min at 100 × *g* and 4 °C).
18. To improve the yield, wash beads twice with two bed volumes of elution buffer. Incubate for 10 min at 1,100 rpm and 16 °C before collecting the eluates by centrifugation.
19. Combine eluates of POI-tagged and control cells when purifying and analyzing protein complexes following the *AP-PM* strategy.

3.2.3 Tryptic In-Solution Digestion and Sample Preparation for LC-MS Analysis

1. Add four volumes of ice-cold 100 % acetone to the eluates and incubate for at least 2 h at −20 °C to precipitate the proteins.
2. Spin down the precipitated proteins at 16,000 × *g* for 10 min. Remove the supernatant and dry the pellet at room temperature.
3. Resuspend the pellet in 10 μl of urea buffer and transfer the solution to a microcentrifuge tube.
4. Add TCEP to a final concentration of 10 mM and incubate for 30 min at 37 °C to reduce disulfide bonds.
5. Add iodoacetamide to a final concentration of 50 mM and incubate for 30 min at room temperature in the dark to alkylate free thiol groups.
6. Add DTT to a final concentration of 20 mM to stop the alkylation reaction.
7. For the protein digest, add 375 ng of trypsin in 25 μl of 20 mM NH_4HCO_3 and incubate overnight at 37 °C.
8. Remove the liquid in vacuo using a SpeedVac and reconstitute the peptides in 40 μl of 0.1 % (v/v) TFA.

9. Spin down insolubilized material (5 min at 16,000 × *g*).
10. Transfer the supernatant containing the peptide mixture to a glass insert.

3.2.4 LC-MS/MS and Data Analysis

1. Assemble glass vials, inserts containing the peptides, caps, and septa and load the peptide mixture onto the C18 μ-precolumn. Preconcentrate and wash peptides with 0.1 % (v/v) TFA for 15 min at a flow rate of 30 μl/min.
2. Switch precolumn in line with the RP nano LC column and elute peptides with a linear gradient ranging from 5 to 42 % solvent B in 150 min and 40–95 % B in 5 min at a flow rate of 300 nl/min (*see* **Note 40**). Wash the column with 95 % solvent B for further 5 min and re-equilibrate with 5 % solvent B for 15 min.
3. Acquire high resolution MS/MS data in data-dependent mode enabling fragmentation of ≥6 of the most intense peptide ions per duty cycle by collision induced dissociation.
4. Use general mass spectrometric settings optimized for the MS instrument employed.
5. Download the MaxQuant software package (*see* **Note 41**).
6. Use MaxQuant default settings (referring to version 1.3.0.5) with the following exceptions: minimum of 1 unique peptide for protein identification; use unique peptides only and a minimum ratio count of 1 for peptide and protein quantification; enable "Match between runs" for biological replicates and "Second peptides" (*see* **Note 42**).
7. Specific interaction partners exhibit protein abundance ratios significantly higher than one while co-purified contaminants derived from both POI-tagged and control cells exhibit abundance ratios of approx. 1. To determine candidate interaction partners based on multiple biological replicates, it is recommended to analyze the data using statistics. For example, calculate the mean $\log_{10}$ protein ratios as well as the *p*-value (*t*-test) for each protein and visualize the data in a 2D plot (Fig. 2b). Proteins with a *p*-value of ≤0.05 and a protein ratio higher than a distinct threshold value are then classified as specific interaction partners (*see* **Notes 43–45**).
8. Integration of data obtained in AP-AM and AP-PM experiments allows for the definition of stable core components and transient interaction partners (Fig. 2b). Core components are classified as specific interaction partners in both AP-AM and AP-PM datasets. Transiently associated proteins, however, typically exhibit high abundance ratios only in AP-PM experiments. Due to exchange of labeled and unlabeled proteins attached to the bait during the purification process employing the AP-AM strategy, transient interaction partners show abundance ratios of approx. 1 leading to misinterpretation as contaminants (*see* **Note 46**).

4 Notes

1. The protocol described here for disruption of *ARG4* and *LYS1* is based on the use of a yeast strain auxotrophic for histidine, leucine, and uracil. However, it can also be applied to strains with deviating auxotrophies. In this case, different plasmids need to be used for gene disruption and marker rescue.
2. In case the strain of choice is neither his⁻ nor leu⁻, use either plasmid pUG6 with a geneticin resistance marker, pUG66 with a phleomycin resistance marker, or pUG72 with the *K. lactis* URA3 marker (Table 1A).
3. If the strain to be used is not ura⁻ or when choosing the URA3 marker for *ARG4* and *LYS1* gene disruption, pSH47 can be substituted by pSH63 with the TRP1 marker or pSH65 with a phleomycin resistance marker (Table 1C).
4. The oligonucleotides consist of two parts: a ~46 bp sequence homologous to either the 5′ or the 3′ end of the coding region of *ARG4* or *LYS1* and a sequence that can anneal to both pUG27 and pUG73. When using different plasmids for gene disruption, the sequences of the oligonucleotides need to be modified accordingly.
5. Sequences of the primers for the confirmation of each ORF are provided on the Web site of the yeast deletion project [4]: http://www-sequence.stanford.edu/group/yeast_deletion_project/downloads.html.
6. Yeast cells can be transformed by heat-shock [35] or by electroporation [36]. Transformation by electroporation is more efficient but requires an electroporator and electroporation cuvettes.
7. Store DO mixes at 4 °C in glass flasks sealed with Parafilm to prevent hydration; liquid media can be stored at room temperature and agar plates at 4 °C.
8. The protocol described here for the SILAC AP-MS-based characterization of membrane protein complexes can be applied to any yeast strain auxotrophic for lysine and arginine. However, we recommend using POI-tagged and control strains with otherwise identical genetic background to exclude possible effects resulting from different genotypes.
9. Stable isotope-coded amino acids can be obtained from Cambridge Isotope Laboratories (Andover, MA, USA), Eurisotop (Saarbrücken, Germany), Sigma-Aldrich, or Silantes (Munich, Germany).
10. In this chapter, we describe a generic protocol for the growth of yeast cultures. Depending on the requirements and nature of the POI, however, the protocol as well as the composition

of the medium may be adjusted. Components that can be used as amino acid source for the cells, such as yeast extract, must be omitted. Cultivate cells for at least five cell doublings to ensure complete incorporation of the heavy amino acids into the proteome.

11. It may be necessary to add unlabeled proline to the medium if conversion of heavy arginine (Arg10) to heavy proline (Pro6) is observed; the amount needs to be determined experimentally.
12. Prepare inhibitors according to the manufacturer's instruction. Stock solutions (100–1,000×) can be stored at −20 °C.
13. Digitonin is a mild, non-ionic detergent frequently used for the solubilization of membrane proteins and the purification of native protein complexes. However, depending on the experimental requirements and the properties of the POI, other detergents may be better suited for the purification. Therefore, it is recommended to test different detergents and to determine the concentration yielding best results before initiating a SILAC AP-MS study. It is also recommended to test the compatibility of the detergent selected with the entire protocol (TEV protease and in-solution trypsin digest; LC/MS analysis).
14. Protease inhibitors may inhibit TEV protease as well as tryptic protein digest. In case addition of protease and/or other inhibitors (e.g., phosphatase inhibitors) to the elution buffer is required, make sure they do not impair the activity of the TEV protease and trypsin.
15. Human IgG-coupled Sepharose beads are commercially available (e.g., from GE Healthcare) or can be prepared as described [37].
16. For sample preparation and subsequent LC-MS analysis, use only HPLC grade solvents and reagents as well as water of Milli-Q purity.
17. The protocol described in this chapter including the analysis of mass spectrometric raw data relies on the acquisition of high resolution MS data and the use of Orbitrap instruments.
18. The MaxQuant software suite including the integrated search engine Andromeda is freely available. MaxQuant supports data acquired with XCalibur, a software specific for instruments from Thermo Scientific such as the Orbitrap family.
19. An *S. cerevisiae* protein sequence database is provided with the MaxQuant software suite or can be downloaded from the Web site www.yeastgenome.org/download-data (*Saccharomyces* Genome Database).
20. For PCR conditions, follow the guidelines specified for the DNA polymerase of choice. When using the Phusion High-Fildelity DNA polymerase, perform PCRs using 0.4 units of

the enzyme, 0.2 mM of each dNTP, 0.5 μM of each primer, and 1–10 ng of template DNA in HF buffer in a final volume of 20 μl. Apply the following thermocycling conditions: 30 s initial denaturation at 98 °C; 35 cycles of: 10 s at 98 °C, 30 s at annealing temperature (T_a), and 20–30 s per kilobase of PCR product at 72 °C; final extension: 10 min at 72 °C; cooling down to 4 °C. When long chimeric oligonucleotides are used, only the part homologous to the plasmid anneals in the first round(s) of PCR. It is therefore recommended to use a T_a of 55 °C for the first five and a T_a of 60 °C for the remaining 30 cycles. Check PCR products by agarose gel electrophoresis. Expected sizes are 1.5 kb for pUG27, 2.5 kb for pUG73, and 2.2 kb for pYM8.

21. The efficiency of co-transforming yeast cells with both PCR products may be low and may result in yeast cells with deficiency in the biosynthesis of arginine or lysine only. We therefore recommend selecting additionally for single transformants as well (*Δarg4* on SC-his and *Δlys1* on SC-leu plates). In case co-transformation results in single transformants only, perform a second transformation to disrupt *LYS1* in *Δarg4* cells or *ARG4* in *Δlys1* cells.
22. In case plasmids carrying marker cassettes other than HIS and LEU are used for gene disruption, the selection media for transformants needs to be modified accordingly.
23. According to our experience, heat-shock transformation may lead to cells with respiration deficiency. We therefore test the growth of single colonies on a medium containing a non-fermentable carbon source such as YPG and select for respiration competence.
24. For direct PCR, resuspend yeast cells in 50 μl Zymolyase 20T solution (60 units/ml), incubate for 30 min at 37 °C, heat for 10 min at 95 °C, spin down cell debris, and use 1 μl of the supernatant for the PCR. Perform the PCR as described in **Note 20**.
25. In our experience, direct PCR works best using a clone grown on YPD.
26. Since the PCR product for *Δarg4::loxP-HisMX-loxP* is only 65 bp longer than the *ARG4* PCR product, it is crucial to perform a control PCR of wild type cells to allow for precisely assessing the correct integration of the marker cassette.
27. If a plasmid other than pSH47 with a different marker cassette is used for marker rescue, the selection medium for transformants needs to be adjusted accordingly.
28. To evaluate the labeling efficiency and assess potential arginine-to-proline conversion, grow yeast in heavy arginine- and lysine-containing medium according to the protocol described in

Subheading 3.2.1; a culture volume of 50–100 ml is sufficient. Harvest cells, extract proteins, subject an aliquot of the protein extract (~2 μg) to tryptic in-solution digest (*see* Subheading 3.2.3) and analyze the peptide mixture by LC-MS (*see* Subheading 3.2.4). Process mass spectrometric raw data with MaxQuant as described in 3.2.4 with the following modifications: include heavy proline (Pro6) as variable modification and disable the option "re-quantify." Use the data listed in the evidence.txt result file to determine the extent to which heavy arginine and lysine are incorporated into the proteins as well as the degree of arginine-to-proline conversion. Remove the hits derived from the reverse and contaminant database.

The incorporation of the heavy amino acids is calculated as follows:

$$\text{Incorporation}\,(\%) = \frac{\text{ratio}\,(H\,/\,L)}{\text{ratio}\,(H\,/\,L)+1}\times 100$$

Peptides for which a ratio H/L was not calculated by MaxQuant show either complete or no incorporation of heavy amino acids. These peptides need to be included into the calculation as well: in case of complete incorporation (MS Intensity $L=0$; Intensity $H>0$), set the incorporation to 100 %; in case of no incorporation (Intensity $L>0$; Intensity $H=0$), set the incorporation to 0 %. Calculate the mean incorporation across all peptides. The incorporation efficiency is limited by the purity of the SILAC amino acids used (typically ≥98 %). To obtain accurate relative quantitative data, the incorporation should be at least 95 %.

To assess the degree of arginine-to-proline conversion, calculate the percentage of Pro6-containing peptides among all identified peptides. To minimize arginine-to-proline conversion (<5 %), the amount of heavy arginine in the medium may be reduced and/or unlabeled proline may be added. However, the presence of a large excess of unlabeled proline in the medium may lead to the reverse metabolic conversion of proline to unlabeled arginine and may also affect other metabolic pathways and cellular processes. Thus, the amounts of heavy arginine and unlabeled proline need to be titrated carefully. Further methods suited to counteract arginine-to-proline conversion are described in the introduction of this chapter.

29. The expected product size for ORF:TEV-PA-kanMX6 is calculated as follows: the size of the expected PCR product of the wild type gene plus 2,163 bp of the TEV-PA-kanMX6 insert minus 3 bp (stop codon) minus the distance of the part of the ORF_TPA_rev primer sequence homologous to the ORF from the stop codon (i.e., 0–200 bp). If the sequences of the primers used for confirmation of the ORF were derived from the

Web site of the yeast deletion project, the size of the PCR product of the wild type gene can be found there as well [4]: http://www-sequence.stanford.edu/group/yeast_deletion_project/downloads.html.

30. We further recommend confirming correct expression and subcellular localization of the POI-TEV-PA fusion protein by western blot analysis using an antibody against the Protein A-tag. If the C-terminal TEV-PA tag compromises correct expression and subcellular localization or affects the function of the POI (*see* **Note 31**), the tag needs to be attached to the N-terminus of the POI. However, N-terminal tagging requires a cloning strategy different from the one described in this chapter.
31. For example, if deletion of the gene coding for the POI shows a growth defect under specific conditions, check whether this phenotype is rescued in the strain expressing the fusion protein.
32. The culture volume needed to purify sufficient amounts of a given protein complex depends on the expression level of the POI as well as its susceptibility to membrane extraction and should be determined before starting with a SILAC experiment.
33. For interaction data of high reliability, we strongly recommend to perform at least three independent biological replicates. We further suggest including a label-swap experiment to avoid potential artifacts arising from the labeling strategy.
34. To avoid induction of stress caused by starvation, it is important to supply sufficient amounts of glucose during cultivation by frequently adding fresh medium.
35. We recommend controlling the efficiency of the protein purification by taking an aliquot of the sample at each purification step and analyzing the samples by immunoblotting with an antibody against the POI or the Protein A tag.
36. For efficient resuspension of the pellet, use a Dounce homogenizer.
37. The detergent-to-protein ratio is critical for effective solubilization of membrane protein complexes and needs to be determined for individually for each protein and detergent by varying both the detergent and the protein concentration.
38. If possible, reduce deceleration force of the centrifuge to prevent the sedimented Sepharose beads from whirling up.
39. The presence of large amounts of TEV protease in the eluate significantly compromises quantitative MS analysis of the yeast proteins and, thus, identification of specific protein interaction partners when performing protein in-solution digest. We therefore recommend binding of the TEV protease to Ni-NTA agarose before collecting the eluate. However, it should be

tested first if the POI exhibits a strong binding to Ni-NTA agarose, too. In this case, we recommend performing SDS-PAGE of the eluates to separate the proteins followed by proteolytic in-gel digestion.

40. The gradient (slope, duration) may need to be adjusted to the RP nano LC column used to identify maximum numbers of peptides and proteins.
41. Information about the computational requirements for successfully setting up MaxQuant are provided under http://maxquant.org/requirements.htm.
42. For more details about the use and applicability of MaxQuant, refer to Chapter 24 of this edition. A further source for valuable information and recommendations as well as for troubleshooting is the MaxQuant Google group (https://groups.google.com/forum/#!forum/maxquant-list).
43. MaxQuant reports normalized and non-normalized data for the protein abundance ratios [33]. The decision which kind of ratio to use for downstream data analysis depends on the nature of the individual dataset. If the data show a normal (or Gaussian) distribution, the normalized ratios are to be used. However, interaction proteomics data may deviate from a Gaussian distribution due to an enrichment of a large number of specific interaction partners and/or the absence of a large pool of co-purified contaminants. In these cases, the non-normalized protein ratios should be used for further data analysis.
44. Quantitative data derived from AP-PM experiments often exhibit a wider distribution due to higher experimental variations (Fig. 2b). As a consequence, the significance threshold for classifying proteins as specific interaction partners can be expected to be higher in such experiments.
45. Threshold values can be calculated using statistical tools, e.g., outlier analysis by boxplots [24], significance B for data showing a normal distribution [33], or determination of a significance line corresponding to a defined FDR for interaction partners [38].
46. Note that the SILAC AP-MS approach does not allow for discriminating between direct and indirect interactions.

Acknowledgments

This work was supported by the Deutsche Forschungsgemeinschaft, Forschergruppe 1905 and the Excellence Initiative of the German Federal and State Governments (EXC 294 BIOSS).

References

1. Goffeau A, Barrell BG, Bussey H et al (1996) Life with 6000 genes. Science 274:563–567
2. Kolkman A, Slijper M, Heck AJ (2005) Development and application of proteomics technologies in Saccharomyces cerevisiae. Trends Biotechnol 23:598–604
3. Botstein D, Fink GR (2011) Yeast, an experimental organism for 21st century biology. Genetics 189:695–704
4. Giaever G, Chu AM, Ni L et al (2002) Functional profiling of the Saccharomyces cerevisiae genome. Nature 418:387–391
5. Huh WK, Falvo JV, Gerke LC et al (2003) Global analysis of protein localization in budding yeast. Nature 425:686–691
6. Gavin AC, Bosche M, Krause R et al (2002) Functional organization of the yeast proteome by systematic analysis of protein complexes. Nature 415:141–147
7. Ghaemmaghami S, Huh WK, Bower K et al (2003) Global analysis of protein expression in yeast. Nature 425:737–741
8. Ho Y, Gruhler A, Heilbut A et al (2002) Systematic identification of protein complexes in Saccharomyces cerevisiae by mass spectrometry. Nature 415:180–183
9. Gavin AC, Aloy P, Grandi P et al (2006) Proteome survey reveals modularity of the yeast cell machinery. Nature 440:631–636
10. Krogan NJ, Cagney G, Yu H et al (2006) Global landscape of protein complexes in the yeast Saccharomyces cerevisiae. Nature 440:637–643
11. Oeljeklaus S, Meyer HE, Warscheid B (2009) New dimensions in the study of protein complexes using quantitative mass spectrometry. FEBS Lett 583:1674–1683
12. Jiang H, English AM (2002) Quantitative analysis of the yeast proteome by incorporation of isotopically labeled leucine. J Proteome Res 1:345–350
13. Pratt JM, Robertson DH, Gaskell SJ et al (2002) Stable isotope labelling in vivo as an aid to protein identification in peptide mass fingerprinting. Proteomics 2:157–163
14. Ong SE, Blagoev B, Kratchmarova I et al (2002) Stable isotope labeling by amino acids in cell culture, SILAC, as a simple and accurate approach to expression proteomics. Mol Cell Proteomics 1:376–386
15. Ong SE, Kratchmarova I, Mann M (2003) Properties of 13C-substituted arginine in stable isotope labeling by amino acids in cell culture (SILAC). J Proteome Res 2:173–181
16. Gruhler A, Olsen JV, Mohammed S et al (2005) Quantitative phosphoproteomics applied to the yeast pheromone signaling pathway. Mol Cell Proteomics 4:310–327
17. Bendall SC, Hughes C, Stewart MH et al (2008) Prevention of amino acid conversion in SILAC experiments with embryonic stem cells. Mol Cell Proteomics 7:1587–1597
18. Bicho CC, de Lima Alves F, Chen ZA et al (2010) A genetic engineering solution to the "arginine conversion problem" in stable isotope labeling by amino acids in cell culture (SILAC). Mol Cell Proteomics 9:1567–1577
19. Ong SE, Mann M (2005) Mass spectrometry-based proteomics turns quantitative. Nat Chem Biol 1:252–262
20. Van Hoof D, Pinkse MW, Oostwaard DW et al (2007) An experimental correction for arginine-to-proline conversion artifacts in SILAC-based quantitative proteomics. Nat Methods 4:677–678
21. Wang X, Huang L (2008) Identifying dynamic interactors of protein complexes by quantitative mass spectrometry. Mol Cell Proteomics 7: 46–57
22. Fang L, Wang X, Yamoah K et al (2008) Characterization of the human COP9 signalosome complex using affinity purification and mass spectrometry. J Proteome Res 7:4914–4925
23. Mousson F, Kolkman A, Pijnappel WW et al (2008) Quantitative proteomics reveals regulation of dynamic components within TATA-binding protein (TBP) transcription complexes. Mol Cell Proteomics 7:845–852
24. Oeljeklaus S, Reinartz BS, Wolf J et al (2012) Identification of core components and transient interactors of the peroxisomal importomer by dual-track stable isotope labeling with amino acids in cell culture analysis. J Proteome Res 11:2567–2580
25. David C, Koch J, Oeljeklaus S et al (2013) A combined approach of quantitative interaction proteomics and live-cell imaging reveals a regulatory role for ER reticulon homology proteins in peroxisome biogenesis. Mol Cell Proteomics 12:2408–2425
26. Gebert N, Gebert M, Oeljeklaus S et al (2011) Dual function of Sdh3 in the respiratory chain and TIM22 protein translocase of the mitochondrial inner membrane. Mol Cell 44: 811–818
27. Gebert M, Schrempp SG, Mehnert CS et al (2012) Mgr2 promotes coupling of the mitochondrial presequence translocase to partner complexes. J Cell Biol 197:595–604
28. Stroud DA, Oeljeklaus S, Wiese S et al (2011) Composition and topology of the endoplasmic

reticulum-mitochondria encounter structure. J Mol Biol 413:743–750

29. von der Malsburg K, Muller JM, Bohnert M et al (2011) Dual role of mitofilin in mitochondrial membrane organization and protein biogenesis. Dev Cell 21:694–707
30. Qiu J, Wenz LS, Zerbes RM et al (2013) Coupling of mitochondrial import and export translocases by receptor-mediated supercomplex formation. Cell 154:596–608
31. Spencer F, Ketner G, Connelly C et al (1993) Targeted recombination-based cloning and manipulation of large DNA segments in yeast. Methods 5:161–175
32. Bradford MM (1976) A rapid and sensitive method for the quantitation of microgram quantities of protein utilizing the principle of protein-dye binding. Anal Biochem 72: 248–254
33. Cox J, Mann M (2008) MaxQuant enables high peptide identification rates, individualized p.p.b.-range mass accuracies and proteome-wide protein quantification. Nat Biotechnol 26:1367–1372
34. Cox J, Neuhauser N, Michalski A et al (2011) Andromeda, a peptide search engine integrated into the MaxQuant environment. J Proteome Res 10:1794–1805
35. Knop M, Siegers K, Pereira G et al (1999) Epitope tagging of yeast genes using a PCR-based strategy, more tags and improved practical routines. Yeast 15:963–972
36. Thompson JR, Register E, Curotto J et al (1998) An improved protocol for the preparation of yeast cells for transformation by electroporation. Yeast 14:565–571
37. Piechura H, Oeljeklaus S, Warscheid B (2012) SILAC for the study of mammalian cell lines and yeast protein complexes. Methods Mol Biol 893:201–221
38. Hubner NC, Bird AW, Cox J et al (2010) Quantitative proteomics combined with BAC TransgeneOmics reveals in vivo protein interactions. J Cell Biol 189:739–754
39. Gueldener U, Heinisch J, Koehler GJ et al (2002) A second set of loxP marker cassettes for Cre-mediated multiple gene knockouts in budding yeast. Nucleic Acids Res 30:e23
40. Gueldener U, Heck S, Fielder T et al (1996) A new efficient gene disruption cassette for repeated use in budding yeast. Nucleic Acids Res 24:2519–2524

Chapter 4

Whole Proteome Analysis of the Protozoan Parasite *Trypanosoma brucei* Using Stable Isotope Labeling by Amino Acids in Cell Culture and Mass Spectrometry

Olivera Cirovic and Torsten Ochsenreiter

Abstract

The single-celled protozoan *Trypanosoma brucei spp.* is the causative agent of human African trypanosomiasis and nagana in cattle. Quantitative proteomics for the first time has allowed for the characterization of the proteome from several different life stages of the parasite (Butter et al., Mol Cell Proteomics 12:172–179, 2013; Gunasekera et al., BMC Genomics 13:556, 2012; Urbaniak et al., PloS One 7(5):e36619, 2012). To achieve this, stable isotope labeling by amino acids in cell culture (SILAC) (Ong et al., Mol Cell Proteomics 1:376–386, 2002) was adapted to *T. brucei spp.* cultures. *T. brucei* cells grown in standard media with dialyzed fetal calf serum containing heavy isotope-labeled amino acids (arginine and lysine) show efficient incorporation of the labeled amino acids into the whole cell proteome (8–12 divisions) and no detectable amino acid conversions. The method can be applied to both of the major life stages of the parasite and in combination with RNAi or gene knockout approaches.

Key words *Trypanosoma brucei*, SILAC, Mass spectrometry, Cell culture, Cell differentiation

1 Introduction

T. brucei spp. belongs to the group of Kinetoplastidae that, despite many peculiarities, has been instrumental in the discovery and understanding of many basic biological principles including RNA editing, trans-splicing, GPI anchoring, and immune evasion [5–9]. *T. brucei spp.* has a digenetic life cycle that alternates between an insect and a mammalian host. Gaining insights into the protein dynamics of the parasite during the cell differentiation that is necessary for the transition between the hosts is of great interest in parasitology and cell biology. Quantitative proteomics using SILAC in *T. brucei* is based on the principles described previously [10, 11]. For the comparison of two *T. brucei* proteomes, the cells are grown under identical conditions, except that one culture medium contains regular amino acids and the second heavy isotope-labeled lysine and arginine ($^{13}C_6$, 99 %; Fig. 1).

Bettina Warscheid (ed.), *Stable Isotope Labeling by Amino Acids in Cell Culture (SILAC): Methods and Protocols*, Methods in Molecular Biology, vol. 1188, DOI 10.1007/978-1-4939-1142-4_4, © Springer Science+Business Media New York 2014

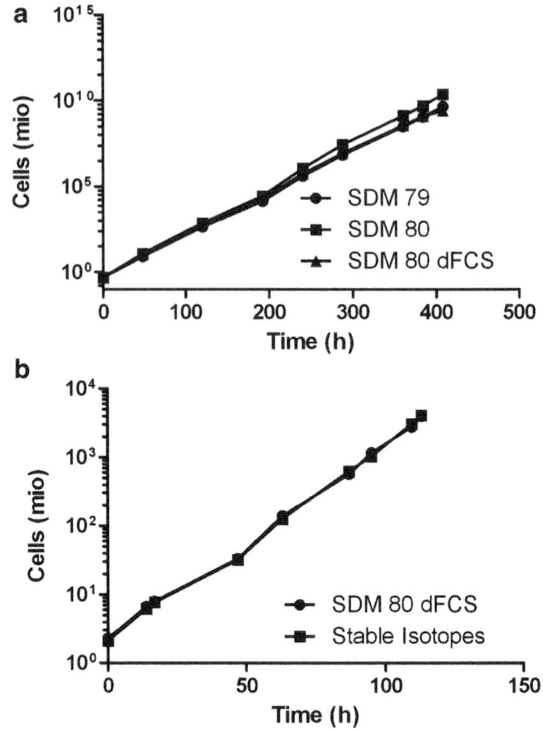

Fig. 1 Comparison of growth rate of *T. brucei* cells in different culture media (**a**). Growth of procyclic cells in SDM-79/80 with regular FCS and SDM-80 with dialyzed FCS (dFCS) (**b**). Growth of procyclic cells in SDM-80 with dialyzed FCS and regular amino acids or cells in SDM-80 with dialyzed FCS and heavy isotope-labeled amino acids (arginine, lysine) (Figure from Gunasekera et al. [2] with permission)

After incorporation (>98 %) of the heavy isotope-labeled amino acids into the total cellular proteome (Fig. 2), the two cell populations are mixed, cells are lysed, and proteins are either fractionated on SDS-PAGE or directly digested using trypsin. Separation of the proteins on SDS-PAGE is a simple and efficient method to decrease the complexity of the samples to be analyzed by mass spectrometry (MS). In general, fractionation by SDS-PAGE leads to higher coverage of the proteome. The use of SILAC in combination with MS is not limited to the characterization of different life stages but also allows for an unbiased approach to characterize proteome changes in RNAi-, overexpression-, or knockout cell lines.

In this chapter we describe the use of SILAC for quantitative proteomics in the two major life stages of *T. brucei*. The detailed description includes the preparation of SILAC media, the procedure for growing *T. brucei* cells in the appropriate media, and the preparation of cell lysates for SDS-PAGE fractionation.

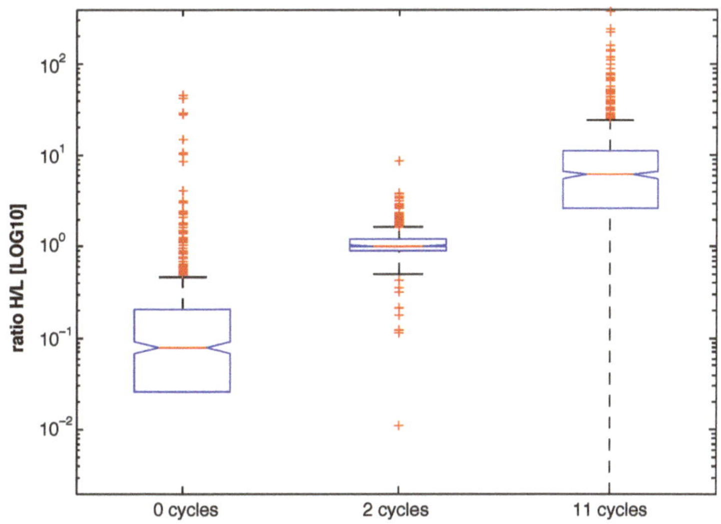

Fig. 2 Labeling efficiency of procyclic *T. brucei* whole cell proteome after 0, 2, and 11 cell division cycles. The box plots depict the mean ratios of heavy/light peptides from the whole proteome measured after 0, 2, and 11 cell doublings [Figure from Gunasekera et al. [2], with permission]

2 Materials

2.1 T. brucei Cell Culture

Trypanosomes are grown in the standard culture media before they are adapted to the SILAC media: procyclic form (PCF) in SDM-79 medium [12] and bloodstream form (BSF) in HMI-9 medium [13]. Both SDM-79 and HMI-9 media are supplemented with 10 % (v/v) of heat-inactivated non-dialyzed FCS.

Media for the SILAC experiments for both life stages of trypanosomes are prepared to contain all the components of the standard growth media except for L-lysine and L-arginine. In the "light" version of the SILAC media, regular L-lysine and L-arginine are added, whereas in the "heavy" version, heavy isotope-labeled L-lysine and L-arginine are added. Both "light" and "heavy" versions of the media are supplemented with 10 % (v/v) of heat-inactivated dialyzed fetal calf serum (FCS; 10,000 Da molecular weight cutoff).

1. SILAC medium for the PCF trypanosomes: SDM-80 medium [14] with glucose, without L-lysine and L-arginine: 1 mM NaH_2PO_4, 116 mM NaCl, 0.8 mM $MgSO_4$, 5.4 mM KCl, 1.8 mM $CaCl_2$, 26.2 mM $NaHCO_3$, 30.7 mM HEPES, 23.9 mM MOPS, 4 mM pyruvate, 1 % (v/v) MEM vitamin solution 100×, 5.2 mM proline, 5.9 mM threonine, 0.58 mM L-methionine, 0.68 mM L-phenylalanine, 0.75 mM L-tyrosine,

0.1 mM L-cysteine, 0.2 mM L-histidine, 0.4 mM L-isoleucine, 0.76 mM L-leucine, 0.05 mM L-tryptophan, 0.4 mM L-valine, 2.25 mM L-alanine, 0.1 mM L-asparagine, 0.1 mM L-aspartic acid, 0.09 mM L-glutamic acid, 0.49 mM L-serine, 0.1 mM glycine, 1.28 mM taurine, 0.46 mM glutamine, 0.2 mM β-mercaptoethanol, 0.1 mM hypoxanthine, 0.017 mM thymidine, 0.008 mM hemin, 5.55 mM D-glucose, 100,000 U/l penicillin, 100 mg/l streptomycin, 10 % heat-inactivated dialyzed FCS (*see* **Notes 1–4**).

- For the "light" version of SDM-80, add 0.4 mM lysine and 1.1 mM arginine.
- For the "heavy" version of SDM-80, add 0.4 mM $^{13}C_6$-lysine (99 %) and 1.1 mM $^{13}C_6$-arginine (99 %).

2. SILAC medium for the BSF trypanosomes (modified HMI-9 medium): IMDM medium without L-lysine and L-arginine, with 4 mM L-glutamine, 1 mM hypoxanthine, 50 μM bathocuproine disulfonate, 1.5 mM L-cysteine, 1 mM sodium pyruvate, 0.16 mM thymidine, 0.2 mM β-mercaptoethanol, 100,000 U/l penicillin, 100 mg/l streptomycin, 10 % heat-inactivated dialyzed FCS.
   - For the "light" version of the modified HMI-9 medium, add 0.798 mM L-lysine and 0.4 mM L-arginine.
   - For the "heavy" version of the modified HMI-9 medium, add 0.798 mM $^{13}C_6$l-lysine (99 %) and 0.4 mM $^{13}C_6$-L-arginine (99 %).
3. Steritop™ Filter Units (Millipore).

2.2 Protein Extraction

1. Phosphate-buffered saline (PBS): 137 mM NaCl, 2.7 mM KCl, 10 mM Na_2HPO_4, 1.8 mM KH_2PO_4 (pH 7.4).
2. 2× SDS loading sample buffer [15]: 100 mM Tris–HCl, pH 6.8, 4 % (w/v) sodium dodecyl sulfate (SDS), 20 % (v/v) glycerol, and 0.002 % (w/v) bromophenol blue. Before boiling the samples, add β-mercaptoethanol to a final concentration 10 % (v/v) or 100 mM dithiothreitol (DTT).

2.3 Sodium Dodecyl Sulfate Polyacrylamide Gel Electrophoresis (SDS-PAGE) Stock Solutions and Buffers (See Note 5)

1. Acrylamide–bis-acrylamide solution (29:1).
2. 10 % (w/v) SDS.
3. 1.5 M Tris–HCl, pH 8.8.
4. 0.5 M Tris–HCl, pH 6.8.
5. 10 % (w/v) ammonium persulfate (APS).
6. *N*,*N*,*N*′,*N*′-tetramethylethylenediamine (TEMED).
7. Running buffer (1×): 25 mM Tris-base, 192 mM glycine, 0.1 % SDS.

8. Coomassie staining solution: 0.25 % (w/v) Coomassie Brilliant Blue R250, 10 % (v/v) acetic acid, 40 % (v/v) methanol.
9. Destaining solution for Coomassie: 10 % (v/v) acetic acid, 25 % (v/v) methanol.

2.4 Excision of Protein Bands

1. Scalpel.
2. 20 % (v/v) absolute ethanol.

3 Methods

Trypanosomes are grown under identical conditions in two different versions of SILAC medium, "light" and "heavy". The cells grown in the SILAC medium containing heavy isotope-labeled amino acids are mixed in equal numbers with cells grown in the SILAC medium with normal amino acids. Cells are lysed and resolved on SDS-PAGE. Subsequently, each lane of the SDS gel containing the fractionated cell lysate is cut into 10–20 pieces. Each piece is cut into small cubes (1 × 1 mm), which then are further processed to digest the contained proteins with trypsin and eventually analyze the peptides by mass spectrometry.

3.1 Heat Inactivation of Dialyzed and Non-dialyzed Fetal Calf Serum

1. Thaw FCS at 37 °C.
2. Heat-inactivate thawed FCS at 56 °C for 30 min.
3. Store at 4 °C for short-term use or at −20 °C for long-term use (aliquoted).

3.2 Preparation of SDM-80 Medium for SILAC

1. To prepare 1 L of SDM-80 medium, add all ingredients to 700 ml of deionized water.
2. Allow stirring until it is completely dissolved (around 1 h).
3. Add water to 900 ml.
4. Adjust the pH to 7.3 with NaOH.
5. Filter medium by applying vacuum to Steritop™ filter unit.
6. Add 100 ml of heat-inactivated dialyzed FCS under sterile conditions (*see* **Note 6**).
7. Store at 4 °C.

3.3 Preparation of Modified HMI-9 Medium for SILAC

1. To prepare 1 L of modified HMI-9 medium, add all ingredients to 800 ml of IMDM medium without L-lysine and L-arginine.
2. Allow stirring until it is completely dissolved (around 1 h).
3. Add IMDM without L-lysine and L-arginine to 900 ml.
4. Filter medium by applying vacuum to Steritop™ filter unit.
5. Add 100 ml of heat-inactivated dialyzed FCS under sterile conditions.
6. Store at 4 °C.

3.4 Cell Culture: Procyclic Form of Trypanosomes

1. Collect procyclic cells grown in SDM-79 medium with 10 % non-dialyzed FCS by centrifugation at $1{,}400 \times g$ for 8 min (*see* **Note 7**).
2. Discard supernatant and add 5 ml of SDM-80 ("light" or "heavy") with 10 % of heat-inactivated dialyzed serum.
3. Centrifuge at $1{,}400 \times g$ for 8 min.
4. Discard supernatant and resuspend the cells in SDM-80 ("light" or "heavy") at a density of 2×10^6 cells/ml (*see* **Note 8**).
5. Grow cells in fully capped culture flasks at 27 °C without CO_2 in SDM-80 ("light" or "heavy").
6. Keep a density of 2×10^6 cells/ml, split the culture every day in order to keep logarithmic growth (Fig. 1, *see* **Note 9**).
7. For cells grown in the "light" medium, collect the samples for protein extraction after several days of stable logarithmic growth.
8. For cells grown in the "heavy" medium, collect the samples for protein extraction after 2 cell division cycles (doubling time 8 h) and 10 cell division cycles (Fig. 2, *see* **Note 10**).

3.5 Cell Culture: Bloodstream Form of Trypanosomes

1. Collect bloodstream cells grown in HMI-9 medium with 10 % non-dialyzed FCS by centrifugation at $1{,}400 \times g$ for 8 min.
2. Discard the supernatant and add 5 ml of modified HMI-9 ("light" or "heavy" version) with 10 % of heat-inactivated dialyzed serum.
3. Centrifuge at $1{,}400 \times g$ for 8 min.
4. Discard the supernatant and resuspend the cells in the "light" or the "heavy" version of modified HMI-9 at a density of 2×10^5 cells/ml.
5. Grow cells in loosely capped culture flasks at 37 °C with 5 % CO_2 in the "light" or the "heavy" version of modified HMI-9.
6. Keep the density of 2×10^5 cells/ml, split the culture every day in order to keep logarithmic growth.
7. For cells grown in the "light" medium, collect the samples for protein extraction after several days of stable logarithmic growth.
8. For cells grown in the "heavy" medium, collect the samples for protein extraction after 2 cell division cycles (doubling time 6–7 h) and 12 cell division cycles.

3.6 Cell Lysis

1. Collect separately the cells from "heavy" and "light" media by centrifugation at $1{,}400 \times g$ for 8 min.
2. Remove the supernatants and resuspend each pellet in 1 ml of cold PBS. Transfer the cells into a microcentrifuge tube.

3. Centrifuge at 1,400 × *g* for 8 min in a benchtop microcentrifuge.
4. Remove the supernatants and resuspend each pellet in 1× SDS loading buffer:

   PCF cells (from "light" or "heavy" medium): 5×10^5 cells/μl.

   BSF cells (from "light" or "heavy" medium): 2.5×10^5 cells/μl.
5. Boil samples at 95 °C for 5 min.
6. Store at −80 °C.

3.7 Preparing Protein Samples for Mass Spectrometry

Protein samples are prepared as described previously [11].

1. Thaw the cell lysates from cells grown in the "light" and the "heavy" medium.
2. For PCF cell lysates, mix 5×10^6 cells grown in the "light" medium with 5×10^6 cells grown in the "heavy" medium.
3. For BSF cell lysates, mix 2.5×10^6 cells grown in the "light" medium with 2.5×10^6 cells grown in the "heavy" medium.
4. Vortex vigorously.
5. Load the samples onto a 10 % acrylamide (1.00 mm) gel and run the SDS-PAGE.
6. Stain the proteins using the Coomassie staining solution for 15 min.
7. Destain the gels.
8. Wash the gel with water for a few hours.
9. Put the gel onto a transparent surface and excise gel lanes with a clean scalpel.
10. Cut each gel lane into 10–20 bands.
11. Cut excised bands in several 1 × 1 mm cubes.
12. Transfer the gel pieces into microcentrifuge tubes containing 200 μl of 20 % (v/v) ethanol.
13. Spin down gel pieces in a benchtop microcentrifuge.
14. Store at 4 °C until MS analysis.

4 Notes

1. All the components can be prepared as 100–1,000× stock solutions, filter-sterilized and kept at 4 °C.
2. Preparation of 100 mM hypoxanthine stock solution: dissolve 1.36 g of hypoxanthine in 100 ml of 1 M sodium hydroxide, heat to 45 °C.
3. Antibiotics (besides penicillin and streptomycin) are added to the media if strains with selection markers are grown for the SILAC experiment.

4. Originally, SDM-80 medium is prepared without glucose [12]. Here, we are referring to SDM-80, but glucose is added to a final concentration of 5.55 mM.
5. All the buffers for SDS-PAGE can be handmade or commercially provided. Also, gels can be hand-casted or pre-casted. When handmade buffers and hand-casted gels are used, pay attention to the chemical purity of the components used in order to reduce chemical background [11].
6. For the testing purposes, prepare small quantities of SILAC medium with "light" versions of L-lysine and L-arginine and 10 % dialyzed FCS. Always include control culture: cells grown in "light" SILAC medium with 10 % non-dialyzed FCS. Growth rate of cells grown in SILAC medium with dialyzed FCS should be similar to those with non-dialyzed FCS. Only after cells are stably growing in "light" SILAC medium with dialyzed FCS, culture in "heavy" SILAC medium can be started.
7. In our experiments, we used for labeling the procyclic strain AnTat 1.1 [16] and the bloodstream strain "New York single marker" a LISTER 427 cell line [17].
8. Sometimes, trypanosomes are not adapting well to media with dialyzed serum. There are several options to overcome this problem: maintain the cultures at higher densities than usual ($2–3 \times 10^6$ cells/ml for PCF, 2×10^5 cells/ml for BSF); add a small amount of non-dialyzed serum to the medium; test dialyzed serum from another manufacturer.
9. Growth curves comparing procyclic *T. brucei* cell growth in three different culture media [SDM-79 medium with non-dialyzed FCS, SILAC medium with "light" ($^{12}C_6$-lysine and $^{12}C_6$-arginine) or "heavy" ($^{13}C_6$-lysine and $^{13}C_6$-arginine) amino acids and dialyzed FCS] are shown in the article Gunasekera et al. [2].
10. In order to test the incorporation of the heavy isotope-labeled lysine and arginine in the parasite proteome, cells grown in the "heavy" media are collected after 2 cell division cycles (early incorporation sample) and at the end of the SILAC experiment (late incorporation sample, 10 cell division cycles for PCF, 12 division cell cycles for BSF). Very often it is difficult to collect cells at these precise division times. For the early incorporation sample, instead of 2 division cycles cells can be collected also after 3 division cycles. End of the SILAC experiment should not be before 10 division cycles for PCF, and 12 division cycles for BSF, but cells certainly can be collected later, up to 20 division cycles, for both life stages. Lysates of the cells grown in the "light" media are used as a control.

Acknowledgements

The authors would like to thank Kapila Gunasekera and Roman Trikin for critically reading the manuscript. The research was supported by the Swiss National Science Foundation.

References

1. Butter F, Bucerius F, Michel M et al (2013) Comparative proteomics of two life cycle stages of stable isotope-labeled *Trypanosoma brucei* reveals novel components of the parasite's host adaptation machinery. Mol Cell Proteomics 12:172–179
2. Gunasekera K, Wüthrich D, Braga-Lagache S et al (2012) Proteome remodelling during development from blood to insect-form *Trypanosoma brucei* quantified by SILAC and mass spectrometry. BMC Genomics 13:556
3. Urbaniak MD, Guther ML, Ferguson MA (2012) Comparative SILAC proteomic analysis of Trypanosoma brucei bloodstream and procyclic lifecycle stages. PLoS One 7(5):e36619
4. Ong SE, Blagoev B, Kratchmarova I et al (2002) Stable isotope labeling by amino acids in cell culture, SILAC, as a simple and accurate approach to expression proteomics. Mol Cell Proteomics 1:376–386
5. Benne R, Van den Burg J, Brakenhoff JP et al (1986) Major transcript of the frame- shifted coxII gene from trypanosome mitochondria contains four nucleotides that are not encoded in the DNA. Cell 46:819–826
6. Borst P, Cross GA (1982) Molecular basis for trypanosome antigenic variation. Cell 29: 291–303
7. Ferguson MA, Cross GA (1984) Myristylation of the membrane form of a Trypanosoma brucei variant surface glycoprotein. J Biol Chem 259:3011–3015
8. de Almeida ML, Turner MJ, Stambuk BB et al (1988) Identification of an acid-lipase in human serum which is capable of solubilizing glycophosphatidylinositol-anchored proteins. Biochem Biophys Res Commun 150:476–482
9. Sutton RE, Boothroyd JC (1986) Evidence for trans splicing in trypanosomes. Cell 47: 527–535
10. Ong SE, Mann M (2006) A practical recipe for stable isotope labeling by amino acids in cell culture (SILAC). Nat Protoc 1: 2650–2660
11. Shevchenko A, Tomas H, Havlis J et al (2006) In-gel digestion for mass spectrometric characterization of proteins and proteomes. Nat Protoc 1:2856–2860
12. Brun R, Jenni L, Tanner M et al (1979) Cultivation of vertebrate infective forms derived from metacyclic forms of pleomorphic Trypanosoma brucei stocks. *Short communication*. Acta Trop 36:387–390
13. Hirumi H, Hirumi K (1989) Continuous cultivation of Trypanosoma brucei blood stream forms in a medium containing a low concentration of serum protein without feeder cell layers. J Parasitol 75:985–989
14. Lamour N, Rivière L, Coustou V et al (2005) Proline metabolism in procyclic Trypanosoma brucei is down-regulated in the presence of glucose. J Biol Chem 280:11902–11910
15. Laemmli UK (1970) Cleavage of structural proteins during the assembly of the head of bacteriophage T4. Nature 227:680–685
16. Le Ray D, Barry JD, Easton C et al (1977) First tsetse fly transmission of the "AnTat" serodeme of Trypanosoma brucei. Ann Soc Belg Med Trop 57:369–381
17. Wirtz E, Leal S, Ochatt C et al (1999) A tightly regulated inducible expression system for conditional gene knock-outs and dominant-negative genetics in Trypanosoma brucei. Mol Biochem Parasitol 99:89–101

Chapter 5

Stable Isotope Labeling by Amino Acids in Cultured Primary Neurons

Guoan Zhang, Katrin Deinhardt, and Thomas A. Neubert

Abstract

Cultured primary neurons are a well-established model for the study of neuronal function. Conventional stable isotope labeling with amino acids in cell culture (SILAC) requires nearly complete metabolic labeling of proteins and therefore is difficult to apply to cultured primary neurons, which do not divide in culture. Here we describe a protocol that utilizes a multiplex SILAC labeling strategy for primary cultured neurons. In this strategy, two different sets of heavy amino acids are used for labeling cells for the different experimental conditions. This allows for a straightforward SILAC quantitation using partially labeled cells because the two cell populations are always equally labeled.

Key words SILAC, Primary neurons, Mass spectrometry, Proteomics, Quantitation

1 Introduction

Stable isotope labeling with amino acids in cell culture (SILAC) involves cell culture in media containing "light" (natural) or "heavy" isotope-containing amino acids [1]. It is important to obtain a high degree of label incorporation because incomplete labeling will skew the SILAC ratio in favor of the light protein [2]. To ensure nearly complete labeling, it is generally required to maintain cells in SILAC media for at least five cell divisions so that even proteins with zero turnover rate will be highly labeled (>97 %) by dilution alone [1, 3].

Primary neurons are widely used as a very important model in neuroscience because in general their functional properties more closely resemble the in vivo state than those of transformed cell cultures. Because the primary neurons do not divide in culture, it is difficult to use the standard light/heavy SILAC labeling because of the issue of incomplete labeling. To overcome this difficulty, a method has been reported in which the SILAC ratio is corrected for incomplete labeling by monitoring the label incorporation of every protein [4, 5]. However, this strategy has several obvious

Bettina Warscheid (ed.), *Stable Isotope Labeling by Amino Acids in Cell Culture (SILAC): Methods and Protocols*, Methods in Molecular Biology, vol. 1188, DOI 10.1007/978-1-4939-1142-4_5, © Springer Science+Business Media New York 2014

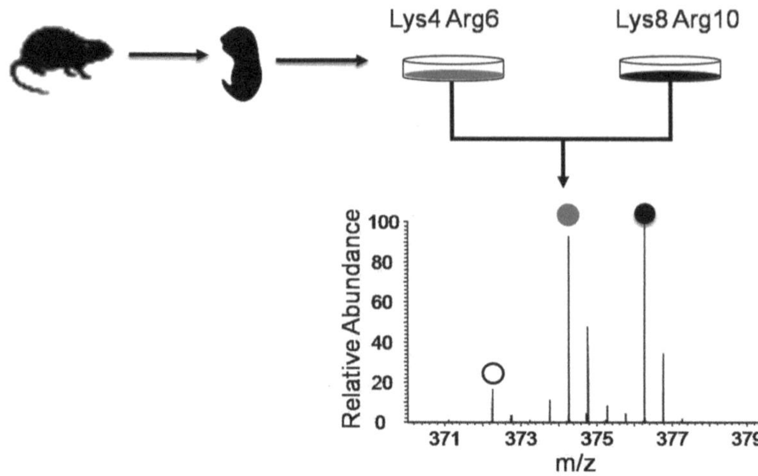

Fig. 1 Schematic of multiplex SILAC labeling for primary cultured neurons. Two different versions of stable isotope-coded amino acids (medium and heavy, represented by *gray* and *black colors*, respectively) were used for labeling. This strategy does not require complete labeling because the medium and heavy amino acids are incorporated at the same rate. An MS spectrum of a representative peptide (LLASLVK from Trim28) is shown as an example. (The *gray* and *black circles* denote the medium and heavy peptide signals, respectively, while the *open circle* denotes the light signal). The MS spectrum was acquired using an LTQ-Orbitrap mass spectrometer

disadvantages. First, each SILAC analysis requires a parallel analysis to measure the label incorporation for every single protein. In addition to the extra cost and effort, it is difficult to obtain the label incorporation for every protein because it requires the protein be quantified in two analyses. A considerable proportion of the SILAC protein ratios cannot be corrected because for analysis of complex protein mixtures by liquid chromatography–tandem mass spectrometry (LC–MS/MS), protein identification is typically only partially overlapped between replicate analyses [2]. Moreover, the correction step introduces additional variation that compromises the accuracy of quantitation.

To circumvent these problems, we employ a multiplex SILAC labeling strategy for primary neurons (Fig. 1) [6–8]. Instead of using light and heavy labeling amino acids to distinguish the two experimental conditions, we use two different sets of heavy amino acids, D_4-lysine/$^{13}C_6$-arginine (Lys4/Arg6) and $^{13}C_6{}^{15}N_2$-lysine/$^{13}C_6{}^{15}N_4$-arginine (Lys8/Arg10).

Because the two cell populations incorporate the two sets of amino acids at the same rate, they are always equally labeled. SILAC quantitation is done using the signals of the medium (Lys4/Arg6) and heavy (Lys8/Arg10) labeled peptides, and the unlabeled peptides can be ignored. This allows for a straightforward and accurate SILAC quantitation using partially labeled cells.

Here we describe a protocol for SILAC culture of primary cortical neurons from embryonic rat brain.

2 Materials

2.1 Equipment and Labware

1. Horizontal laminar flow hood for dissections.
2. Laminar flow biological hood for cell culture.
3. Humidified culture incubator at 37 °C and 5 % CO_2.
4. Dissecting microscope.
5. Dissecting tools (we use scissors and Dumont #5 and #5/45 forceps from Fine Science Tools).
6. Centrifuge to accommodate 15 ml tubes.
7. Vacuum-driven filter units (0.45 μm pore size, 250 ml).
8. Water bath 37 °C.
9. 70 μm cell strainers.
10. 10 cm and 6 cm sterile petri dishes.
11. Tissue culture dishes (size depends on number of cells in the experiment).
12. Glass Pasteur pipettes, fire-polished to three different pore sizes, each about 50 % the size of the previous opening (approx. 1 mm, 0.5 mm, and 0.2 mm in diameter).
13. Hemocytometer.

2.2 Animal and Reagents

1. Timed-pregnant female (Sprague–Dawley) rat, embryonic day E18.
2. Ca–Mg-free Hanks' balanced salt solution (HBSS).
3. Neurobasal medium.
4. Neurobasal medium deficient of arginine and lysine (custom-made from Invitrogen).
5. B-27 supplement 50×.
6. 5-Fluorouridine/uridine (5-FU; 10 mM each as stock solution).
7. L- Glutamine, 200 mM stock.
8. D_4-lysine, $^{13}C_6$-arginine, $^{13}C_6{}^{15}N_2$-lysine and $^{13}C_6{}^{15}N_4$-arginine.
9. Penicillin–streptomycin (5,000 U of penicillin and 5,000 μg of streptomycin per ml).
10. 0.5 % Trypsin–EDTA (Invitrogen).
11. Fetal bovine serum (FBS).

2.3 SILAC Media and Other Solutions

1. Coating solution: Dissolve poly-D-lysine in cell culture water and dilute to 0.1 mg/ml. Filter with filter units (0.45 μm).
2. Dissection solution: Add glucose stock (37 %, sterile) to HBSS (1:100).

3. SILAC media: Prepare 1,000× stock solutions for labeling amino acids: 398 mM for arginine and 798 mM for lysine (in water). Add B-27 supplement (1:50), L-glutamine (1:400) and penicillin–streptomycin (1:100) to neurobasal medium deficient of arginine and lysine. Split the medium equally into two parts. To half of the medium add the Arg6 and Lys4 stocks (1:1,000). To the other half add the Arg10 and Lys8 stocks (1:1,000). Filter both media using 0.45 μm filter units. The SILAC media can be store at 4 °C for up to 4 weeks.

3 Methods

It is important that all tissue culture steps be performed in a sterile environment. Therefore, dissections are carried out in a horizontal laminar flow hood, using a dissecting microscope. Tissue dissociation and cell culture are performed in a vertical flow hood.

3.1 Dissection

1. In the vertical flow hood, coat culture plates with poly-D-lysine for at least 1 h, then wash with water and let dry. Coated and dried plates can be stored at 4 °C for a week or at −20 °C for a month.
2. Sacrifice pregnant female rat using CO_2, followed by cervical dislocation. Remove uterus, place it into a 10 cm sterile petri dish filled with dissection solution, and move to horizontal flow hood.
3. Isolate E18 fetuses and collect them in fresh dissection solution. Make sure to rinse off remaining blood.
4. Isolate embryonic brains and place them into a 6 cm sterile petri dish filled with dissection solution. Using a dissecting microscope, dissect out cortices and remove the meninges (*see* **Notes 1** and **2**).
5. Collect isolated cortices in dissection solution and move to vertical flow hood.
6. Split cortices to collect 6–8 embryos worth of tissue per 15 ml conical tube in 3 ml of dissection solution. Add 300 μl of trypsin–EDTA and incubate at 37 °C for 8 min to disrupt cell–cell contacts (*see* **Note 3**).
7. Add 1 ml of FBS to quench the trypsin and spin cells for 3 min at 300 × *g*. Remove supernatant and resuspend cells in 1 ml of regular neurobasal medium.
8. Triturate cells with fire-polished glass Pasteur pipettes, starting with the largest opening, followed by the medium and the smallest opening. Triturate about six to eight times with each pipette. This will dissociate tissue into a single cell suspension (*see* **Notes 4** and **5**).

9. Pool all dissociated cells, filter through a cell strainer to remove non-dissociated larger chunks.
10. Count cells using a hemacytometer. Typical yields are ~15 million cells per embryo for cortical neurons.

3.2 SILAC Cell Culture

1. Suspend the cells well by pipetting and split equally into two 15 ml tubes. Spin cells down for 3 min at 300×*g*. Carefully remove the supernatant.
2. Resuspend cells in SILAC media at required density.
3. Add 5-FU to the culture media (5 μM each final concentration) (*see* **Note 6**).
4. Plate cells in culture dishes (*see* **Notes 7** and **8**).
5. Place cells in a humidified incubator at 37 °C and 5 % CO_2.
6. Check cells daily under a microscope. Make sure cells grow equally well in the two SILAC media (*see* **Note 9**).
7. Exchange half of the culture media for fresh media every 3 days. Add 5-FU to fresh media before each medium exchange (*see* **Note 10**).
8. Grow neurons for about 10 days before proceeding to cell treatment/lysis/MS analysis (*see* **Notes 11–15**).

4 Notes

1. This protocol describes the methods using rat embryonic cortical cultures. Other neuronal cultures, e.g., from genetically modified mice, can be used as required.
2. During all steps, make sure the tissue is always submerged in dissection solution and thus does not dry.
3. It is important that the tissue is not incubated in trypsin for more than 8–10 min as this leads to neuronal damage and reduces the health of the cultures.
4. The diameter of the fire-polished Pasteur pipettes (*see* Subheading 2.1, **item 12**) is important, as too large a diameter will not efficiently dissociate the tissue, while too small a diameter can damage the cells.
5. Be careful to avoid excess bubble formation when triturating, as the increase in surface tension can rupture the cells. If tissue or cells start forming sticky clumps, add 10 μg/ml of DNase to the trituration; this should dissolve clumps.
6. Addition of 5-FU inhibits growth of dividing cells such as glia cells. Typically in our cultures less than 1 % of the cell population is non-neuron cells.

7. Many protocols require neurons be plated in medium containing FBS. However, we found that plating neurons in FBS-free medium does not affect cell survival as compared to FBS-containing medium.
8. For SILAC experiments that require a lot of material, we plate cells at a density of up to 15 million/10 cm culture dish. We found that at this density, cell growth is normal compared to lower densities and there is no need for more frequent medium change.
9. It is critical to maintain equal growth between the two SILAC-labeled populations because unequal growth will cause differences in label incorporation levels in the two cell populations and eventually lead to errors in SILAC quantitation. To ensure this, make every effort to handle the two cell populations in the same way as much as possible.
10. Cultured neurons secrete growth factors that are important for growth and maintenance of neurons. Refreshing only half of the medium allows for a certain amount of these growth factors to be retained.
11. Label incorporation increases rapidly in the first several days and slows down after 6 or 7 days in culture. *See* Fig. 2 for an illustration of label incorporation versus culture time.
12. Although significant label incorporation occurs within only 2 days of culture (average incorporation >50 %), longer labeling times are preferred to increase the amount of labeled proteins if permitted by the experimental design. We normally label cells for 10 days, which should result in >90 % average label incorporation.
13. We do not recommend labeling time less than 2 days. Low incorporation level will decrease the number of quantifiable proteins. Also, accuracy of quantitation will be compromised due to decreased signal-to-noise ratio and partial overlapping of isotopic distributions between light and medium signals.
14. Different proteins are labeled at considerably different rates. For example, nuclear proteins such as histones are labeled at much slower rates than typical proteins. After 10 days of labeling, the incorporation is usually around 30–40 % for histones. This needs to be taken into consideration when performing experiments that target slow-turnover proteins.
15. Arginine to proline conversion [1, 9] typically occurs in SILAC labeling of cortical neurons at a low level (<5 %). We have found that this does not significantly affect protein quantitation. However, occasionally high levels of conversion (>30 %) were observed for reasons that are not well understood. In such a case, the conversion can be inhibited by adding 300 mg/L

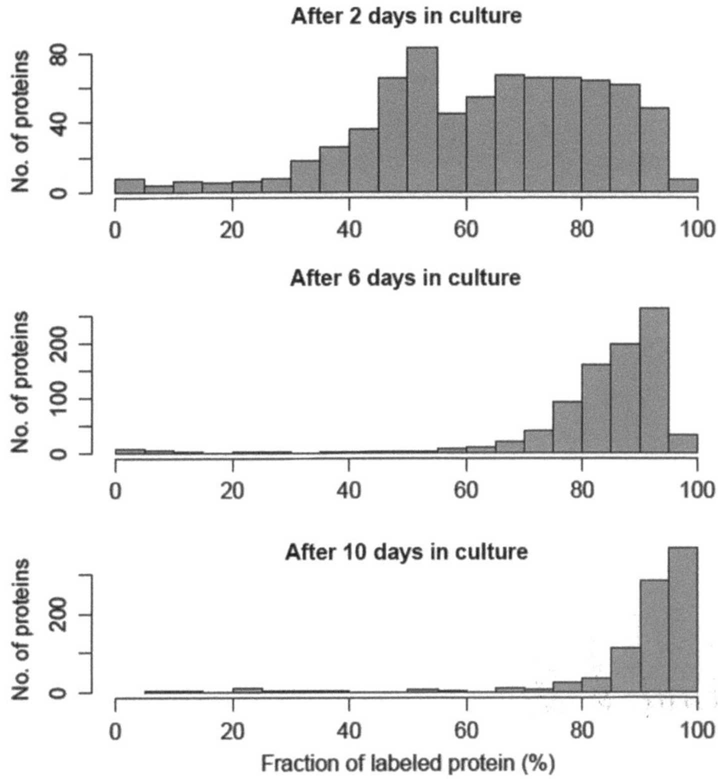

Fig. 2 Label incorporation after SILAC labeling of primary cultured cortical neurons for 2, 6, and 10 days. The histograms are plotted based on 746, 844, and 862 quantified proteins from the 2, 6, and 10 day-neurons, respectively. The MS data were acquired using an LTQ-Orbitrap mass spectrometer [7]

L-proline into the SILAC culture medium. For other types of neurons, the conversion level may be different. We also caution that for quantitation of proline-containing peptides, the effect of this conversion on quantitation may be more pronounced [9].

Acknowledgement

This work was supported by NIH NINDS grant P30 NS050276 to TAN and the 100 Women in Hedge Funds foundation.

References

1. Ong SE, Blagoev B, Kratchmarova I et al (2002) Stable isotope labeling by amino acids in cell culture, SILAC, as a simple and accurate approach to expression proteomics. Mol Cell Proteomics 1:376–386
2. Elias JE, Haas W, Faherty BK et al (2005) Comparative evaluation of mass spectrometry platforms used in large-scale proteomics investigations. Nat Methods 2:667–675
3. Ong SE, Mann M (2006) A practical recipe for stable isotope labeling by amino acids in cell culture (SILAC). Nat Protoc 1:2650–2660
4. Spellman DS, Deinhardt K, Darie CC et al (2008) Stable isotopic labeling by amino acids in

cultured primary neurons: application to brain-derived neurotrophic factor-dependent phosphotyrosine-associated signaling. Mol Cell Proteomics 7:1067–1076

5. Liao L, Park SK, Xu T et al (2008) Quantitative proteomic analysis of primary neurons reveals diverse changes in synaptic protein content in fmr1 knockout mice. Proc Natl Acad Sci U S A 105:15281–15286
6. Selbach M, Schwanhausser B, Thierfelder N et al (2008) Widespread changes in protein synthesis induced by microRNAs. Nature 455: 58–63
7. Zhang G, Deinhardt K, Chao MV et al (2011) Study of neurotrophin-3 signaling in primary cultured neurons using multiplex stable isotope labeling with amino acids in cell culture. J Proteome Res 10:2546–2554
8. Zhang G, Neubert TA, Jordan BA (2012) RNA binding proteins accumulate at the postsynaptic density with synaptic activity. J Neurosci 32: 599–609
9. Van Hoof D, Pinkse MW, Oostwaard DW et al (2007) An experimental correction for arginine-to-proline conversion artifacts in SILAC-based quantitative proteomics. Nat Methods 4:677–678

Chapter 6

SILAC and Alternatives in Studying Cellular Proteomes of Plants

Annemarie Matthes, Karin Köhl, and Waltraud X. Schulze

Abstract

Quantitative proteomics by metabolic labeling has a high impact on the growing field of plant systems biology. SILAC has been pioneered and optimized for plant cell culture systems allowing for SILAC-based quantitative experiments in specialized experimental setups. In comparison to other model organisms, the application of SILAC to whole plants is challenging. As autotrophic organisms, plants under their natural growth conditions can hardly be fully labeled with stable isotope-coded amino acids. The metabolic labeling with inorganic nitrogen is therefore the method of choice for most whole-plant physiological questions. Plants can easily metabolize different inorganic nitrogen isotopes. The incorporation of the labeled inorganic nitrogen then results in proteins and metabolites with distinct molecular mass, which can be detected on a mass spectrometer. In comparative quantitative experiments, similarly as in SILAC experiments, treated and untreated samples are differentially labeled by nitrogen isotopes and jointly processed, thereby minimizing sample-to-sample variation. In recent years, heavy nitrogen labeling has become a widely used strategy in quantitative proteomics and novel approaches were developed for metabolite identification. Here we present a typical hydroponics setup, the workflow for processing of samples, mass spectrometry and data analysis for large-scale metabolic labeling experiments of whole plants.

Key words SILAC, Plant quantitative proteomics, Hydroponics, Metabolic labeling, Inorganic nitrogen, *Arabidopsis thaliana*

1 Introduction

1.1 Applicability of SILAC to Plants

SILAC (stable isotope labeling of amino acids in cell cultures) is a widely applied metabolic labeling strategy in quantitative proteomics. It is based on distinguishing and quantifying peptides by the mass shift introduced to one proteome by specific amino acids containing nonradioactive isotopes. Amino acids consist of carbon, hydrogen, nitrogen, sulfur, and oxygen. All these atoms have natural nonradioactive isotopes. Most common in nature are their light forms (^{1}H, ^{12}C, ^{14}N, ^{16}O, and ^{32}S). The heavy isotopes (^{2}H, ^{13}C, ^{18}O, ^{34}S) are rather of low abundance. The artificially enriched incorporation of those heavy atoms in biological compounds results in the heavy amino acids used in SILAC. Arginine and lysine

Bettina Warscheid (ed.), *Stable Isotope Labeling by Amino Acids in Cell Culture (SILAC): Methods and Protocols*, Methods in Molecular Biology, vol. 1188, DOI 10.1007/978-1-4939-1142-4_6, © Springer Science+Business Media New York 2014

are established as the preferred amino acids for stable isotope labeling [1], as the specificity of enzymatic digestion with trypsin results in an isotope label of one of these basic amino acids at the C-terminus of the digested peptides. Only the peptide corresponding to the original C-terminus is missing this label. A complete tryptic digestion therefore results in peptide mixtures with a mass shift of exactly one incorporated lysine or arginine per labeled and unlabeled peptide pairs. This makes data-dependent acquisition of spectra and quantitative data analysis rather straightforward.

In the original SILAC experiments, two cell cultures with differential treatment—one grown on light amino acids the other on the heavy amino acids—were compared [1]. For relative quantitation of the corresponding peptides from both cell cultures, ratios of peak intensities were calculated from the extracted ion chromatograms of co-eluting labeled and unlabeled ions. Due to the simple data analysis workflows in SILAC experiments compared to other quantitation strategies, SILAC is now widely applied to study various biochemical and biological aspects (reviewed in ref. [2]). Particularly, in the recent years, SILAC has been applied beyond the original pairwise comparisons of two conditions, to also study small molecule–protein interactions [3], micro-RNA targets [4], and protein translation [5].

The use of SILAC has soon been expanded to other organisms, such as auxotrophic bacteria [6] or yeast strains [7], *Drosophila* [8], *Caenorhabditis* [9], and mouse [10].

For the plant kingdom, amino acid labeling with heavy arginine has been successfully applied for *Chlamydomonas* strains deficient of arginine biosynthesis to study adaptive responses to iron deficiency [11] and to characterize proteins regulated by anaerobic responses [12]. In addition, amino acid labeling was used in quantitative immunoprecipitations combined with knockdown strains to find specific interaction partners to chloroplast proteins [13].

In contrast to algae, full SILAC labeling of the higher plant model organism *Arabidopsis thaliana* has not been achieved. In contrast to other organisms, plants are able to synthesize their whole set of amino acids autotrophically by several pathways downstream of photosynthesis and the assimilation of inorganic nitrogen. Moreover, plants have a very efficient system for the nitrogen recycling from organic breakdown products in the cell. Even in plant cell culture systems, this leads to several challenges to the use of SILAC in higher plants. So far, two reports for the use of SILAC in higher plant cell cultures are available. Gruhler et al. [14] labeled cell cultures with $^{2}H_{3}$-leucine, $^{13}C_{6}$-arginine and $^{2}H_{4}$-lysine under constant light conditions but in the presence of ammonium as an inorganic nitrogen source in the growth medium. The maximum labeling efficiency of arginine under these conditions was about 80 % and this imposed particular challenges to the data analysis [14]. The second report from Schütz et al. [15] elegantly used also the incorporation of lysine in cell cultures, again in the presence of ammonium in the growth medium. To cope with the incomplete labeling, two sets of cell cultures were labeled with either the medium-heavy Lys4 and or the heavy Lys8, representing the treated and the control sample. Since both sets of cell cultures should have the same incomplete labeling efficiency, differential effects between the two culture sets could be derived from calculating the ratio between heavy and medium-heavy ion intensities [15]. When grown in the light, the labeling efficiency reached only 58 %, which is too low for efficient quantitative coverage. However, when the cell cultures were grown in the dark, an incorporation of the stable isotope-labeled lysine of up to 91 % could be reached. Unfortunately, dark grown plant cell cultures for most plant physiological studies do not represent a particularly meaningful condition, but for specialized projects this system can well be applied.

It might be still possible to improve the protocols for the use of SILAC also for labeling of whole plants. Possibly also mutants in amino acid transporters or the amino acid biosynthesis [16, 17] or photorespiration mutants [18] in combination with feeding the plant with labeled amino acids as sole nitrogen source [17, 19] could be tested. However, these mutants might again result in a plant system that hardly represents the typical physiological properties and therefore would have limited applicability.

1.2 ^{15}N Metabolic Labeling as an Alternative in Plants

Due to the inefficient labeling, high interconversion between amino acids (e.g., arginine to proline) and labeling experimental setups that do not necessarily represent the typical physiological properties of the plant organism, plant researchers have established an alternative metabolic labeling system involving full ^{15}N-labeling with inorganic nitrogen sources. The inorganic nitrogen can be supplied to plants as nitrate nitrogen and ammonium nitrogen and is readily incorporated into amino acids and proteins. In fact, the labeling of organic substances by inorganic ^{15}N salts has already been used for a long time in studying metabolites (e.g., by NMR [20]). Its application to large-scale proteomics has evolved only over the last decade [21].

In contrast to SILAC labeling, in which only atoms of single amino acids are isotope-labeled, labeling on basis of ^{15}N inorganic salts results in isotope label incorporation to every single nitrogen atom of the proteins. This has consequences for the comparison of the light and the heavy (^{15}N) labeled peptide [22]. The normal light peptide isotope clusters consists of a monoisotopic peak and additional satellite peptide peaks with higher m/z ratios, mainly caused by the different degree of incorporated natural ^{13}C atoms (Fig. 1a). This introduces a mass shift of 1 amu/z between the different isotope peaks. A fully ^{15}N-labeled peptide would have the same shape as the ^{14}N peptide only shifted to a higher 1 amu/z per incorporated ^{15}N atom. Since different amino acids contain

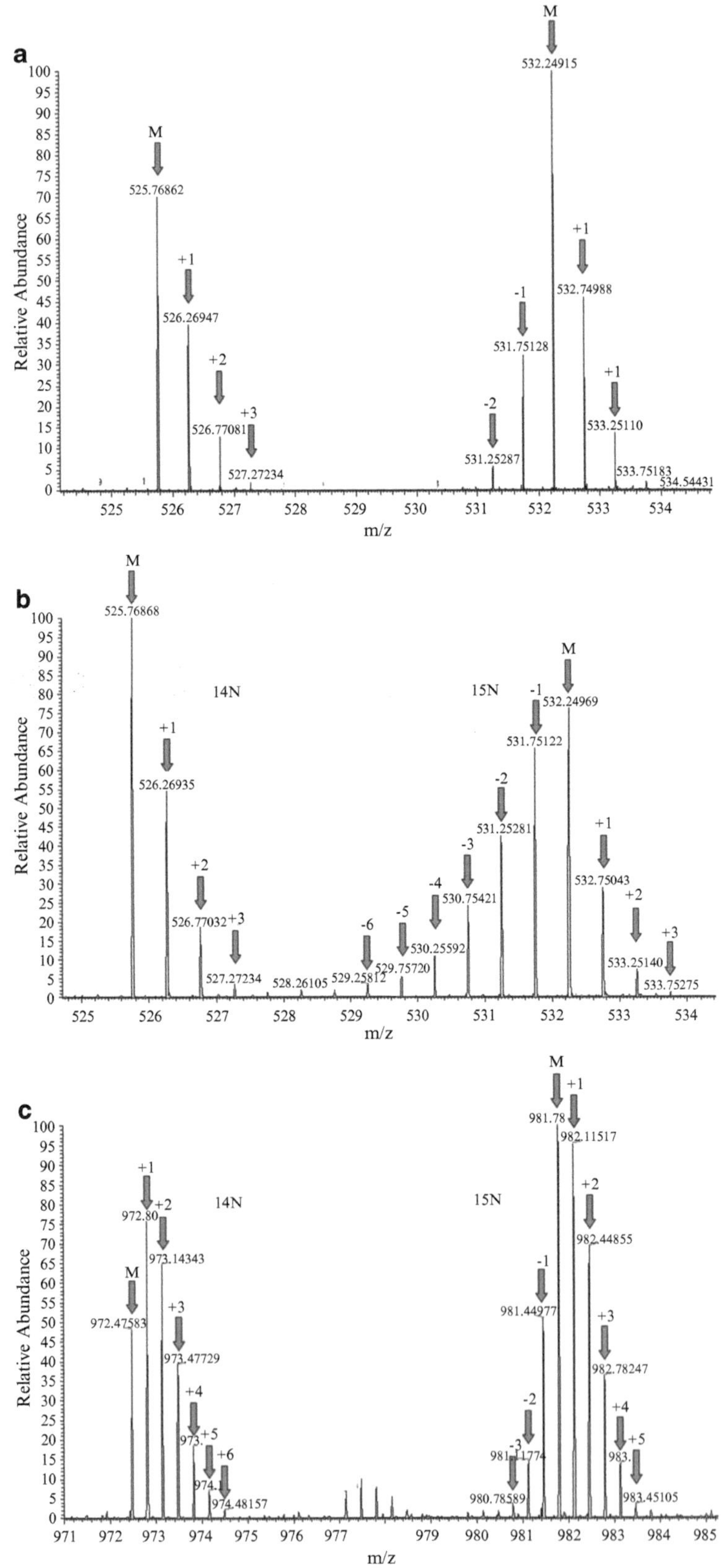
a
Relative Abundance
m/z
M
525.76862
+1
526.26947
+2
526.77081
+3
527.27234
-2
531.25287
-1
531.75128
M
532.24915
+1
532.74988
+1
533.25110
533.75183
534.54431
b
14N
15N
M
525.76868
+1
526.26935
+2
526.77032
+3
527.27234
528.26105
529.25812
-6
-5
529.75720
-4
530.25592
-3
530.75421
-2
531.25281
-1
531.75122
M
532.24969
+1
532.75043
+2
533.25140
+3
533.75275
c
14N
15N
M
972.47583
+1
972.80
+2
973.14343
+3
973.47729
+4
973.
+5
974.
+6
974.48157
-3
980.78589
-2
981.11774
-1
981.44977
M
981.78
+1
982.11517
+2
982.44855
+3
982.78247
+4
983.
+5
983.45105

different amounts of nitrogen atoms, the mass shift is therefore dependent of the amino acid composition of the peptide and differs between various peptides. Moreover, ^{15}N inorganic salts usually come only in about 98 % purity grade, with still residual ^{14}N atoms. In peptide mass spectra, this results in satellite peaks with lower masses than the ^{15}N monoisotopic peak, dependent on the number of incorporated ^{14}N atoms, and is additionally overlain by the ^{13}C isotope (Fig. 1b).

In plants, the residual amount of ^{14}N can be additionally increased by ^{14}N coming from seed storage nitrogen, particularly in young seedlings. The different isotope clusters of the ^{15}N and the ^{14}N peptide has to be taken into consideration in the quantitative data analysis [22]. Corrective actions have to be taken, e.g., for long peptides with a high number of nitrogen atoms (Fig. 1c). This is a major difference to SILAC where a constant mass shift is introduced to the peptides and the isotopic envelope of the peptide ion is not changed.

1.2.1 Metabolic Labeling of Plants in Various Culture Systems

Depending on the plant species, a combination of different nitrogen-containing salts should be used to achieve optimal growth. For example, *Arabidopsis* prefers a nitrate and ammonium combination [23], but high ammonium concentrations can even be toxic. Plants can be cultured very easily in liquid or solid culture media with the supply of these inorganic nitrogen sources. Quantitative proteomic studies have been carried out using inorganic ^{15}N-labeling of plant tissues from various culture systems such as cell cultures [24, 25], and whole plants grown in hydroponic systems [26, 27], axenic in liquid culture [28–30] or on ^{15}N-supplied agar [31–33] and even a close to natural plant labeling system involving a soil-like matrix was developed [34]. All SILAC studies involving higher plants were done on cell culture systems [15, 35].

1.2.2 Experiments Involving Full ^{15}N-Labeling

Experimental designs involving ^{15}N-labeling from inorganic salts in plants basically are done in analogy to metabolic labeling experiments with SILAC in other organisms. Either the labeled or unlabeled proteome serves as internal standard to which the other

Fig. 1 Full scan spectra of the ^{14}N and ^{15}N isotope clusters at different labeling efficiencies and for different peptides. (**a**) Peptide FEGDTLVNR ($C_{45}H_{71}N_{13}O_{16} + H_2O$; charge state 2), the ^{15}N peptide is labeled 97.65 %. (**b**) Peptide FEGDTLVNR ($C_{45}H_{71}N_{13}O_{16} + H_2O$; charge state 2) at incomplete ^{15}N-labeling from a 2 week old plant germinated from ^{14}N seeds and grown in ^{15}N hydroponics. (**c**) Peptide spectra of the ^{14}N and ^{15}N peptide VGADISVVGYDDTEDSSCYIPPLTTIK ($C_{125}H_{194}N_{28}O_{45}S_1 + H_2O$; charge state 3). The monoisotopic peak is the highest peak in the ^{15}N isotope cluster but not in the ^{14}N isotope cluster. M: monoisotopic peak; +x: Satellite peak with increasing contents of ^{13}C; −x: Satellite peak due to increased incorporation of ^{14}N

(treated) sample is compared. Intensity ratios of the heavy and light forms of co-eluting peptide ions in the combined sample reflect relative abundance ratios of each peptide in treatment and control. Variations in experimental strategies range from simple pair wise [26] comparisons to reciprocal labeling experiments [30, 33, 36–39] or the use of a universal standard sample [40].

Such ^{15}N-labeling experiments have been used to study several aspects in plant stress physiology, such as the effect of elicitors on protein phosphorylation [36]. Whole plant studies were carried out to characterize phosphorylation changes upon abscisic acid treatment [30], and protein abundance changes were monitored during leaf senescence [32] and heat shock responses [41]. A study of plant adaptation to osmotic stress revealed new insights for the role of mitochondrial proteins [33].

1.2.3 Experiments Involving Partial ^{15}N-Labeling

Partial ^{15}N-labeling uses subtle changes in the isotopomer envelope of ions in partially labeled peptides to quantify the changes in protein abundance [29]. The described algorithm enables comparison of relative abundances of labeled and unlabeled proteins, although the degree of labeling was only to about 5 %. This approach was applied to compare protein composition of light and dark grown plants [29]. A thorough assessment of the use of full versus partial ^{15}N-labeling revealed that both strategies are in general comparable with regards to precision and accuracy.

While partial labeling is more challenging with respect to automated identification of labeled and unlabeled peptide pairs, and in quantifying the change in isotope cluster distribution, it allowed for quantification of more peptides across the whole dynamic range [29]. To make use of partial metabolic labeling in quantitative proteomics, the isotopic envelope of the ^{15}N-labeled form must diverse sufficiently from the unlabeled counterpart, so robust quantification of relative changes are possible. One of the main reasons for developing the partial metabolic labeling approach is the decrease of the costs for labeled nutrients and its potential application also to soil-grown plants.

1.2.4 Applications in Pulse-Chase Experiments to Study Protein Synthesis Rates

An interesting application of the partial ^{15}N-labeling strategies is in pulse-chase experiments to study protein stability and protein turnover. In an elegant experiment using pulse-labeling with ^{15}N inorganic salts and analysis of intact peptide ions by MALDI-TOF, the dynamics of photosystem II assembly were studied [42]. Pulse-labeling using full metabolic ^{15}N-labeling is challenging due to difficult interpretation of fragment spectra of partially labeled peptides and due to difficult pairing of isotope envelopes. Therefore, in the study mentioned above, proteins were purified on two-dimensional gels prior to mass spectrometric analysis, and the peptide identification was carried out based on the unlabeled ion.

The complication of peptide identification for partially labeled ions may be a reason why pulse-labeling with full ^{15}N-labeling on a larger proteome-wide scale still has not been carried out. However, recently, promising approaches using ^{15}N pulse-labeling across more than one protein have been pioneered for mice [43] and have recently started to be used also in plants to study protein turnover rates in *Ostreococcus* [44] and in *Arabidopsis* [45].

2 Materials

2.1 Hydroponic System for Metabolic Labeling of Whole Arabidopsis Plants

1. Grey polypropylene box (Allibert) with volume of about 5 L and open top (*see* **Note 1**).
2. Sowing tray with a perforated bottom (e.g., Piki-Saat 80, Wisauplast, Wisau, Germany), fitting on top of the plastic box like a lid.
3. Rockwool-cressmats (Grodan, Denmark) (*see* **Note 2**).
4. Saran wrap.
5. Algae foil: thin black polyethylene foil with pinprick holes available from suppliers for commercial plant growers (*see* **Note 3**).
6. Optional: Topagar with Gibberellic acid (0.1 Topagar, 3.5 ng/ml Gibberellic acid 4+7) for germination (*see* **Note 4**).
7. Arabidopsis seeds.
8. Graduated cylinder, 5 L.
9. Glass pipettes, 5 ml.
10. Nutrient solution medium according to Loque et al. [46] (Table 1).

The chemicals should have the purest grade. Stock solutions can be prepared and can be stored at room temperature for several months.

11. Tweezers.

2.2 Protein Isolation

1. The chemicals for your protein isolation protocol.
2. The chemicals for your protein concentration determination of your choice.

2.3 Protein Precipitation Especially for Low Protein Amount or Small Volume

Use for the preparation of all solutions H_2O bidest.

1. Glycogen solution (20 μg/μl).
2. Ethanol, 100 % (high grade).
3. Sodium acetate, 2.5 M pH 5; pH is adjusted with acetic acid/KOH.

Table 1
Composition of a typical nutrient solution used for metabolic labeling of plants [46]

Chemical	Concentration in nutrient solution (μM)	Stock solution (M)
NH_4NO_3 (^{15}N-ammonium-^{15}N nitrate > 98 %)	1,000	1
$CaCl_2$	250	1
FeEDTA	100	0.1
KH_2PO4	1,000	1
$MgSO_4$	1,005	1
H_3BO_3	100	0.1
$CuSO_4$	1.5	0.0015
KCl	50	0.05
$MnSO_4$	10	0.01
Na_2MoO_4	0.1	0.0001
Na_2O_3Si	100	0.1
$ZnSO_4$	2	0.002

Here, ammonium nitrate is the only nitrogen source and can be supplied in ^{15}N-enriched form. For labeling with amino acids, the labeled amino acids need to be added to the medium at concentrations up to 300 μM

2.4 LysC–Trypsin In-Solution Digestion

All the chemicals are of the highest purity grade.

Use for the preparation of all solutions H_2O bidest.

1. Urea.
2. Dithiothreitol (DTT).
3. Iodoacetamide.
4. Ammonium bicarbonate.
5. LysC.
6. Trypsin, sequencing grade.
7. Denaturation buffer: 6 M urea adjusted with 50 mM ammonium bicarbonate to pH 8.
8. Reduction buffer: 1 M DTT in 50 mM ammonium bicarbonate.
9. Alkylation buffer: 200 mM iodoacetamide in 50 mM ammonium bicarbonate. This buffer is prepared immediately before use as iodoacetamide is light sensitive.

2.5 Desalting of the Samples by C_{18} StageTips

Use for the preparation of all solutions H_2O bidest.

1. Several 200 μl pipette tips.
2. Empore™ C18 solid phase extraction disks.
3. Glacial acetic acid.
4. Acetonitrile hypergrade for LC-MS.
5. Trifluoroacetic acid for peptide synthesis.
6. Solution A: 0.5 % (v/v) acetic acid.
7. Solution B: 80 % (v/v) ACN, 0.5 % acetic acid.
8. Resuspension solution: 0.5 % (v/v) trifluoroacetic acid, 2.5 % (v/v) acetonitrile.

2.6 MS-Analysis

Use for the preparation of all solutions H_2O bidest.

1. Acetonitrile hypergrade for LC-MS.
2. Isopropanol HPLC grade.
3. Trifluoroacetic acid for peptide synthesis.
4. Solution A: 0.5 % acetic acid and 5 % isopropanol (*see* **Note 5**). Solution B: 0.5 % acetic acid, 5 % isopropanol, 80 % acetonitrile.
5. Nano-HPLC (Thermo Fisher Scientific).
6. Chromolith® CapRod® RP-18e 150-0.1 monolithic analytical column (Merck).
7. LTQ-Orbitrap (Thermo Fisher Scientific).

2.7 Data Analysis

1. Xcalibur (Thermo Fisher Scientific).
2. DTA Supercharge Peak converter included in the MSQuant software.
3. Mascot Peptide identification software (at least version 2.2).
4. MSQuant quantitation software (at least version 1.5).

3 Methods

3.1 Hydroponic System for Metabolic Labeling of Whole Arabidopsis Plants

The method described in the following is very useful for large scale experiments, since several hundred plants can be grown in parallel with minor effort.

3.1.1 Sowing of the Seeds and Plant Cultivation

1. Prepare a clean working surface for the following steps (*see* **Note 6**).
2. Mount the sowing tray on top of the 5 L plastic box. Fit the rock-wool inside the tray and wet it with nutrient solution (Table 1).

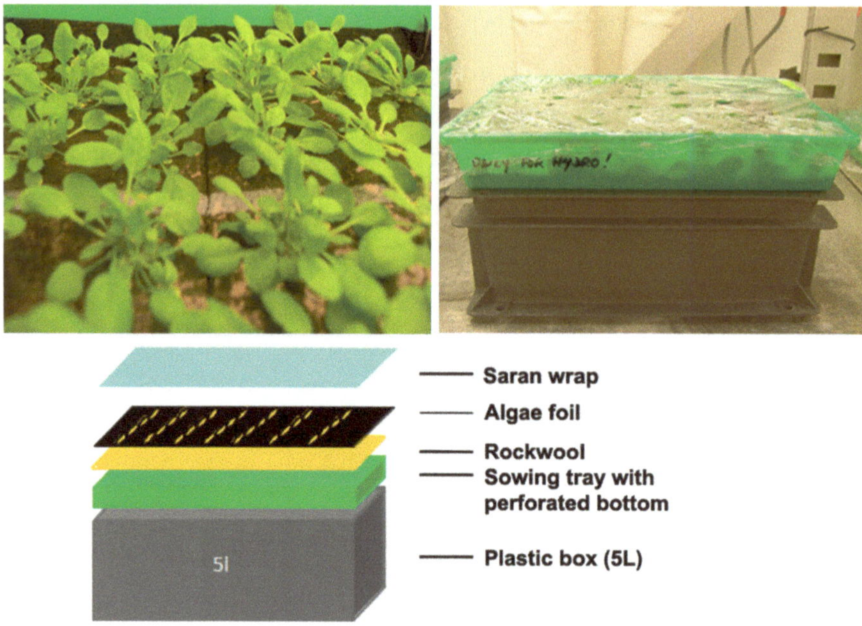

Fig. 2 Hydroponic system for metabolic labeling of plants (*top*). *Arabidopsis* plants grown for metabolic labeling and view of the experimental setup (*bottom*). Schematic diagram of the hydroponic system for metabolic labeling of whole plants

Make sure the rockwool is equally wet throughout the sowing tray, as otherwise the germination and seedling growth can be influenced. A schematic drawing of the hydroponic culture setup is shown in Fig. 2.

3. Punch holes into the algae foil. The size of the holes should be large enough to position several seeds inside the hole (about 5 mm). The distance of the holes should be at least 3 cm to allow one plant to grow up to full size and even maybe produce seeds.
4. Spread the algae foil on top of the wet rockwool.
5. For germination, preincubate the seeds for 5 min in cold (4 °C) 0.1 % agar containing 3.5 ng/ml gibberellic acid 4 + 7 (*see* **Note 4**).
6. At least 2 seeds should be sown per hole to make sure that a seeding establishes in every hole. Seedlings can only be transferred very carefully from one hole to another in the first 2 weeks after germination. After 2–3 weeks, the germinated seedlings should be individualized to prevent growth competition. Therefore, the surplus of plants is removed with tweezers.
7. Cover the sowing tray with saran wrap until the seeds have fully developed their cotyledons. To gradually reduce humidity to ambient conditions, holes are punched into the wrapping film before removing it completely after 3 weeks (*see* **Note 7**).

8. After 1 week, fill the plastic box with nutrient solution with either ^{14}N or ^{15}N. Be careful, as the humidity under the saran wrap will increase afterwards.
9. Exchange the respective ^{15}N or ^{14}N nutrient solution every week. It is also possible in the beginning, when the plants are small, to change the medium only every second weeks, but then the pH of the medium and possible growth of other organisms in the medium should be controlled.
10. The plants are fully labeled after 3 weeks, even when unlabeled ^{14}N seeds were used.
11. To obtain ^{15}N-labeled seeds from the labeled plants, the exchange of the nutrient solution is stopped after seed set. The plants can then be bagged while the seeds are drying.

3.1.2 Processing of the Plant Material

1. Harvest the $^{14}N/^{15}N$ plant material separately.
2. Processing of the fresh or frozen material is carried out according to the particular experimental design. Protein extraction or organelle enrichment protocols have been described [47–49].
3. Mix the ^{14}N plant material with the ^{15}N plant material as soon as possible. In case of the comparison of two developmental stages, labeled and unlabeled plant material is mixed based on fresh weight (assuming equal protein extraction efficiencies). In case of biochemical treatments of one of the plant samples it might be necessary to mix the sample based on total protein content after the treatment or even derived parameters, such as chlorophyll content in chloroplast preparations.
4. Determine the protein concentration.
5. Precipitate the proteins in case of low protein concentration.

3.1.3 Protein Precipitation Especially for Low Protein Amount and Small Volume

1. Add 1 μl of glycogen to the sample (as carrier).
2. Add 3 volumes of high grade ethanol (use 5 volumes if sample is in high salt, i.e., urea).
3. Add 40 μl per ml of total volume of 2.5 M NaOAc, pH 5.
4. Let reaction stand at room temperature for 2–3 h.
5. Spin at 13,000 ×*g* in Eppendorf centrifuge.
6. Discard supernatant.

3.1.4 LysC–Trypsin Digestion [50]

For the whole procedure, *see* **Notes 8** and **9**.

1. The protein pellets are diluted in the denaturing solution. Keep the volume of the solution low to increase digestion efficiency.
2. Optional: Reduce the peptide disulfide bonds by DTT. Add DTT solution to a final concentration of 4 mM to the sample. Incubate for 30 min at RT (*see* **Note 10**).

3. Optional: The reduced disulfide bonds are blocked by iodoacetamide. Add the iodoacetamide to a final concentration of 20 mM. Incubate for 30 min in the dark (*see* **Note 10**).
4. Add LysC to the protein solution. The ratio of LysC–protein (w/w) should be 1:100.
5. Incubate for 4 h at RT.
6. Dilute the sample with ammonium bicarbonate buffer to a final urea concentration of 1.6 M.
7. Add trypsin in a ratio trypsin–protein 1:50 and incubate over night at RT.
8. Acidify the sample with TFA to a final TFA concentration of 0.2 %.
9. Process the sample immediately when possible. Otherwise store the samples at −20 °C after desalting.

3.1.5 Desalting the Samples [51]

Electrospray ionization is sensitive to salts. Therefore, the samples have to be desalted.

1. Place one Empore disk on a clean surface (for example a petri dish).
2. Take the flat tipped syringe needle and punch out 2–3 little disks and eject them into the 200 μl pipette tip. Fix them properly at the bottom of the tip.
3. Prepare as many tips as you need for your samples.
4. Condition the StageTip with 50 μl B and centrifuge at 14,000 × *g*.
5. Equilibrate the StageTip two times with 100 μl solution A and centrifuge each time for 1 min at 14,000 × *g*.
6. Load the sample and centrifuge for 15 min at 400 × *g* (*see* **Note 11**).
7. The samples are desalted by washing two times with 100 μl solution A. The washing solution is removed by centrifuging the samples 2 min at 1000 × *g*.
8. Concentrate the samples in a speed vacuum centrifuge until the samples are semidry (ca. 1 μl left) (*see* **Note 12**).
9. Add final solution to a volume of 15 μl for 25 μg of protein.

3.2 Mass Spectrometry Analysis

1. Mount your HPLC column on the LC-MS/MS setup according to the manufacturer's instruction.
2. Separate the peptides by reversed phase chromatography with a linear gradient running solution A from 95 % decreasing to 40 % in 2 h and an elution step for 10 min.
3. Run the LTQ-Orbitrap with the following settings: resolution of 30,000 (full width at half maximum) in the Orbitrap; data dependent acquisition: top 5 sequencing in the mass

Table 2
Quantitation software that is free to use but may require specific input formats

Software	Website	Features	Formats	Reference
ASAP Ratio	tools.proteomecenter.org/wiki/index.php?title=Software:ASAPRatio	^{15}N labeling, ICAT, label-free; possibility to self-define mass tags	mzXML, pepXML, DTAselect	[57]
Census	fields.scripps.edu/census/index.php	^{15}N labeling, ICAT, ITRAQ, label-free; possibility to self-define mass tags	mzXML, pepXML, DTAselect	[58]
MaxQuant	maxquant.org	SILAC, label-free; possibility to self-define mass tags; NOT for ^{15}N labeling	Instrument raw files (Thermo)	[59–61]
MSQuant	msquant.sourceforge.net	^{15}N labeling, ICAT, label-free; possibility to self-define mass tags	Mascot output in htm-format instrument raw files (Thermo, ABI, Waters)	[52]
Proteomatic				[62]
XPRESS	tools.proteomecenter.org/wiki/index.php?title=Software:XPRESS	ICAT	mzXML, pepXML, DTAselect	[63]

All except for MaxQuant are also compatible with full ^{15}N-labeling

range from 300 to 1,500 in the LTQ. The collisional energy is 35 %. The settings for the spray are: spray voltage 1.5 kV, capillary temperature 200 °C, capillary voltage 23 V, tube lens voltage 180 V.

3.3 Data Analysis Workflow in Metabolic Labeling of Plants

After data acquisition on the mass spectrometer, data has to be processed by several steps before resulting in protein identifications and quantitative protein values. In many cases, the identification process and the quantitation process are carried out independently and are linked later at the level of individual spectra. Identification by database search is aided by various software packages such as SEQEST (Thermo Finnigan), Mascot (Matrix Science), X!Tandem (thegpm.org), OMSSA (NCBI), and Phenyx (GeneBio), and most of these allow for definition of search parameters for full ^{15}N-labeling. Also for quantitation, a number of freely available programs can be used (Table 2), that allow for automation of most of these tasks. Generic formats such as mzData, mzXML, or pepXML allow for instrument vendor-independent use of software algorithms.

In the following, we describe a data analysis workflow based on LTQ-Orbitrap raw data processed for Mascot database search and quantitation with MSQuant.

1. Conversion of the raw data into a peak list: The precise settings for each converter are dependent on the mass spectrometer used. Example of Peak list generation from Thermo raw files with DTASupercharge (compatible with a later Mascot Search):

   Preprocess settings: run extract_msn.exe (newer version ExtractMSn.exe; Daemon 2.3 and earlier versions do not recognize this and have to be renamed. *See*: http://www.matrixscience.com/help/instruments_xcalibur.html); Postprocess settings: mgf file generation; deletion of DTA files is recommended, as lots of files are generated that are not needed after the peak list file has been generated. Batch processing of several files can be set in the "Automation" menu.

2. Peptide and protein identification: The peptide search is done according the usual settings with extension to ^{15}N peptide identification. In Mascot for example one should additionally select the respective quantitation mode, such as "^{15}N Metabolic (MD)". An example of typical parameters for Mascot used for data from an Orbitrap mass spectrometer:

   Max. missed cleavages: 2, decoy database: yes; fixed modifications: carbamidomethyl (if alkylation with iodoacetamide was done); variable modifications: oxidation (M); peptide charge: 2+ and 3+; monoisotopic: yes; peptide tolerance: 10 ppm; MS/MS ion search: yes; MS/MS tolerance +/− 0.8 Da; quantitation: ^{15}N metabolic (MD); instrument: ESI-TRAP. Mascot search results are saved in html format.

3. Quantification using MSQuant [52]: Associate raw data with the peptide search results, take suitable score threshold settings for the experiment and let the program process and quantify the data. Similar to the previous data processing steps the precise settings depend on which methods/machines you used and which kind of information is needed from the data.

MSQuant is compatible with Mascot. Therefore, many options are linked to the settings in the Mascot search, and information and given in the search result (e.g., score thresholds). The following options settings are recommended for typical pairwise comparisons: selection of bold red (refers to the color code of Mascot assigned to unambiguously identified proteins), parenthesized (includes duplicated peptides identifications, i.e., from different modification states for quantification), checked red only (includes top-ranking peptides assigned to individual lower abundant proteins). In the peptide filter option charge ranges, sequence length of the peptides, mass range and score thresholds can be selected based on the significance scores in the Mascot search result.

Modification filters can additionally be set in the "peptide filter". For peptide recalibration select "frequency" if using an Orbitrap mass spectrometer. A quantitation mode allowing for full ^{15}N-quantitation needs to be defined and selected, and the respective peak list generator and raw file type must be specified [53].

3.4 Statistical Analysis

Good quality spectra will be averaged by MSQuant to yield abundance ratios of labeled and unlabeled forms for a peptide, and peptides will be averaged to obtain protein ratios. The peptide ratios can be derived from two values: (1) the ion intensities of the monoisotopic and the +1 peak or (2) the extracted ion chromatogram. The values from the extracted ion chromatogram are preferable, as they are independent of the peptide amino acid content. The program does not correct for long peptides and their complex isotope clusters. So if one would like to use the ion intensities, one may have to apply respective correction factors [22]. If the standard deviation of a protein ratio is high, it is advised to check the quantification of the single peptides manually in MSQuant. There might be cases in which the program finds presumably corresponding peaks by assigning spectra of two different peptides. Also the assignment of peptides to proteins is challenging, since many peptide sequences can match more than one protein. For accurate quantitation, it is important to only consider those peptides in the quantitation that are unique to a particular protein, so-called proteotypic peptides [54]. In any case, replicated experiments are necessary to draw meaningful biological conclusions. The final step in the data analysis workflow is the statistical analysis, which can be aided by using StatQuant [55] or cRacker [56].

4 Notes

1. It is not recommended to reuse boxes that had previously been used for ^{14}N hydroponic medium, as these boxes could already have absorbed ^{14}N to their surface, and this could lead to a source of ^{14}N contamination.
2. Usually there is no inorganic nitrate or ammonium nitrogen in the rockwool, but we still recommend to test this by soaking some rockwool in water and checking for nitrate with nitrate test strips (Merck).
3. The algae foil prevents algae from growing on the rockwool without formation of condensation water below the film.
4. This is an optional step and also incubation of seeds in Topagar without gibberellic acid or even just in water is possible. However, seed germination is more uniform in presence of

the hormone. Furthermore, seeds in agar can be sown with a pipette.

5. Isopropanol is used to stabilize the spray.
6. Any contamination with soil should be avoided, as it will introduce Cyanobacteria or blue-green algae to the system. These algae can lead to an incomplete labeling as a result of nitrogen fixation and thus contamination of the system with ^{14}N ammonium.
7. As a rule, there should be visible mild condensation on the saran wrap in the first days up to 2 weeks, but formation of water drops should be avoided. If the plants develop glassy leaves, the humidity is far too high, and humidity should slowly be reduced by punching more and more holes into the saran wrap. In contrast, too fast changes from high humidity to low humidity can cause necrotic speckles on the leaves. A proper humidity control is the most important point for uniform seed germination and plant development. Avoid moving the hydroponic cultures as this may lead to roots sticking to the bottom of the sowing tray.
8. To avoid keratin contaminations, work is done under a clean fume hood or the sterile bench. The lab coat should fully cover the skin. To avoid Latex contaminations, use nitrile gloves.
9. The solutions used for protein digestion can cause different peptide modifications, which can later on interfere with the data analysis. For example, urea can cause carbamylation, iodoacetamide can lead to over-alkylation, and alkaline buffers can result in deamidation or glutamine cyclization. The extent of these modifications depends highly on the temperature and the incubation time. Therefore, the reduction step should be done at room temperature, the alkylation step should not exceed the 30 min and the digestion time should be kept as short as possible.
10. The reduction and the alkylation could be skipped if there is sufficient protein coverage of the peptides without cysteine containing peptides.
11. For good peptide recovery, it is important to centrifuge at a low speed. At this low speed, it takes sometimes longer to sufficiently spin the liquid through.
12. The samples should not be completely dried as this causes difficulties in resolubilizing the peptides for mass spectrometric analysis. Ideally, the drying process is stopped when there is about 1–2 μl of liquid left. Do not interrupt the drying process and ideally store samples in dry state. In the dehydration process, the samples undergo a strong pH gradient to more acidic, which can cause hydrolysis of the peptides.

References

1. Ong SE, Blagoev B, Kratchmarova I et al (2002) Stable isotope labeling by amino acids in cell culture, SILAC, as a simple and accurate approach to expression proteomics. Mol Cell Proteomics 1:376–386
2. Bantscheff M, Schirle M, Sweetman G et al (2007) Quantitative mass spectrometry in proteomics: a critical review. Anal Bioanal Chem 389:1017–1031
3. Ong SE, Schenone M, Margolin AA et al (2009) Identifying the proteins to which small-molecule probes and drugs bind in cells. Proc Natl Acad Sci U S A 106:4617–4622
4. Selbach M, Schwanhäusser B, Thierfelder N et al (2008) Widespread changes in protein synthesis induced by microRNAs. Nature 455:58–63
5. Schwannhäusser B, Gossen M, Dittmar G et al (2009) Global analysis of cellular protein translation by pulsed SILAC. Proteomics 9:205–209
6. Soufi B, Kumar C, Gnad F et al (2010) Stable isotope labeling by amino acids in cell culture (SILAC) applied to quantitative proteomics of Bacillus subtilis. J Proteome Res 9:3638–3646
7. Gruhler A, Olsen JV, Mohammed S et al (2005) Quantitative phosphoproteomics applied to the yeast pheromone signaling pathway. Mol Cell Proteomics 4:310–327
8. Cuomo A, Bonaldi T (2010) Systems biology "on-the-fly": SILAC-based quantitative proteomics and RNAi approach in Drosophila melanogaster. Methods Mol Biol 662:59–78
9. Larance M, Bailly AP, Pourkarimi E et al (2011) Stable-isotope labeling with amino acids in nematodes. Nat Methods 8:849–851
10. Krüger M, Moser M, Ussar S et al (2008) SILAC mouse for quantitative proteomics uncovers kindlin-3 as an essential factor for red blood cell function. Cell 134:353–364
11. Naumann B, Busch A, Allmer J et al (2007) Comparative quantitative proteomics to investigate the remodeling of bioenergetic pathways under iron deficiency in Chlamydomonas reinhardtii. Proteomics 7:3964–3979
12. Therashima M, Specht M, Naumann B et al (2010) Characterizing the anaerobic response of Chlamydomonas reinhardtii by quantitative proteomics. Mol Cell Proteomics 9:1514–1532
13. Heide H, Nordhues A, Drepper F et al (2009) Application of quantitative immunoprecipitation combined with knockdown and cross-linking to Chlamydomonas reveals the presence of vesicle-inducing protein in plastids 1 in a common complex with chloroplast HSP90C. Proteomics 9:3079–3089
14. Gruhler A, Schulze WX, Matthiesen R et al (2005) Stable isotope labeling of Arabidopsis thaliana cells and quantitative proteomics by mass spectrometry. Mol Cell Proteomics 4:1697–1702
15. Schütz W, Hausmann N, Krug K et al (2011) Extending SILAC to proteomics of plant cell lines. Plant Cell 23:1701–1705
16. Svennerstam H, Ganeteg U, Näsholm T (2008) Root uptake of cationic amino acids by Arabidopsis depends on functional expression of amino acid permease 5. New Phytol 180:620–630
17. Hirner A, Ladwig F, Stransky H et al (2006) Arabidopsis LHT1 is a high-affinity transporter for cellular amino acid uptake in both root epidermis and leaf mesophyll. Plant Cell 18:1931–1946
18. Somerville CR, Ogren WL (1980) Photorespiration mutants of Arabidopsis thaliana deficient in serine-glyoxylate aminotransferase activity. Proc Natl Acad Sci U S A 77:2684–2687
19. Forsum O, Svennerstam H, Ganeteg U et al (2008) Capacities and constraints of amino acid utilization in Arabidopsis. New Phytol 179:1058–1069
20. Ippel JH, Pouvreau L, Kroef T et al (2004) In vivo uniform (15)N-isotope labelling of plants: using the greenhouse for structural proteomics. Proteomics 4:226–234
21. Arsova B, Kierszniowska S, Schulze WX (2012) The use of heavy nitrogen in quantitative proteomics experiments in plants. Trends Plant Sci 17:102–112
22. Gouw JW, Tops BBJ, Mortensen P et al (2008) Optimizing identification and quantitation of ^{15}N-labeled proteins in comparative proteomics. Anal Chem 80:7796–7803
23. Marschner H (1998) Mineral nutrition of higher plants. Academic, London
24. Engelsberger WR, Erban A, Kopka J et al (2006) Metabolic labeling of plant cell cultures with $K_{15}NO3$ as a tool for quantitative analysis of proteins and metabolites. Plant Methods 2:1–11
25. Lanquar V, Kuhn L, Lelièvre F et al (2007) ^{15}N-metabolic labeling for comparative plasma membrane proteomics in Arabidopsis cells. Proteomics 7:750–754
26. Bindschedler LV, Palmblad M, Cramer R (2008) Hydroponic isotope labelling of entire

plants (HILEP) for quantitative plant proteomics; an oxidative stress case study. Phytochemistry 69:1962–1972

27. Laganowsky A, Gomez SM, Whitelegge JP et al (2009) Hydroponics on a chip: anaysis of the Fe deficient Arabidopsis thylakoid membrane proteome. J Proteomics 72:397–415
28. Nelson CJ, Huttlin EL, Hegeman AD et al (2007) Implications of ^{15}N-metabolic labeling for automated peptide identification in Arabidopsis thaliana. Proteomics 7:1279–1292
29. Huttlin EL, Hegeman AD, Harms AC et al (2007) Comparison of full versus partial metabolic labeling for quantitative proteomic analysis in Arabidopsis thaliana. Mol Cell Proteomics 6:860–881
30. Kline KG, Barrett-Wilt GA, Sussman MR (2010) In planta changes in protein phosphorylation induced by the plant hormone abscisic acid. Proc Natl Acad Sci U S A 107:15986–15991
31. Guo G, Li N (2011) Relative and accurate measurement of protein abundance using (15) N stable isotope labeling in Arabidopsis (SILIA). Phytochemistry 72:1028–1039
32. Hebeler R, Oeljeklaus S, Reidegeld KA et al (2008) Study of early leaf senescence in Arabidopsis thaliana by quantitative proteomics using reciprocal ^{14}N/15N labeling and difference gel electrophoresis. Mol Cell Proteomics 7:108–120
33. Skirycz A, Memmi S, De Bodt S et al (2011) A reciprocal ^{15}N-labeling proteomic analysis of expanding Arabidopsis leaves subjected to osmotic stress indicates importance of mitochondria in preserving plastid functions. J Proteome Res 10:1018–1029
34. Schaff JE, Mbeunkui F, Blackburn K et al (2008) SILIP: a novel stable isotope labeling method for in planta quantitative proteomic analysis. Plant J 56:840–854
35. Gruhler A, Kratchmarova I (2008) Stable isotope labeling by amino acids in cell culture (SILAC). Methods Mol Biol 424:101–111
36. Benschop JJ, Mohammed S, O'Flaherty M, Heck AJ, Slijper M, Menke FL (2007) Quantitative phospho-proteomics of early elicitor signalling in Arabidopsis. Mol Cell Proteomics 6:1705–1713
37. Kierszniowska S, Seiwert B, Schulze WX (2009) Definition of Arabidopsis sterol-rich membrane microdomains by differential treatment with methyl-ß-cyclodextrin and quantitative proteomics. Mol Cell Proteomics 8: 612–623
38. Kierszniowska S, Walther D, Schulze WX (2009) Ratio-dependent significance thresholds in reciprocal ^{15}N-labeling experiments as a robust tool in detection candidate proteins responding to biological treatment. Proteomics 9:1916–1924
39. Keinath NF, Kierszniowska S, Lorek J et al (2010) PAMP-induced changes in plasma membrane compartmentalization reveal novel components of plant immunity. J Biol Chem 285:39140–39149
40. Mühlhaus T, Weiss J, Hemme D et al (2011) Quantitative shotgun proteomics using a uniform ^{15}N-labeled standard to monitor proteome dynamics in time course experiments reveals new insights into the heat stress response of Chlamydomonas reinhardtii. Mol Cell Proteomics 10:M110.004739
41. Palmblad M, Mills DJ, Bindschedler LV (2008) Heat-shock response in Arabidopsis thaliana explored by multiplexed quantitative proteomics using differential metabolic labeling. J Proteome Res 7:780–785
42. Nowaczyk MM, Hebeler R, Schlodder E et al (2006) Psb27, a cyanobacterial lipoprotein, is involved in the repair cycle of photosystem II. Plant Cell 18:3121–3131
43. Zhang Y, Reckow S, Webhofer C et al (2011) Proteome scale turnover analysis in live animals using stable isotope metabolic labeling. Anal Chem 83:1665–1672
44. Martin SF, Munagapati VS, Salvo-Chirnside E et al (2012) Proteome turnover in the green alga Ostreococcus tauri by time course ^{15}N metabolic labeling mass spectrometry. J Proteome Res 11:476–486
45. Li L, Nelson CJ, Solheim C et al (2012) Determining degradation and synthesis rates of Arabidopsis proteins using the kinetics of progressive ^{15}N labeling of two-dimensional gel-separated protein spots. Mol Cell Proteomics 11:M111.010025
46. Loqué D, Tillard P, Gojon A et al (2003) Gene expression of the NO3- transporter NRT1.1 and the nitrate reductase NIA1 is repressed in Arabidopsis roots by NO2-, the product of NO3- reduction. Plant Physiol 132:958–967
47. Carroll AJ, Heazlewood JL, Ito J et al (2008) Analysis of the Arabidopsis cytosolic ribosome proteome provides detailed insights into its components and their post-translational modification. Mol Cell Proteomics 7:347–369
48. Marmagne A, Salvi D, Rolland N et al (2006) Purification and fractionation of membranes for proteomic analyses. Methods Mol Biol 323:403–420
49. Lilley KS, Dupree P (2007) Plant organelle proteomics. Curr Opin Plant Biol 10:594–599
50. Glatter T, Ludwig C, Ahrné E et al (2012) Large-scale quantitative assessment of different

in-solution protein digestion protocols reveals superior cleavage efficiency of tandem Lys-C/ trypsin proteolysis over trypsin digestion. J Proteome Res 11:5145–5156

51. Rappsilber J, Ishihama Y, Mann M (2003) Stop And Go Extraction tips for matrix-assisted laser desorption/ionization, nanoelectrospray, and LC/MS sample pretreatment in proteomics. Anal Chem 75:663–670
52. Mortensen P, Gouw JW, Olsen JV et al (2010) MSQuant, an open source platform for mass spectrometry-based quantitative proteomics. J Proteome Res 9:393–403
53. Gouw JW, Krijgsveld J, Heck AJ (2010) Quantitative proteomics by metabolic labeling of model organisms. Mol Cell Proteomics 9:11–24
54. Ong SE, Mann M (2005) Mass spectrometry-based proteomics turns quantitative. Nat Chem Biol 1:252–262
55. van Breukelen B, van den Toorn HW, Drugan MM et al (2009) StatQuant: a post-quantification analysis toolbox for improving quantitative mass spectrometry. Bioinformatics 25:1472–1473
56. Zauber H, Schulze WX (2012) Proteomics wants cRacker: Automated standardized data analysis of LC/MS derived proteomic data. J Proteome Res 11:5548–5555
57. Li XJ, Zhang H, Ranish JR et al (2003) Automated statistical analysis of protein abundance ratios from data generated by stable isotope dilution and tandem mass spectrometry. Anal Chem 75:6648–6657
58. Park SK, Venable JD, Xu T et al (2008) A quantitative analysis software tool for mass spectrometry-based proteomics. Nat Methods 5:319–322
59. Cox J, Mann M (2008) MaxQuant enables high peptide identification rates, individualized p.p.b.-range mass accuracies and proteome-wide protein quantification. Nat Biotechnol 26:1367–1372
60. Cox J, Matic I, Hilger M et al (2009) A practical guide to the MaxQuant computational platform for SILAC-based quantitative proteomics. Nat Protoc 4:698–705
61. Cox J, Neuhauser N, Michalski A et al (2011) Andromeda: a peptide search engine integrated into the MaxQuant environment. J Proteome Res 10:1794–1805
62. Specht M, Kuhlgert S, Fufezan C et al (2011) Proteomics to go: Proteomatic enables the user-friendly creation of versatile MS/MS data evaluation workflows. Bioinformatics 27: 1183–1184
63. Han DK, Eng J, Zhou H et al (2001) Quantitative profiling of differentiation-induced microsomal proteins using isotope-coded affinity tags and mass spectrometry. Nat Biotechnol 19:946–951

Chapter 7

In Vivo Stable Isotope Labeling by Amino Acids in *Drosophila melanogaster*

Matthias D. Sury*, Jia-Xuan Chen*, and Matthias Selbach

Abstract

The fruit fly *Drosophila melanogaster* is one of the most widely used and well-studied model organisms in biology and therefore a promising tool for quantitative proteomics. Here, we describe a method to label *D. melanogaster* with stable isotope labeled amino acids in vivo. Feeding flies with heavy lysine labeled yeast cells leads to virtually complete heavy labeling already in the first filial generation. The approach is simple, fast, and cost-effective, which makes SILAC flies an attractive model system for the emerging field of in vivo quantitative proteomics.

Key words *Drosophila melanogaster*, SILAC, In vivo labeling, Yeast, Protein extraction

1 Introduction

Studies in *D. melanogaster* have so far mainly focused on genetic aspects. However, recent advances in mass spectrometry have led to an increased interest in quantitative proteomic analysis of *D. melanogaster* [1]. SILAC in the fly system can be performed with cell lines grown in culture [2, 3]. While highly useful, these cell culture models cannot appropriately reflect all relevant regulatory mechanisms of multicellular eukaryotes in vivo. As an alternative, we therefore labeled entire flies by feeding larvae on a diet of SILAC labeled yeast [4]. SILAC flies generated in this way enable in vivo proteome-wide quantification with higher precision than label free methods. Our data shows that feeding flies with heavy lysine labeled yeast leads to almost complete labeling in the first filial generation and can be used to distinguish sex-specific proteins in vivo (Fig. 1). To avoid potential arginine to proline conversion, labeling with heavy lysine was chosen.

Metabolic labeling by heavy lysine of *D. melanogaster* is simple, fast, and cost-effective. Together with other animals in the SILAC

*Matthias D. Sury and Jia-Xuan Chen have equally contributed to this Chapter.

Bettina Warscheid (ed.), *Stable Isotope Labeling by Amino Acids in Cell Culture (SILAC): Methods and Protocols*, Methods in Molecular Biology, vol. 1188, DOI 10.1007/978-1-4939-1142-4_7, © Springer Science+Business Media New York 2014

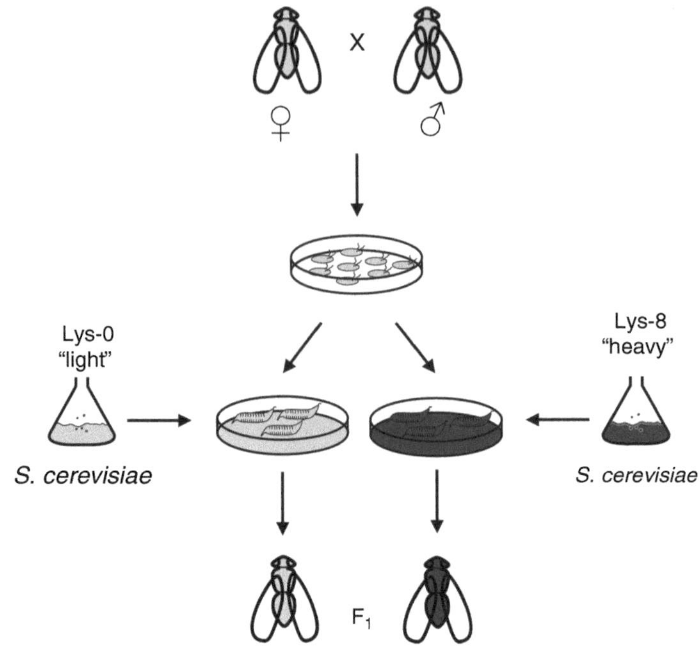

Fig. 1 Workflow of stable isotope labeling with amino acids in *D. melanogaster*. Embryos from a mixed male and female *D. melanogaster* population are collected. Larvae are fed with "light" $^{12}C_6{}^{14}N_2$ L-lysine (Lys-0) or "heavy" $^{13}C_6{}^{15}N_2$ L-lysine (Lys-8) labeled *S. cerevisiae*. Adult F1 subpopulations are mixed and analyzed by LC-MS/MS. Pairs of identical peptides with different stable-isotope compositions can be distinguished by the mass spectrometer based on their mass difference of 8 Da

zoo [5–8], the fly is an attractive model system to gain new insights into biological processes in vivo.

2 Materials

2.1 D. melanogaster Standard Culture

1. *D. melanogaster* standard medium: 8.5 g/l agar, 76.6 g/l cornmeal, 81.6 g/l malt extract, 40.8 g/l molasses, 18.0 g/l brewer's yeast, 10.0 g/l soy flour, 0.45 % (v/v) propionic acid, 0.16 % (v/v) methyl 4-hydroxybenzoate (Nipagin) in 95 % EtOH (*see* **Notes 1** and **2**).
2. Overhead stirrer.
3. Peristaltic dispenser.
4. *D. melanogaster* culture tubes: polystyrene flat bottom tubes (68 ml), ceaprene foam plugs.

2.2 S. cerevisiae SILAC Labeling

1. Yeast drop out minimal medium for SILAC labeling [9]: 1.7 g/l yeast nitrogen base (without amino acids, without ammonium sulfate), 20 g/l D-glucose, 5 g/l ammonium sulfate, 200 mg/l adenine hemisulfate, 20 mg/l uracil, 100 mg/l

Tyr, 10 mg/l His, 60 mg/l Leu, 10 mg/l Met, 60 mg/l Phe, 40 mg/l Trp, 100 mg/l Arg, 30 mg/l Lys-$^{12}C_6{}^{14}N_2$ (Lys-0), or 30 mg/l Lys-$^{13}C_6{}^{15}N_2$ (Lys-8) (*see* **Note 3**).

2. Lysine auxotrophic yeast strain [10]: SUB62/DF5 (MATalpha lys2-801 leu2-3/112 ura3-52 his3-delta200 trp1-1) (*see* **Note 4**).
3. Orbital shaker.
4. Spectrophotometer.

2.3 Embryo Collection

1. Apple juice agar solution: 22.5 g/l agar, 25 % (v/v) apple juice, 25 g/l sucrose, 0.16 % (v/v) Nipagin in 95 % EtOH (*see* **Note 5**).
2. Petri dishes (100 mm × 15 mm).
3. Yeast from *Saccharomyces cerevisiae* Type 2 (Sigma-Aldrich, St. Louis, MO, USA).
4. Embryo collection container (Fig. 2a).
5. Embryo wash unit (Fig. 2b).
6. CO_2 blowgun and CO_2 flypad (Genesee Scientific, San Diego, CA, USA).
7. Paint brush.
8. Plastic wrap.

2.4 D. melanogaster Labeling

1. *Drosophila* labeling medium: 60 % (w/v) labeled SUB62 yeast (wet mass), 320 mM sucrose, 0.3 mM ampicillin, 0.1 % (v/v) Nipagin in 95 % EtOH, 0.5 % propionic acid, and 2.5 % phosphoric acid.
2. Labeling container (Fig. 3).
3. Petri dish.
4. Medical grade cotton wool.
5. Disinfected tissue paper.
6. Toothpick.
7. Plastic cup.
8. Paint brush.
9. Plastic wrap.

2.5 Protein Extraction

1. Modified radioimmunoprecipitation assay (RIPA) buffer: 50 mM Tris–HCl (pH 7.4), 150 mM NaCl, 1 % NP-40, 0.25 % Na-deoxycholate, 1 mM EDTA, 0.1 % SDS, and 1X protease inhibitor cocktail complete.
2. Microcentrifuge tubes.
3. Tissue homogenizer (Ultra-Turrax, Staufen, Germany).
4. Ultrasonic bath.
5. Coomassie Plus (Bradford) Protein Assay.

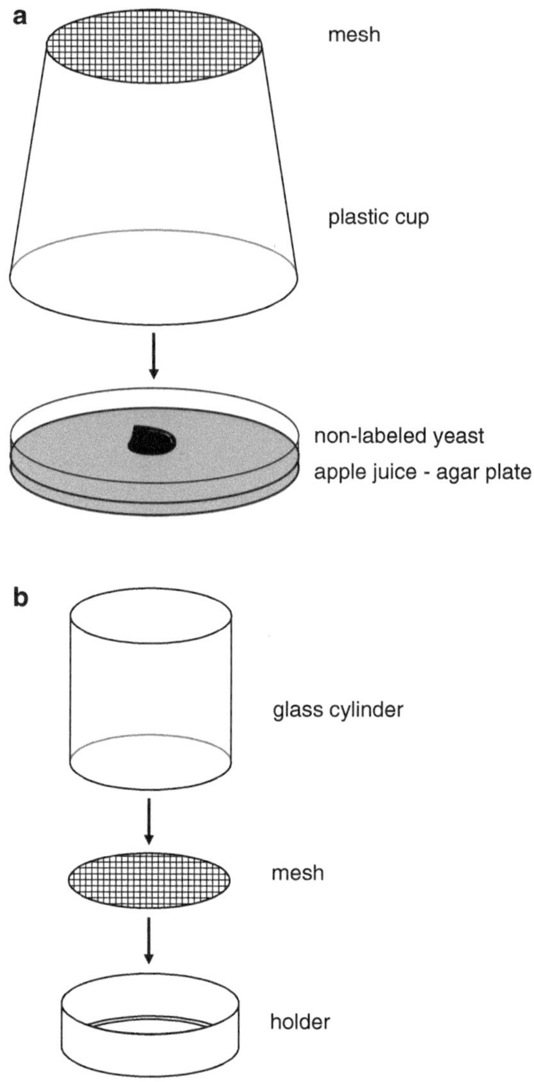

Fig. 2 (**a**) Setup of the embryo collection container. A petri dish containing 10 ml apple juice agar and a layer of non-labeled yeast is capped with a plastic cup. For optimal aeration, the bottom of the cup is replaced with a fine mesh. The cup is attached to the petri dish via plastic wrap. (**b**) Setup of the embryo wash unit. To collect embryos from the apple juice agar plate, embryos are filtered through a 150 μm metal mesh placed between a glass cylinder and a plastic holder

3 Methods

3.1 D. melanogaster Standard Culture

1. To prepare 1 l of culture medium, boil agar in 800 ml water until agar is dissolved, then add cornmeal by constantly stirring.
2. Dissolve malt extract in 100 ml water.

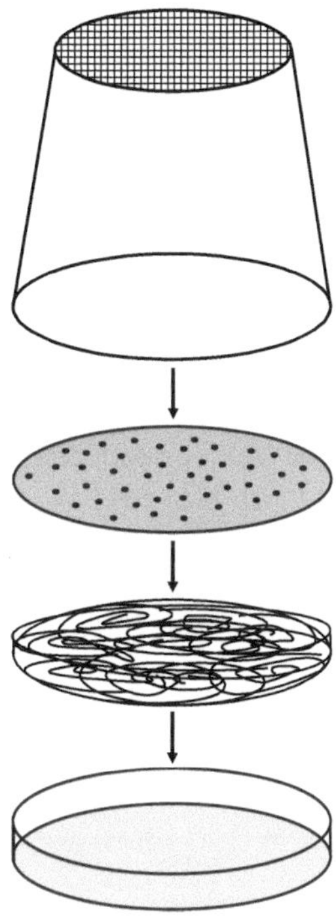

Fig. 3 Setup of the labeling container. Cotton wool soaked with labeling medium is placed in a petri dish and covered with a perforated tissue paper. Embryos are dispersed with a paint brush on the tissue paper. The petri dish is capped with a plastic cup with its bottom replaced with a fine mesh. The labeling container is sealed with plastic wrap

3. Warm up molasses in 100 ml water and mix with dissolved malt extract, soy flour, and yeast. Add this mixture to the heated agar–cornmeal solution and return to a boil.
4. Cool down to 60 °C and then add Nipagin and propionic acid. Stir for additional 15 min.
5. Dispense 25 ml culture medium per culture tube by using a peristaltic dispenser. Store tubes at 4 °C until use (*see* **Note 6**).
6. Keep flies in standard culture tubes at 18 °C, ~60 % relative humidity (*see* **Note 7**) and in a 12 h light-dark cycle. Passage flies every 8 weeks.

3.2 S. cerevisiae SILAC Labeling

1. Transfer a small clone of the unlabeled lysine auxotrophic SUB62 strain to a 100 ml Erlenmeyer flask containing 5 ml of yeast labeling medium.
2. Incubate at 30 °C overnight with orbital shaking at 250 rpm.
3. Transfer 10 μl of the liquid culture to another Erlenmeyer flask containing 5 ml of yeast labeling medium. Culture the pre-culture as described in **step 2**.

4. Transfer 125 μl of the second pre-culture into 125 ml of yeast labeling medium in a 1-l baffled Erlenmeyer flask.
5. Incubate at 30 °C for ~24 h with orbital shaking at 250 rpm. Dilute 100 μl of the liquid culture 1:10 in water. Measure the optical density (OD) of the diluted culture at 600 nm with a spectrophotometer.
6. When the OD600 of the diluted liquid culture reaches 0.7–0.8, harvest the SILAC yeasts by centrifuging at 1,900×*g* for 10 min at RT.
7. Remove the supernatant and store the SILAC yeast pellet at −20 °C (*see* **Note 8**).

3.3 Embryo Collection

1. To prepare 1 l of apple juice agar solution, dissolve 22.5 g agar in 750 ml water.
2. Autoclave the agar solution.
3. When the autoclaved agar solution has cooled down to ~60 °C, keep it in a water bath at 60 °C.
4. Pre-warm 250 ml of apple juice to 60 °C and add sucrose and Nipagin.
5. Thoroughly mix the apple juice into the autoclaved agar solution.
6. Dispense the apple juice agar solution into the petri dishes (10 ml/dish).
7. Once the apple juice agar plates have cooled and hardened, store them at 4 °C until use.
8. Add a thick paste (~2 g) of non-labeled yeast in the center of the apple juice agar plate. The yeast serves as a food source during egg deposition (*see* **Note 9**).
9. To anesthetize the flies, pierce the blowgun into the culture tube and carefully blow CO_2 into the tube. Wait until the flies do not move anymore. Transfer the flies to a CO_2 flypad and sort them according to their sex. Transfer 40–50 male and female flies to an embryo collection container. Seal container with plastic wrap and incubate at 25 °C overnight (*see* **Notes 10** and **11**).
10. Pierce the blowgun into the embryo collection container and anesthetize the flies with CO_2. To collect the embryos, remove the apple juice agar plate and rinse embryos extensively with water. By using a paint brush, filter the embryos through the embryo wash unit to get rid of the yeast (*see* **Note 12**).
11. After working with embryos, freeze all devices overnight at −20 °C to avoid uncontrolled hatching of *D. melanogaster*.

3.4 D. melanogaster Labeling

1. Place a thin layer of medical grade cotton wool onto a 100 mm×15 mm petri dish.
2. Dispense 10 ml of the *Drosophila* labeling medium evenly onto the cotton wool. Cover the cotton with a thin piece of

disinfected tissue paper. Make small holes through the paper using a toothpick. This allows the larvae to have easy access to the labeling medium (*see* **Notes 13** and **14**).

3. Transfer the embryos onto the tissue paper. Couple the dish to a plastic cup using plastic wrap. Incubate at 25 °C until hatching of adult flies (*see* **Note 15**).
4. If depleted, additional *Drosophila* labeling medium should be added during larval stage.
5. Transfer hatched flies to a new embryo collection container attached to an apple juice agar plate. Provide ~6 g/day SILAC labeled yeast (wet mass) until the flies reach the desired age for collection.
6. At this stage, hatched adult flies are almost completely labeled. However, they can be kept in an embryo collection container to collect fully labeled embryos.
7. For further processing, snap freeze flies or embryos in dry ice or liquid nitrogen and store at −80 °C.

3.5 Protein Extraction

1. Transfer whole flies or embryos to a 2-ml microcentrifuge tube containing ice-cold modified RIPA buffer (150 μl of modified RIPA buffer per ~10 flies).
2. Homogenize the whole flies vigorously for ~20 s on ice using a tissue homogenizer.
3. Put the fly homogenate for 3 min in an ultrasonic bath filled with ice-cold water.
4. Centrifuge at 16,000 × *g* for 10 min at 4 °C. Transfer the supernatant to a new microcentrifuge tube.
5. Determine the protein concentration by using the Bradford protein assay.
6. Store protein extracts at −80 °C until use.
7. Check labeling efficiency by mass spectrometry.

4 Notes

1. To prepare a 16 % (w/v) Nipagin solution, dissolve 16 g methylparaben in 100 ml EtOH (95 %). To get a final concentration of 0.16 %, dilute 10 ml Nipagin solution (16 %) in 990 ml *D. melanogaster* food.
2. Other *Drosophila* standard medium recipes can be found on the website of the Bloomington Drosophila Stock Center at Indiana University (http://flystocks.bio.indiana.edu).
3. Met (10 g/l), Lys (146 g/l), and Arg (84 g/l) stock solutions are prepared in PBS, Tyr (50 g/l), His (10 g/l), and Trp (40 g/l) stock solutions are prepared in 0.5 M NaOH, Leu

(30 g/l), and Phe (30 g/l) stock solutions are prepared in 0.5 M HCl.

4. The SUB62 yeast strain, also called DF5 (Dan Finley 5), is available at the American Type Culture Collection (ATCC), order number ATCC 200912. However, in principle any other lysine auxotrophic strain can be used.
5. Other juices such as grape juice may also be used.
6. To avoid desiccation, coat the culture tubes with moistened tissue papers and store the tubes in a sealed plastic bag at 4 °C.
7. Excessive humidity increases the risk of fungal growth. On the other hand, too low humidity leads to desiccation of the culture medium.
8. Labeling efficiency of SILAC yeast should be almost 100 %. Therefore, check labeling efficiency of SUB62 yeast strain by mass spectrometry before continuing with *D. melanogaster* labeling. Labeling efficiency can be checked by calculating the heavy to light ratio of the 100 most intense proteins [4], for example.
9. Mix the yeast with a small amount of water until it forms a sticky paste.
10. Pay attention that the CO_2 blow is not too strong, otherwise the flies will stick to the culture medium.
11. Female imagines are bigger than males. Furthermore, male flies have a rounded abdomen and strongly pigmented last abdominal tergites.
12. A larger amount of embryos can be collected by using a filter unit connected to a vacuum pump [11].
13. According to our experience, 10 ml of *Drosophila* labeling medium is enough to breed ~150 flies.
14. The tissue paper prevents embryos from drowning in the *Drosophila* labeling medium.
15. It takes ~10 days at 25 °C until the flies reach their adult stage.

Acknowledgements

We would like to thank Erik McShane for critical comments on the manuscript. This work was supported by the Helmholtz Association and the Swiss National Science Foundation.

References

1. Krijgsveld J, Ketting RF, Mahmoudi T et al (2003) Metabolic labeling of C. elegans and D. melanogaster for quantitative proteomics. Nat Biotechnol 21(8):927–931
2. Bonaldi T, Straub T, Cox J et al (2008) Combined use of RNAi and quantitative proteomics to study gene function in Drosophila. Mol Cell 31(5):762–772
3. Hilger M, Bonaldi T, Gnad F et al (2009) Systems-wide analysis of a phosphatase knockdown by quantitative proteomics and phosphoproteomics. Mol Cell Proteomics 8(8):1908–1920
4. Sury MD, Chen JX, Selbach M (2010) The SILAC fly allows for accurate protein quantification in vivo. Mol Cell Proteomics 9:2173–2183
5. Fredens J, Engholm-Keller K, Giessing A et al (2011) Quantitative proteomics by amino acid labeling in C. elegans. Nat Methods 8(10):845–847
6. Kruger M, Moser M, Ussar S et al (2008) SILAC mouse for quantitative proteomics uncovers kindlin-3 as an essential factor for red blood cell function. Cell 134(2):353–364
7. Larance M, Bailly AP, Pourkarimi E et al (2011) Stable-isotope labeling with amino acids in nematodes. Nat Methods 8(10): 849–851
8. Westman-Brinkmalm A, Abramsson A, Pannee J et al (2011) SILAC zebrafish for quantitative analysis of protein turnover and tissue regeneration. J Proteomics 75(2):425–434
9. Gruhler A, Olsen JV, Mohammed S et al (2005) Quantitative phosphoproteomics applied to the yeast pheromone signaling pathway. Mol Cell Proteomics 4(3): 310–327
10. Finley D, Ozkaynak E, Varshavsky A (1987) The yeast polyubiquitin gene is essential for resistance to high temperatures starvation and other stresses. Cell 48(6):1035–1046
11. Wodarz A (2008) Extraction and immunoblotting of proteins from embryos. Methods Mol Biol 420:335–345

Chapter 8

Stable Isotope Labeling for Proteomic Analysis of Tissues in Mouse

Soraya Hölper, Aaron Ruhs, and Marcus Krüger

Abstract

Since the first metabolic labeling experiments with stable isotopes beginning of the last century, several approaches were pursued to monitor protein dynamics in living animals. Today, almost all model organisms from bacteria to rodents can be fully labeled with SILAC (stable isotope labeling of amino acids in cell culture) amino acids. The development of special media and diets containing the labeled amino acids provides an efficient way to metabolically label prokaryotic and eukaryotic organisms. Preferentially, the essential amino acid lysine ($^{13}C_6$-lysine) is used to label mice (*Mus musculus*) and after one generation the natural isotope is fully replaced by the stable $^{13}C_6$-lysine isotope. So far, the SILAC mouse approach has been used to analyze several transgenic and knockout mouse models. Spike-in of labeled proteins into non-labeled samples provides an accurate relative protein quantification method without any chemical modification. Here we describe how to establish a SILAC mouse colony and describe the analysis of skeletal muscle tissue with different metabolic and contractile profiles.

Key words SILAC, Mass spectrometry, Proteomics, Mouse, Skeletal muscle, Stable isotope labeling, In vivo, Quantitative

1 Introduction

The administration of stable isotopes to living animals is a common method for metabolic labeling of proteins. In early tracer experiments, Schoenheimer and Rittenberg used stable isotopes to investigate the dynamic nature of the protein pool in the body [1]. Today, the metabolic labeling with stable isotopes (i.e., ^{15}N, ^{13}C) has been applied to almost all model organisms, including bacteria [2], yeast [3], plants [4], flies [5], worms [6], newts [7], and rodents [8, 9]. Clearly, the complete labeling of those organisms with stable isotopes (i.e., the amino acid ^{13}C-lysine) and the combination with newly developed mass spectrometric methods allows for an in-depth quantitative characterization of proteomes [10, 11]. Although all models are relevant to investigate molecular and biochemical mechanisms, to date the mouse is most often used to

Bettina Warscheid (ed.), *Stable Isotope Labeling by Amino Acids in Cell Culture (SILAC): Methods and Protocols*, Methods in Molecular Biology, vol. 1188, DOI 10.1007/978-1-4939-1142-4_8,

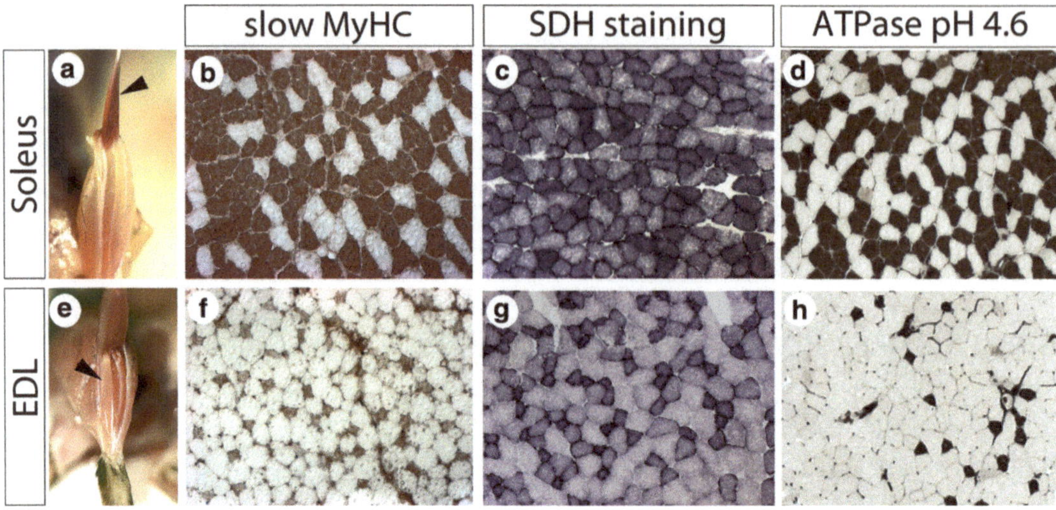

Fig. 1 Three staining methods for slow and fast fibers and spike-in workflow with EDL and soleus. (**a**, **e**) Photographs showing the localization of the Soleus and EDL muscle within the mouse hind limb. (**b–d**, **f–h**) Three different staining methods based on antibody staining against slow myosin heavy chain (MyHC-I), succinate dehydrogenase (SDH) and myosin heavy chain ATPase at pH 4.6 to identify different muscle fiber types

study human diseases including cardiovascular, neurological and metabolic disorders. Recently, it has been shown that besides its contractile function, skeletal muscle plays also a crucial role in metabolic syndromes such as diabetes mellitus type II [12]. Moreover, unbiased quantitative proteome analysis of neuromuscular disorders like amyotrophic lateral sclerosis (ALS) [13], multiple sclerosis [14], and Duchenne muscular dystrophy [15] has identified novel molecules, involved in metabolism, cell signaling, and contractility, which had not been described previously to be associated with those diseases. Thus, these findings based on quantitative mass spectrometry revealed new, interesting targets for clinical studies.

Muscle tissue has been investigated for more than a century. The first description of different types of muscle were based on its color, which ranges from white to red, representing fast and slow contracting fibers. Biochemical methods helped to categorize muscle fibers according to their metabolic and enzymatic activity. For example, the myosin ATPase and succinate dehydrogenase (SDH) activity can be used to delineate slow oxidative and fast glycolytic muscle fibers. Another method is based on the immunohistochemical detection of myosin heavy chain proteins (MyHC). Slow fibers express MyHC-I, whereas fast fibers are positive for MyHC-2A, MyHC-2B, or MyHC-2X (Fig. 1) [16]. In addition, several other proteins, including myosin light chains, troponins, α-actinin, and the calcium-ATPase SERCA, were identified as marker for specific fiber types.

Recently, several studies showed that the SILAC mouse is a powerful tool to perform quantitative proteomics of complex organs, including heart, liver, and brain [8, 17]. Moreover, we have shown that the approach is suitable to investigate skeletal muscle with different metabolic and contractile structures [18]. In this chapter, we describe a protocol for labeling mice with the stable isotope $^{13}C_6$-lysine (termed Lys6) and give an example on how to analyze two different types of skeletal muscles.

2 Materials

For all buffers and solvents listed in this chapter, use ultrapure Milli-Q water and HPLC-grade reagents.

2.1 Diet and Mouse Strain

1. The mouse diet with Lys6 can be purchased from Silantes GmbH, Germany. It contains 1 g of Lys6 per 100 g of mouse diet, which is according to the standard mouse nutritional requirements [19] (*see* **Note 1**).
2. Mouse strain C57BL/6.

2.2 Dissection of Mouse Muscles

1. Anaesthesia with 10 % ketamine (100 mg/kg mouse weight) and 2 % xylazine (10 mg/kg mouse weight) in 0.9 % NaCl solution.
2. Sterile phosphate buffered saline (1× PBS, Gibco).
3. For cardiac perfusion, use a 50 ml syringe and a 22-gauge (G) blunt end needle.
4. Dissecting set including scissors and forceps.
5. Liquid nitrogen for snap-freezing isolated tissue.

2.3 Protein Extraction and In-Solution Digestion of Mouse Muscle Tissue

1. Lysis buffer: 4 % SDS in 0.1 M Tris–HCl, pH 7.6.
2. Ultra-Turraxx (IKA Works) and sonicator.
3. DC Protein Assay (Bio-Rad).
4. Acetone.
5. *GlycoBlue* (Ambion).
6. Urea buffer: 6 M urea, 2 M thiourea, 10 mM Hepes, pH 7.4 (Gibco).
7. Reduction stock solution: 100 mM DTT. Store aliquots at −20 °C.
8. Alkylation stock solution: 550 mM iodoacetamide in 50 mM ammonium bicarbonate (ABC). Aliquots can be stored at −20 °C. Iodoacetamide is light sensitive and should be stored in the dark.
9. Digestion solution: 50 mM ABC.

10. Endopeptidase LysC solution: 0.5 μg/μl LysC (Wako Chemicals USA) in 50 mM ABC.
11. Acidifying solution: TFA.
12. Stage tips: 200 μl tips, Empore™ Octadecyl C18 47 mm Extraction Disk (Supelco).

3 Methods

Initially, SILAC was developed for quantitative proteomic studies in cell culture systems in which cells are preferentially labeled with the amino acids arginine and lysine containing the stable isotopes ^{2}H, ^{13}C, and ^{15}N. In the case of a SILAC experiment, the non-labeled (termed "light") and the labeled (termed "heavy") samples are mixed 1:1 according to their protein concentration and the comparison of light and heavy peak intensities is used for relative protein quantification [20]. In contrast, the SILAC mouse (also termed "heavy") is used as an internal protein standard, which is spiked into the light control and light experimental condition, respectively. A direct comparison of two light conditions can be achieved by dividing the heavy/light of the control ratio and the heavy/light of the experimental ratio (Fig. 2).

The advantage of the "ratio of ratio" approach is that one accounts changes in protein expression levels due to potential differences in diet, mouse strains, age, and gender.

In addition, only one mouse colony has to be generated which serves as a standard for multiple experiments. The versatility of the in vivo SILAC approach in mice has been demonstrated by several other studies [21–24].

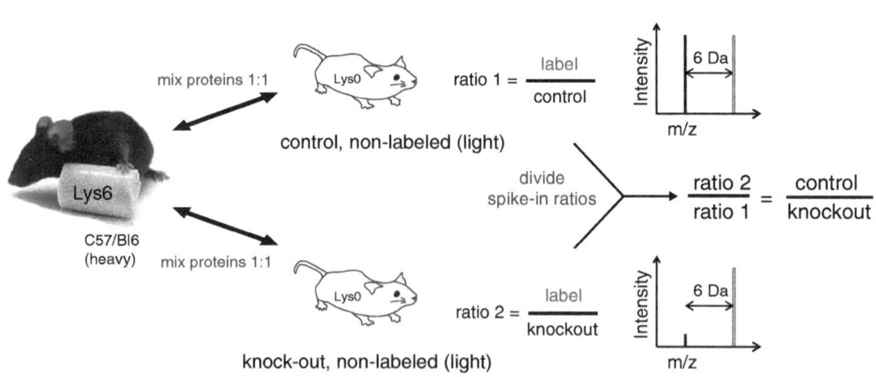

Fig. 2 The SILAC mouse as an internal standard for protein quantification. Lys6-labeled mouse tissue is "spiked-in" into non-labeled (*light*) samples of a control and a knockout mouse. After mass spectrometric measurements, peptides are identified as SILAC pairs and the intensities of the light (Lys0) and heavy (Lys6) peak are divided to achieve relative protein quantification (ratio 1 = *H*/*L* and ratio 2 = *H*/*L*). Calculation of direct ratios between the light samples is obtained by dividing ratio 2/ratio 1, which results in a direct comparison of non-labeled animals. In the presented example, the selected peptide is clearly downregulated in the knockout animal

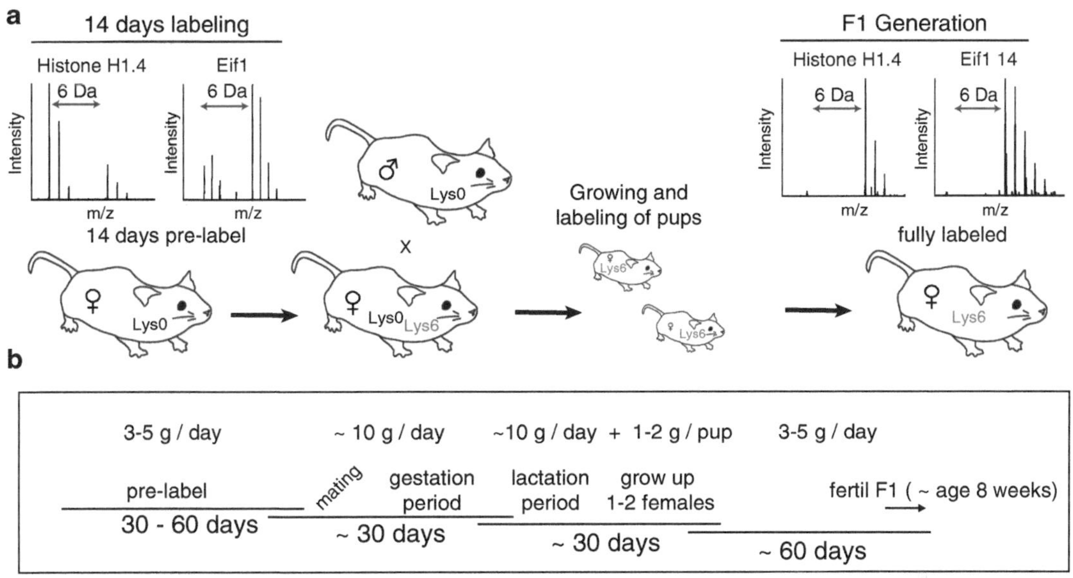

Fig. 3 Generation of a SILAC colony. (**a**) After the initial pre-labeling period, a non-labeled C57BL/6 male is used for mating. The F1 generation shows a Lys6 incorporation of ~96 % in all organs, tissues, and blood. Selected SILAC pairs from a histone and the eukaryotic translation initiation factor show complete labeling in the F1 generation. (**b**) Time schedule of mouse breeding and food consumption

3.1 Mouse Labeling with Lys6

An essential requirement for SILAC based quantitative proteomics is the complete incorporation of stable isotope-labeled amino acids. For SILAC mouse labeling, we used the stable isotope of the essential amino acid lysine, which cannot be synthesized by mammalian cells (*see* **Note 2**).

1. To establish a SILAC mouse colony, use a female mouse of the strain C57BL/6 (*see* **Note 3**). First, the F0 mouse is pre-labeled for approx. 8 weeks. This step is important to increase the Lys6 incorporation rate of the following F1 generation. During the initial pre-labeling period, the mouse is fed with 3–5 g of the SILAC diet (*see* **Note 4**, Fig. 3). The Lys6 incorporation rate can be monitored by blood sampling via the tail or eye vein and should reach ~80 % of Lys6 incorporation. The Lys6 incorporation is calculated as SILAC ratio $= [\text{ratio}(H/L) \times 100]/[\text{ratio}\ (H/L)+1]$.
2. After the pre-labeling period and mating with a non-labeled male, the food amount is increased to 5–10 g during pregnancy. During lactating and weaning periods, the food is once more increased (up to 15 g) depending on the number of pups. It is important to monitor the food uptake and adjust it to the number of pups and the body weight. This ensures to avoid unnecessary use of food (*see* **Note 5**).
3. Incorporation efficiency for organs and tissues of the adult (age >8 weeks) F1 generation should be more than 96 %.

If not, one should continue the breeding to generate an F2 generation.

4. Although food consumption strongly depends on the number of labeled animals, we estimate approx. 1 kg of food to generate and maintain a SILAC mouse generation for approximately 6 months.

3.2 Isolation of Skeletal Muscle Tissue

1. Perform all animal procedures in accordance with institutional guidelines.
2. Prepare a 50 ml syringe/22G blunt end needle with ice-cold 1× PBS. This will be used for cardiac perfusion to wash out the blood from the body (*see* **Note 6**).
3. Anesthetize a C57BL/6 mouse with ketamine and xylazine by intraperitoneal injection (*see* **Note 7**).
4. After anesthesia, expose the heart and perform a left ventricular perfusion with 50 ml of ice-cold 1× PBS. Continue the perfusion until no blood is visible in the liver. After perfusion, sacrifice mice by cervical dislocation.
5. For skeletal muscle dissections, remove the skin of the leg and carefully open the fascia. First peel off the tibialis anterior muscle from the ankle upwards and remove the EDL (*Extensor digitorum longus*, white color). The soleus muscle is lying below the gastrocnemius muscle. Cut the Achilles tendon and then peel off the gastrocnemius muscle to expose the soleus muscle, this is a flat and broad muscle with a red color (*see* **Note 8**) (Fig. 1a, e).
6. Immediately snap-freeze isolated muscle tissue in liquid nitrogen. Samples can be stored at −80 °C. Alternatively, tissue samples can be directly homogenized in lysis buffer.

3.3 Quantitative Proteomic Analysis of Slow and Fast Muscle Tissue

Here, we describe the isolation of the soleus and EDL muscle. The experiment is adapted from an earlier study on mouse skeletal muscle cells [18]. Generally, muscle tissue contains mainly proteins responsible for contraction and energy production. Some of the most abundant sarcomeric proteins, including titin, myosins, and actinin, are shown in the SDS-PAGE stained with colloidal Coomassie (Fig. 4a). As a consequence, mass spectrometric measurements of skeletal muscle tissue usually results in lower numbers of protein identifications compared to other tissues.

3.3.1 Skeletal Muscle Preparation, Protein Isolation, and In-Solution Digestion

1. After isolation of labeled and non-labeled soleus and EDL muscles, respectively, cut the tissues (length ~1 cm) at 4 °C into smaller pieces (1–2 mm) and transfer the tissues into SDS-lysis buffer using 1:10 (w/v) sample to buffer. Use an Ultra-Turrax for complete homogenization and perform a 5 min incubation step at 95 °C. The following lysis procedure is done at room temperature.

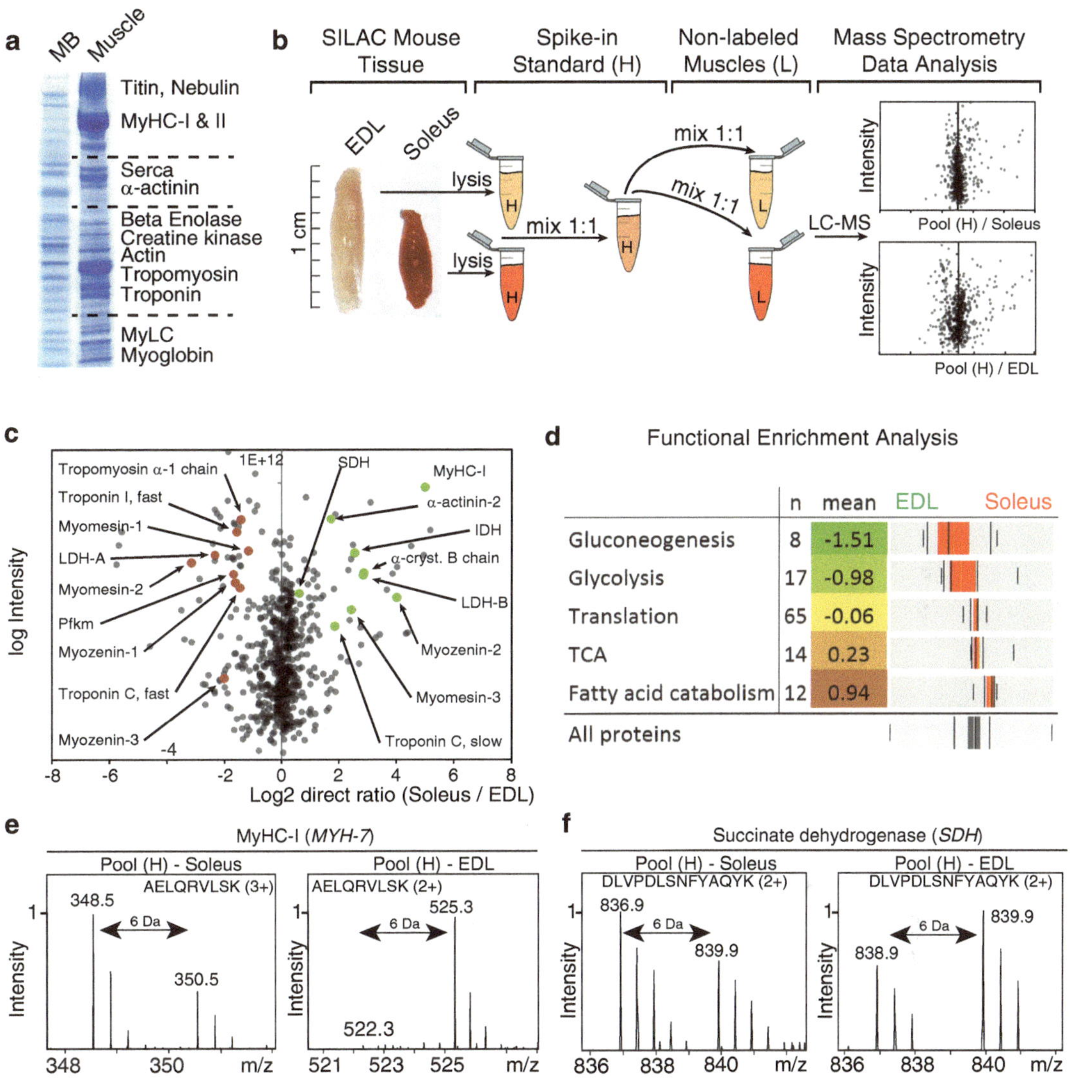

Fig. 4 Experimental workflow. (**a**) SDS-PAGE of myoblast cells and muscle lysate showing the most abundant proteins. (**b**) The EDL and the soleus muscle of the SILAC mouse are equally mixed and are spiked-in as a heavy standard (termed as Pool (H)), which is then mixed 1:1 to the light soleus and light EDL, respectively. The proteins are digested with LysC, analyzed with LC-MS and processed with MaxQuant. (**c**) Direct ratio (soleus–EDL) plotted against the mean log-intensities. Selected proteins for slow fibers (soleus) are marked in *green* and for fast fibers (EDL) are marked in *red*. (**d**) Functional enrichment analysis using the ResA tool. Terms for gluconeogenesis and glycolysis are clearly enriched in the fast EDL muscle. (**e**) SILAC pairs from MyHC-I (gene *MYH7)* and succinate dehydrogenase (SDH)

2. Sonicate the samples to reduce the viscosity and centrifuge the sample for 5 min at 16,000 × *g* and transfer the cleared lysates into a new tube (*see* **Note 9**).
3. Estimate the protein concentration using a standard protocol (for instance Bradford or BCA) according to the manufacturer's instructions.

4. To generate the heavy standard, mix the Lys6-labeled soleus and EDL 1:1 according to the protein concentration. Next, spike-in equal amounts of the heavy standard into the non-labeled soleus and the non-labeled EDL (Fig. 4b).

   In total, we use about 10–20 μg of protein for the in-solution digestion (*see* **Note 10**).

5. For protein precipitation add 4 volumes of ice-cold acetone (−20 °C) and for pellet visualization 1 μl of *GlycoBlue* to the sample. Vortex vigorously and incubate the sample for 1 h at −20 °C. Afterwards, centrifuge for 10 min at 15,000 × *g* at 4 °C. Decant the supernatant and wash the protein pellet with ice-cold 90 % acetone. After centrifugation, dissolve the pellet in urea buffer (*see* **Note 11**).

6. For in-solution digestion, reduce and alkylate proteins by adding DTT to a final concentration of 10 mM and incubate at room temperature for 30 min. Next, incubate samples with a final concentration of 55 mM iodoacetamide for 20 min at room temperature in the dark. Add the endoproteinase LysC at an enzyme to substrate ratio of 1:100 and incubate at room temperature for 2–3 h. Repeat the LysC digestion a second time using the same enzyme to substrate ratio and incubate the sample overnight. Stop the digestion by acidifying with TFA. Finally, desalt and clean peptides with Stop and Go extraction tips (StageTips) [25]. C18-StageTips with the sample can be stored at 4 °C until MS measurements (*see* **Note 12**).

3.4 Mass Spectrometry Measurement and Data Analysis

In principle, any high performance liquid chromatography–mass spectrometry setup can be used for the analysis. For instance, hybrid mass spectrometers like the LTQ-Orbitrap Velos and Q-Exactive mass spectrometers are capable of accurately recording peptide masses and are able to fragment isolated peptides. Detailed methods for liquid chromatography connected to hybrid mass spectrometers are described in [26, 27]. After mass spectrometric measurements, data can be processed with programs such as the MaxQuant software tool based either on Mascot or Andromeda search algorithms (http://maxquant.org/ and www.matrixscience.com) [28].

1. After protein quantification, the direct ratio between the slow and fast muscle can be manually calculated by dividing the heavy/Soleus and heavy/EDL ratio (*see* also Fig. 4e, f). It is recommended to transform the SILAC fold-changes to log2 values. This step ensures that regulated candidates have the same distance from the zero value (equates a 1:1 ratio) (Fig. 4c) (*see* **Note 13**).

2. For a more systematic overview the soleus–EDL dataset based on cellular location and biological function was analyzed. The ResA tool (Resampling Analysis of Arbitrary Annotations) was used to visualize significant enrichments of GO-terms [29].

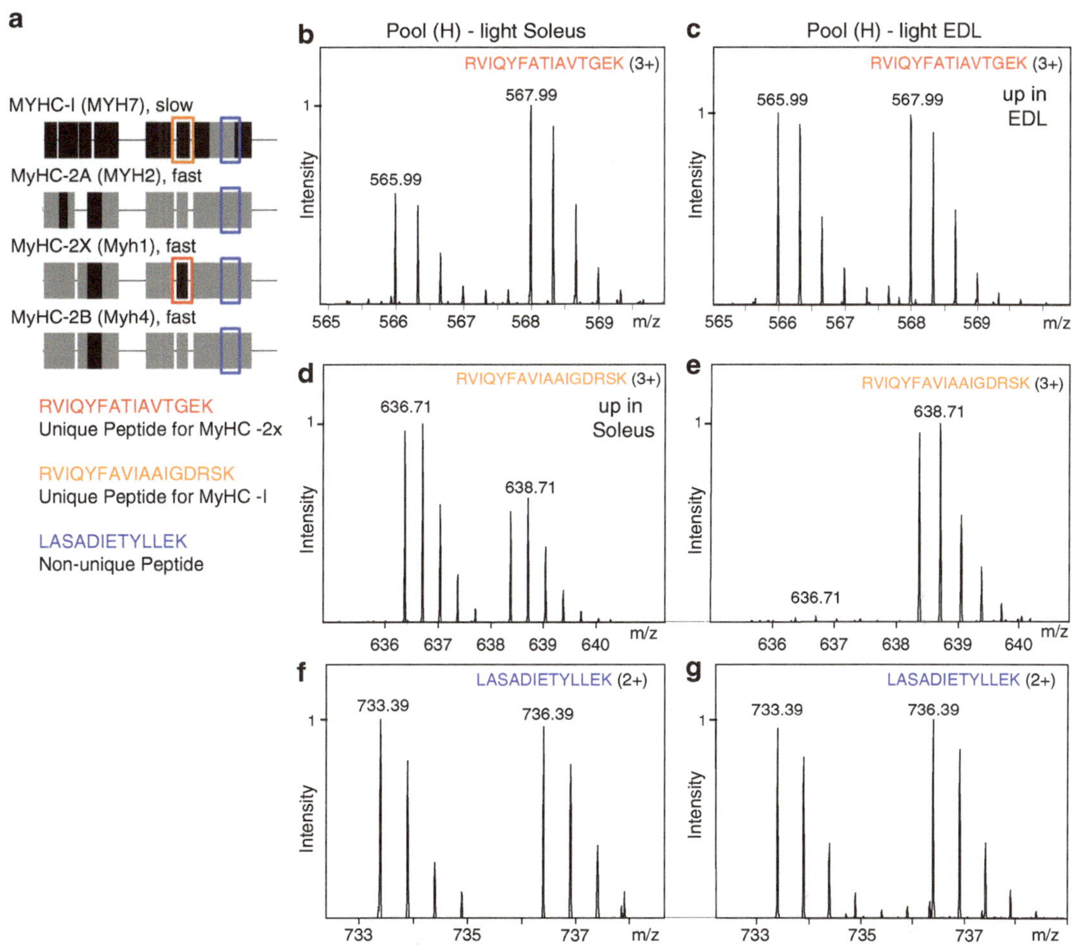

Fig. 5 Quantification on MyHC peptides. (**a**) Schematic overview of MyHC sequences. Black areas represent unique peptides whereas *grey* areas are nonunique peptides. (**b**, **c**) Unique peptide for MyHC-2X; (**d**, **e**) unique peptide for MyHC-I; (**f**, **g**) nonunique peptide, which might belong to several different MyHC proteins

The tool provides unbiased enrichment analysis without choosing a cutoff to define the target dataset (Fig. 4d) (*see* **Note 14**).

3. The identification of unique peptides for MyHC proteins is important to perform correct protein quantification. Examples for unique and nonunique peptides are shown in Fig. 5 (*see* **Note 15**).

4 Notes

1. A detailed list of all components of the diet can be found at www.silantes.com. In principle, any other lysine-free diet can be used for the generation of a SILAC mouse colony. A description of the manual production of the SILAC mouse diet is described in [30]. We recommend having at least 1 % lysine in the diet to achieve optimal growth rates of mice [31].

In case of aberrant food compositions, one should compare this with institutional guidelines.

2. The most commonly used SILAC amino acids for cell culture experiments are lysine and arginine, which has the advantage that after trypsin digestion all peptides have at least one labeled amino acid. But it is recommended to label the mouse with lysine only, since arginine is converted to proline in some cell lines and organisms such as yeast and flies. In case of exclusive lysine labeling, the protease LysC is recommended. However, the combination of lysine and LysC leads to a reduced number of identified peptides during mass spectrometric analysis. In experiments which focus on the enrichment of acetylated or ubiquitinated peptides, the protease trypsin can be used for digestion because the modified lysine is not recognized by the protease trypsin, thus generating miscleaved peptides with at least one labeled lysine.
3. Although C57BL/6 is the most commonly used mouse strain to maintain knockout animals, it is also possible to label other inbred and outbred mouse strains. It is recommended to use a female mouse which delivered and weaned successfully at least one mouse generation. To increase the chance to obtain an F1 generation, 2–3 F0 females can be SILAC-labeled. However, this also increases the cost for the diet.
4. Food consumption varies between different mouse strains, housing conditions, and SILAC diets. Thus, we recommend monitoring the body weight and food uptake during the pre-labeling period.
5. To maintain a SILAC mouse colony, only females have to be fed with the SILAC diet. Males can be taken out as soon as possible. For breeding of the next generation, the presence of the non-labeled mouse should be as short as possible to reduce costs for the Lys6 mouse diet.
6. Perfusion is an optional washing step and not mandatory. A description of the cardiac perfusion is available at http://physics.ucsd.edu/neurophysics/lab/sop%2029.pdf.
7. Anesthesia and perfusion of animals need to follow institutional guidelines and only trained persons are allowed to perform those experiments [32].
8. A comprehensive anatomical description for isolation of skeletal muscle tissue is reported in [33]. Further fast muscles are the tibialis anterior from the hind leg and the extraocular muscle, which controls movements of the eye.
9. Be careful with the sonication step. A duty cycle between 10 and 20 % for a time period of 1–2 min is usually sufficient to fragment the DNA.

10. We recommend an SDS-based lysis buffer to lyse all cellular components. However, the detergent SDS interferes with mass spectrometric analysis and a protein precipitation step is necessary. Alternative methods such as the FASP (filter aided sample preparation) or SDS spin columns can be used for protein isolation and digestion [34, 35].
11. To perform a more in depth proteome analysis, subcellular fractionation or size-exclusion chromatography (SEC) help to separate abundant proteins. In addition, techniques such as in-gel digestion [36], isoelectric focusing of peptides and strong anion exchange chromatography (SAX) [37] can be used to further fractionate the protein samples.
12. For long-term storage, we recommend to elute samples from StageTips and store peptides at −80 °C.
13. For most SILAC experiments, a fold change of >1.5 can be used as a biological cut-off. However, the generation of biological replicates allows for assessing statistically significant outliers.
14. ResA can be accessed at http://resa.mpi-bn.mpg.de. Detailed help on how to use the tool can be found at http://resa.mpi-bn.mpg.de/resa/example.html and in the referenced publication.
15. MyHC proteins are essential components for the contractility of skeletal muscle fibers. So far, 11 different isoforms are described which are differentially expressed between slow, fast, and cardiac muscle tissues. Since some MyHC share common sequence motifs, their identification by shotgun proteomics is problematic.

References

1. Schoenheimer R, Rittenberg D (1935) Deuterium as an indicator in the study of intermediary metabolism. Science 82(2120): 156–157
2. Soufi B, Kumar C, Gnad F et al (2010) Stable isotope labeling by amino acids in cell culture(SILAC) applied to quantitative proteomics of Bacillus subtilis. J Proteome Res 9(7):3638–3646
3. Gruhler A, Olsen JV, Mohammed S et al (2005) Quantitative phosphoproteomics applied to the yeast pheromone signaling pathway. Mol Cell Proteomics 4(3):310–327
4. Schutz W, Hausmann N, Krug K et al (2011) Extending SILAC to proteomics of plant cell lines. Plant Cell 23(5):1701–1705
5. Sury MD, Chen JX, Selbach M (2010) The SILAC fly allows for accurate protein quantification in vivo. Mol Cell Proteomics 9(10): 2173–2183
6. Larance M, Bailly AP, Pourkarimi E et al (2011) Stable-isotope labeling with amino acids in nematodes. Nat Methods 8(10):849–851
7. Looso M, Michel CS, Konzer A et al (2012) Spiked-in pulsed in vivo labeling identifies a new member of the CCN family in regenerating newt hearts. J Proteome Res 11(9): 4693–4704
8. Kruger M, Moser M, Ussar S et al (2008) SILAC mouse for quantitative proteomics uncovers kindlin-3 as an essential factor for red blood cell function. Cell 134(2):353–364
9. McClatchy DB, Dong MQ, Wu CC et al (2007) 15N metabolic labeling of mammalian tissue with slow protein turnover. J Proteome Res 6(5):2005–2010
10. Ong SE, Mann M (2005) Mass spectrometry-based proteomics turns quantitative. Nat Chem Biol 1(5):252–262

11. Mann M, Kelleher NL (2008) Precision proteomics: the case for high resolution and high mass accuracy. PNAS 105(47):18132–18138
12. Beck-Nielsen H (2012) The role of glycogen synthase in the development of hyperglycemia in type 2 diabetes: "To store or not to store glucose, that's the question". Diabetes Metab Res Rev 28(8):635–644
13. Bergemalm D, Forsberg K, Jonsson PA et al (2009) Changes in the spinal cord proteome of an amyotrophic lateral sclerosis murine model determined by differential in-gel electrophoresis. Mol Cell Proteomics 8(6):1306–1317
14. Kroksveen AC, Aasebo E, Vethe H et al (2012) Discovery and initial verification of differentially abundant proteins between multiple sclerosis patients and controls using iTRAQ and SID-SRM. J Proteomics 78:312–325
15. Lewis C, Doran P, Ohlendieck K (2012) Proteomic analysis of dystrophic muscle. Methods Mol Biol 798:357–369
16. Schiaffino S, Reggiani C (2011) Fiber types in mammalian skeletal muscles. Physiological Rev 91(4):1447–1531
17. Walther DM, Mann M (2011) Accurate quantification of more than 4000 mouse tissue proteins reveals minimal proteome changes during aging. Mol Cell Proteomics 10(2): M110.004523
18. Drexler HC, Ruhs A, Konzer A et al (2012) On marathons and sprints: an integrated quantitative proteomics and transcriptomics analysis of differences between slow and fast muscle fibers. Mol Cell Proteomics 11(6): M111.010801
19. Benevenga NJ, Calvert C, Eckhart CD, Fahey GC, Greger JL, Keen CL, Knapka JJ, Magalhaes H, Oftedal OT (1995) Nutrient requirements of the mouse. In: Benevenga NJ, Calvert C, Eckhert CD, Fahey GC, Greger JL (Eds.) Nutrient requirements of laboratory animals. National Academy Press, Washington, DC, pp 80–102
20. Rigbolt KT, Blagoev B (2010) Proteome-wide quantitation by SILAC. Methods Mol Biol 658:187–204
21. Boettger T, Beetz N, Kostin S et al (2009) Acquisition of the contractile phenotype by murine arterial smooth muscle cells depends on the Mir143/145 gene cluster. J Clinic Invest 119(9):2634–2647
22. Dey A, Seshasayee D, Noubade R et al (2012) Loss of the tumor suppressor BAP1 causes myeloid transformation. Science 337(6101): 1541–1546
23. Huang TC, Sahasrabuddhe NA, Kim MS et al (2012) Regulation of lipid metabolism by Dicer revealed through SILAC mice. J Proteome Res 11(4):2193–2205
24. Jordan SD, Kruger M, Willmes DM et al (2011) Obesity-induced overexpression of miRNA-143 inhibits insulin-stimulated AKT activation and impairs glucose metabolism. Nat Cell Biol 13(4):434–446
25. Rappsilber J, Ishihama Y, Mann M (2003) Stop and go extraction tips for matrix-assisted laser desorption/ionization, nanoelectrospray, and LC/MS sample pretreatment in proteomics. Anal Chem 75(3):663–670
26. Shen TL, Noon KR (2004) Liquid chromatography-mass spectrometry and tandem mass spectrometry of peptides and proteins. Methods Mol Biol 251:111–140
27. Michalski A, Damoc E, Hauschild JP et al (2011) Mass spectrometry-based proteomics using Q Exactive, a high-performance benchtop quadrupole Orbitrap mass spectrometer. Mol Cell Proteomics 10(9):M111.011015
28. Cox J, Matic I, Hilger M et al (2009) A practical guide to the MaxQuant computational platform for SILAC-based quantitative proteomics. Nat Protoc 4(5):698–705
29. Ruhs A, Cemic F, Braun T et al (2013) ResA3: a web tool for resampling analysis of arbitrary annotations. PLoS One 8(1):e53743
30. Zanivan S, Krueger M, Mann M (2012) In vivo quantitative proteomics: the SILAC mouse. Methods Mol Biol 757:435–450
31. Bell JM, John AM (1981) Amino acid requirements of growing mice: arginine, lysine, tryptophan and phenylalanine. J Nutrition 111(3): 525–530
32. Flecknell PA (1987) Laboratory animal anesthesia: an introduction for research workers and technicians. Academic Press, San Diego, London
33. Greene E (1968) Anatomy of the rat. Hafner, New York
34. Wisniewski JR, Zougman A, Nagaraj N et al (2009) Universal sample preparation method for proteome analysis. Nat Methods 6(5):359–362
35. Bereman MS, Egertson JD, MacCoss MJ (2011) Comparison between procedures using SDS for shotgun proteomic analyses of complex samples. Proteomics 11(14):2931–2935
36. Vasilj A, Gentzel M, Ueberham E et al (2012) Tissue proteomics by one-dimensional gel electrophoresis combined with label-free protein quantification. Proteome Res 11(7): 3680–3689
37. Wisniewski JR, Zougman A, Mann M (2009) Combination of FASP and StageTip-based fractionation allows in-depth analysis of the hippocampal membrane proteome. J Proteome Res 8(12):5674–5678

Chapter 9

Identification of Novel Protein Functions and Signaling Mechanisms by Genetics and Quantitative Phosphoproteomics in *Caenorhabditis elegans*

Julius Fredens*, Kasper Engholm-Keller*, Jakob Møller-Jensen, Martin Røssel Larsen, and Nils J. Færgeman

Abstract

Stable isotope labeling by amino acids combined with mass spectrometry is a widely used methodology for measuring relative changes in protein and phosphorylation levels at a global level. We have applied this method to the model organism *Caenorhabditis elegans* in combination with RNAi-mediated gene knockdown by feeding the nematode on pre-labeled lysine auxotroph *Escherichia coli*. In this chapter, we describe in details the generation of the *E. coli* strain, incorporation of heavy isotope-labeled lysine in *C. elegans*, and the procedure for a comprehensive global phosphoproteomic experiment.

Key words Quantitative proteomics, Phosphoproteomics, Mass spectrometry, LC-MS/MS, *Caenorhabditis elegans*, Stable isotope labeling by amino acids in cell culture (SILAC), RNAi-compatible *Escherichia coli*, Gene knockdown

1 Introduction

Quantitative mass spectrometry is widely used to examine proteomic changes in biological systems and stable isotope labeling by amino acids in cell culture (SILAC) [1, 2] has contributed to accurate comparison of several proteomes. The concept of SILAC involves labeling of cell cultures by incorporating specific amino acids containing the stable isotopes ^{2}H, ^{13}C, and ^{15}N. By mixing the samples at a very early stage in the sample preparation workflow, no differential sample handling errors will be introduced downstream of this point, thereby minimizing the quantitative variation. The relative levels of the respective proteomes are easily distinguishable and compared in the mass spectrometric analysis due to the specific differences in their peptide masses.

*Julius Fredens and Kasper Engholm-Kelle have equally contributed to this Chapter.

Bettina Warscheid (ed.), *Stable Isotope Labeling by Amino Acids in Cell Culture (SILAC): Methods and Protocols*, Methods in Molecular Biology, vol. 1188, DOI 10.1007/978-1-4939-1142-4_9, © Springer Science+Business Media New York 2014

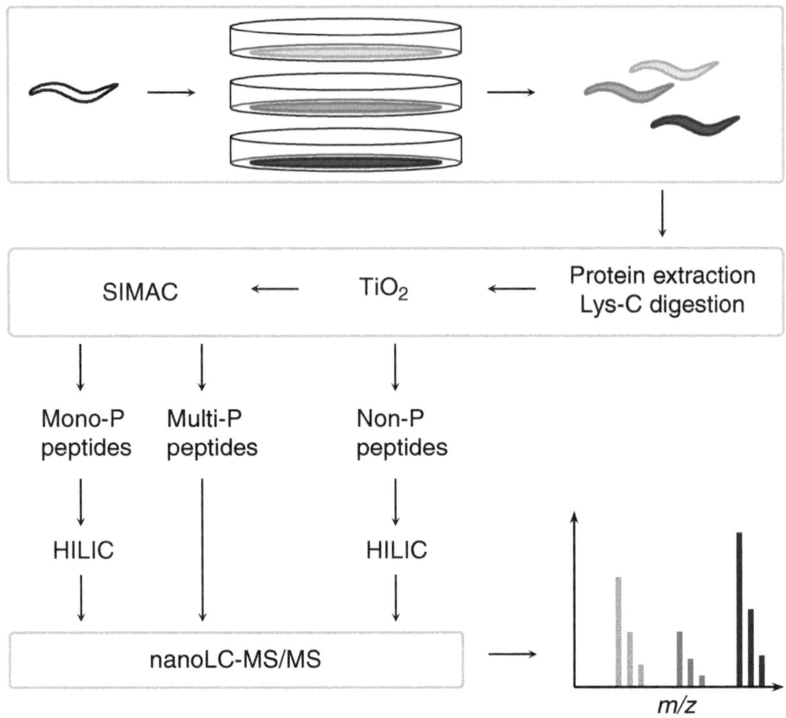

Fig. 1 Flow diagram of labeling, protein preparation, phosphopeptide enrichment, fractionation, and analysis. Lys0, Lys4, or Lys8 is incorporated into the proteome of *C. elegans* over two generations. Proteins are extracted and the three populations are mixed before digestion with Lys-C. Phosphopeptides are enriched using TiO_2 and the mono- and multi-phosphorylated peptides are separated by sequential elution from IMAC (SIMAC). The non-phosphorylated and mono-phosphorylated peptides are fractionated by hydrophilic interaction liquid chromatography (HILIC) and all samples are separated by nanoLC prior to MS/MS analysis

SILAC is increasingly being applied to eukaryotic model organisms, including yeast [1, 3], fruit flies [4], plants [5], mice [6], and nematodes [7, 8]. By a simple procedure, lysine with three different isotopic labels can be completely incorporated into *Caenorhabditis elegans* by feeding the nematodes on pre-labeled lysine auxotroph *Escherichia coli* for just one generation (3 days) (*see* Fig. 1). Furthermore, this *E. coli* strain can be modified to perform efficient simultaneous RNAi in *C. elegans*, which allows convenient gene knockdown. Thus, this method can be used to compare three populations in a single experiment and study proteomic changes upon different treatments, mutations, or knockdowns.

We have recently shown that loss of function or knockdown of the nuclear hormone receptor NHR-49 results in a vast number of regulated proteins [7]. Consistent with previous observations, the majority of the downregulated proteins were involved in lipid metabolism [9] as well as carbohydrate and amino acid metabolism. We have recently extended this methodology, enabling us to

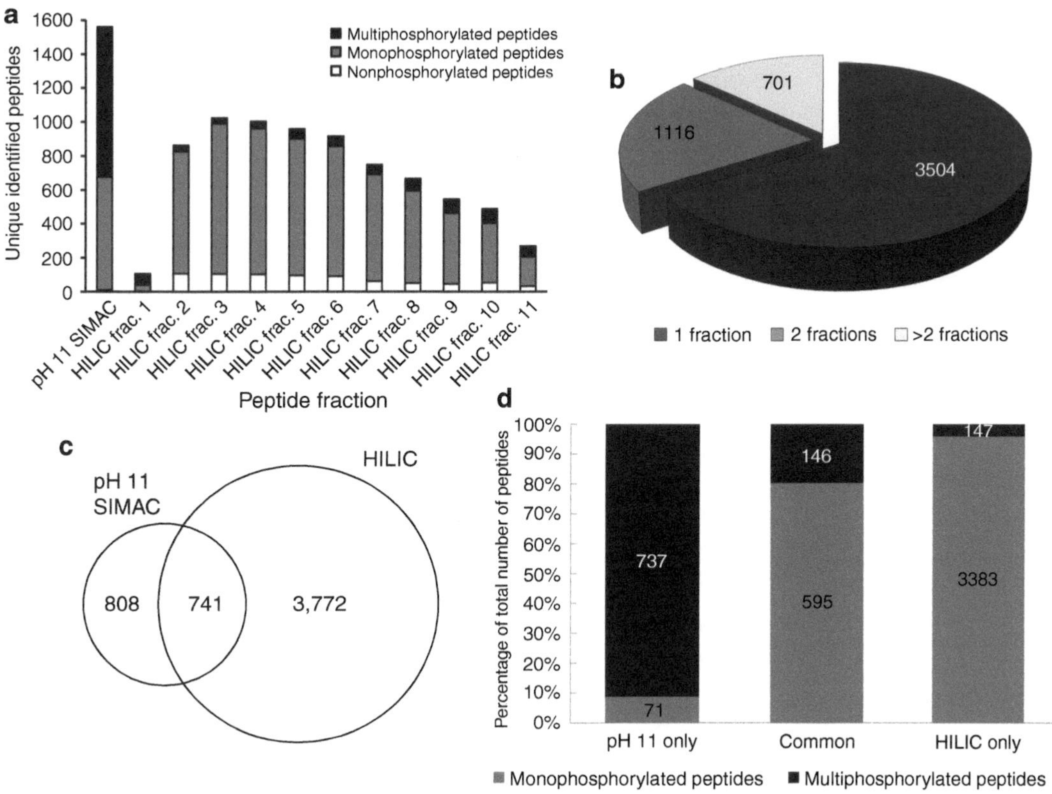

Fig. 2 Enrichment of phosphopeptides from *C. elegans* using TiO_2 chromatography and sequential elution from IMAC. (**a**) Fractionation of phosphopeptides and specificity of the subsequent phosphopeptide enrichment in a technical replicate of a SILAC experiment. Non-phosphorylated peptides are indicated in *white*, mono-phosphorylated in grey, and multi-phosphorylated in *black*. (**b**) Fractionation efficiency of the phosphopeptide separation setup. The pie chart indicates the number of phosphopeptides identified in one (*dark grey*), two (*light grey*), or more than two fractions (*white*). (**c**) A Venn diagram showing the fractionation efficiency of the phosphopeptide separation setup. The phosphopeptide identified in the pH 11 SIMAC multi-phosphopeptide fraction, the HILIC fractions, and the phosphopeptides to two sample subsets are shown. (**d**) Distribution of mono- and multi-phosphorylated peptides in the three subsets of peptide identifications shown in (**c**). Monophosphorylated peptides are shown in *grey*, while multi-phosphorylated peptides are *black*

use three different lysine labels in a single experimental setup. Including fed wild type, starving wild type, and starving AMPK-deficient animals labeled with Lys0, Lys4, and Lys8, respectively, we have identified more than 5,000 proteins with high confidence (false-discovery rate ≤ 1 %) of which almost 4,000 proteins could be quantified in all three states by at least two quantification events (Fredens, Engholm-Keller, Larsen, Færgeman, manuscript in preparation). By phosphopeptide enrichment using TiO_2 chromatography and liquid chromatography-mass spectrometry (LC-MS) [10–12] and sequential elution from immobilized metal affinity chromatography (SIMAC) [13] (*see* Fig. 2), we have identified 3,791 phosphorylation sites with high certainty (localization probability ≥ 75 %) of which 3,577 could be quantified in the same experimental setup (manuscript in preparation).

In this chapter, we describe how an *E. coli* strain is rendered compatible to perform RNA interference, incorporation of heavy isotope-labeled lysine in *C. elegans*, and a comprehensive phosphopeptide enrichment and analysis strategy.

2 Materials

2.1 Generation of an RNAi-Compatible E. coli Strain

For all solutions in this chapter, use analytical grade reagents and ultrapure water. All plates are 10-cm petri dishes. *C. elegans* is incubated at 20 °C and *E. coli* is incubated at 37 °C unless specified otherwise.

1. Lysine auxotroph *E. coli* strain ET505 (F−, λ− *lysA0::Tn10 IN(rrnD-rrnE)1*) (Coli Genetic Stock Center).
2. *E. coli* strain SK7621 (*Δrnc-38*) [14].
3. Lysogeny broth (LB): 1 % tryptone, 0.5 % yeast extract, and 1 % NaCl, pH 7.5 (autoclaved).
4. Use antibiotics at the following concentrations: 12.5 μg/mL tetracycline (stock: 12.5 mg/mL in 96 % ethanol), 50 μg/mL kanamycin (stock: 50 mg/mL in H_2O), and 100 μg/mL ampicillin (stock: 100 mg/mL in H_2O).
5. Agar.
6. 10-cm petri dishes.
7. Maltose.
8. 1 M $MgSO_4$.
9. λDE3 Lysogenization kit (Merck Millipore/Novagen) containing lysates of λDE3 phage, helper phage, selection phage, and T7 tester phage.
10. 10× Phage dilution buffer: 1 M NaCl, 0.2 M Tris–HCl (pH 7.5), 0.1 M $MgSO_4$ (autoclaved).
11. Sterile 50 % glycerol.
12. Tryptone.
13. 100 mM $CaCl_2$.
14. 1 M glucose.
15. P1 phage lysate.
16. Chloroform.
17. 1 M Na-citrate (pH 5.5).
18. PCR tubes.
19. 10× High Fidelity buffer with $MgCl_2$ (Roche).
20. *rnc* forward primer 5′-GCAGAACCATGTATATCAGG-3′.

21. *rnc* reverse primer 5′-GTGGATTTGCCAACGTTCGG-3′.
22. dNTP.
23. High Fidelity Polymerase (Roche).
24. Agarose.
25. Transformation buffer I (TfbI): 30 mM potassium acetate, 100 mM RbCl, 10 mM $CaCl_2$, 50 mM $MgCl_2$, 15 % glycerol. Adjust to pH 5.8 with 0.2 M acetic acid (sterile-filtered).
26. Transformation buffer II (TfbII): 10 mM MOPS, 10 mM RbCl, 75 mM $CaCl_2$, 15 % glycerol, adjust to pH 6.5 with 1 M NaOH (sterile-filtered).

2.2 Labeling of C. elegans

1. Agarose.
2. NaCl.
3. 1 M $CaCl_2$.
4. 1 M $MgCl_2$.
5. 1 M K_2HPO_4/KH_2PO_4 (pH 6.0).
6. 5 mg/mL cholesterol in 96 % ethanol.
7. 1 M IPTG.
8. 10-cm petri dishes.
9. Lysine-free EZ-rich defined medium kit (Teknova).
10. Lysine (Lys0).
11. 96–98 % enriched $^2H_4$-lysine (Lys4).
12. 97–99 % enriched $^{13}C_6{}^{15}N_2$-lysine (Lys8).
13. 5 N NaOH.
14. 5 % NaOCl.
15. S-basal: 0.585 % NaCl, 0.1 % K_2HPO_4, and 0.6 % KH_2PO_4 (autoclaved).
16. 0.9 % NaCl.
17. Glass Pasteur pipettes.

2.3 Protein Extraction and Digestion

1. Complete protease inhibitor with EDTA (Roche).
2. PhosSTOP phosphatase inhibitor cocktail (Roche) (*see* **Note 1**).
3. 2 M Tris–HCl (pH 7.5).
4. 2 M HCl.
5. 10 % SDS.
6. 50 % glycerol.
7. 1 M DTT.
8. Sonicator with 3 mm diameter tip.
9. Benzonase nuclease, purity > 90 % (Merck).

10. Protein determination kit compatible with the SDS buffer, e.g., BCA protein assay kit (Thermo Scientific).
11. Triethyl-ammonia bicarbonate (TEAB).
12. 500 mM iodoacetamide (IAA) in 100 mM (TEAB).
13. Methanol.
14. 6 M urea, 2 M thiourea.
15. Lysyl endopeptidase (Lys-C).
16. Trifluoroacetic acid (TFA).
17. Biochrom 30 amino acid analyzer.

2.4 Phosphopeptide Pre-enrichment

1. Titansphere 5 μm TiO_2 resin (GL Sciences Inc, Tokyo, Japan).
2. Loading buffer: 1 M glycolic acid, 80 % acetonitrile (ACN), 5 % trifluoroacetic acid (TFA).
3. Buffer 1: 80 % ACN, 1 % TFA.
4. Buffer 2: 20 % ACN, 0.2 % TFA.
5. Elution buffer: 1 % aqueous NH_4OH, pH should be 11.3, and not adjusted further.

2.5 Sequential Elution from IMAC Separation of Multiply and Mono-phosphorylated Peptides

1. PhosSelect IMAC resin (Sigma-Aldrich).
2. Gel loader tips.
3. Syringes.
4. Formic acid (FA).
5. POROS Oligo R3 (Life Technologies).
6. Empore extraction C18 disk (3 M).

2.6 Preparation of Non-phosphorylated Peptides

1. Sep-Pak C18 Plus cartridge (Waters).

2.7 Peptide Fractionation by HILIC

1. Agilent 1200 capillary flow HPLC system equipped with a 80-nl UV flow cell and a micro fraction collector (Agilent Technologies).
2. HILIC column consisting of fused silica capillary tubing (0.32 × 180 mm; Polymicro Technologies) and a polyether ether ketone (PEEK) inline microfilter (Upchurch Scientific) packed with 3 μm TSKGel Amide 80 HILIC resin (Tosoh Bioscience).
3. Solvent A: 0.1 % TFA.
4. Solvent B: 90 % ACN, 0.1 % TFA.
5. Mascot Distiller (Matrix Science, London, UK), Proteome Discoverer (ThermoScientific, Bremen, Germany), TransProteomic Pipeline [15], or MaxQuant [16].

3 Methods

3.1 Generation of an RNAi-Compatible E. coli Strain

The ability to fully incorporate a stable isotope-labeled amino acid into the proteins of cell cultures or model organisms is fundamental to all SILAC experiments. To avoid contamination with unlabeled lysine (*see* **Note 2**), *C. elegans* must feed on a lysine auxotroph *E. coli* strain. To enable efficient RNAi-mediated knockdown in *C. elegans*, the *E. coli* strain must contain an inducible T7 polymerase and must not degrade the overexpressed dsRNA. This is achieved, for example, by integration of a DE3 construct and deletion of *rnc*-encoding RNaseIII.

1. Make LB + agar (LA) plates with antibiotics: Mix 15 g of agar and 1 l of LB medium. Autoclave for 20 min and cool to 55 °C. Add appropriate antibiotics and pour 20 mL of LA into each petri dish. Dry at RT and store at 4 °C.
2. Grow lysine auxotroph *E. coli* ET505 in 5 mL of LB + tetracycline with aeration overnight. Dilute 50 μL of this culture in 5 mL of LB + 10 mM $MgSO_4$, 0.2 % maltose, and tetracycline. Incubate with aeration for 1.5–2 h to reach an $OD_{600} = 0.5$.
3. Add 2×10^8 pfu of λDE3 phage lysate to a 1.5-mL tube and dilute 1:10 with 10× Phage dilution buffer. In separate tubes, repeat this with helper phage lysate and selection phage lysate.
4. Add 5 μL of the ET505 culture from **step 2** to one tube and 10 μL to another one. To each tube, add half of the three diluted phage lysates. Mix gently and incubate at RT for 20 min to allow adhesion to the cells. Spread each mixture on an LA plate with tetracycline using a spatula. Allow the plates to dry and incubate overnight (*see* **Note 3**).
5. Pick a few colonies and restreak them on new LA plates with tetracycline and incubate overnight.
6. Make overnight cultures of the ET505(DE3) candidates by adding each colony to 5 mL of LB + tetracycline and incubate with aeration.
7. Prepare frozen stocks in 1.5-mL cryotubes by adding 700 μL of sterile 50 % glycerol and 1 mL overnight culture. Mix gently and store at −80 °C.
8. To test candidate clones for integration of DE3, dilute 50 μL of the overnight culture in 5 mL of LB + 10 mM $MgSO_4$, 0.2 % maltose, and tetracycline to $OD_{600} = 0.5$.
9. Dilute T7 tester phage lysate in 10× Phage dilution buffer to a final concentration of 2,000 pfu/mL in a final volume of 200 μL × number of clones to be tested.
10. In duplicates, mix 100 μL of each culture with 100 μL of diluted T7 tester phage lysate. Incubate at RT for 10 min.

Prepare top agar by mixing 1 g of tryptone, 0.8 g of NaCl, 0.8 g of agar, and 100 mL of H_2O. Autoclave for 15 min and cool to 45 °C. Mix each culture and T7 tester phage with 3 mL top agar. Pour one duplicate onto an LA plate with tetracycline and the other one onto an LA plate with tetracycline and 0.4 mM IPTG. Incubate at RT overnight. The presence of plaques on the plate containing IPTG confirms integration of DE3.

11. For transduction of the deletion insertion *Δrnc-38*, grow donor *E. coli* strain SK7621 in 5 ml of LB + kanamycin with aeration overnight. Dilute 25 μL of this culture in 2.5 mL of LB + 5 mM $CaCl_2$ and 0.2 % glucose. Incubate with aeration for 1 h. Add 100 μL of P1 phage lysate and continue incubation for 2–3 h until the bacterial culture has lysed completely and becomes clear. Transfer 1.8 mL of the lysate to a 2-mL tube. Add 100 μL of chloroform and vortex thoroughly to disrupt surviving cells. Centrifuge at 16,000 × *g* for 2 min and transfer the supernatant to another tube. Add a few drops of chloroform and store at 4 °C.
12. Grow recipient strain ET505(DE3) in 2 mL of LB + tetracycline with aeration overnight. Transfer 1.5 mL of this culture to a tube and centrifuge at 3,300 × *g* for 2 min. Remove the supernatant and resuspend cells in LB + 100 mM $MgSO_4$ and 5 mM $CaCl_2$. Mix 100 μL of ET505(DE3) with 100 μL of the P1 phage lysate from **step 11** in a tube. Prepare negative controls consisting of 100 μL of ET505(DE3) and 100 μL of LB in one tube as well as 100 of P1 phage lysate and 100 μL of LB + 100 mM $MgSO_4$ and 5 mM $CaCl_2$ in another one. Incubate for 30 min.
13. Add 200 μL of 1 M Na-citrate (pH 5.5) and 1 mL of LB to each tube. Incubate for 2 h.
14. Centrifuge tubes at 3,300 × *g* for 2 min. Remove the supernatant and resuspend cells in 100 μL of LB + 100 mM Na-citrate (pH 5.5). Select for positive clones by plating the content of each tube onto an LA plate with tetracycline and kanamycin. Incubate at 30 °C for 2 days.
15. On LA plates with tetracycline and kanamycin, spread 100 μL of 100 mM Na-citrate (pH 5.5) and allow plates to dry. Restreak a few colonies from **step 14** on these plates and incubate at 37 °C overnight.
16. Verify integration of *Δrnc-38* by colony-PCR using primers that bind upstream and downstream of *rnc*. Prepare 100 μL PCR mix on ice: 79 μL of H_2O, 10 μL of 10× High Fidelity buffer with $MgCl_2$, 4 μL of 20 mM forward primer, 4 μL of 20 mM reverse primer, 2.25 μL of 10 mM dNTP, and 1 μL of High Fidelity Polymerase. In separate tubes, dissolve a bit of each clone in 20 μL of H_2O and incubate at 95 °C for 5 min. Use SK7621 and ET505(DE3) as positive and negative control,

respectively. Add 2 μL of a dissolved clone and 23 μL of PCR mix to a PCR tube on ice. Mix gently by pipetting and run the PCR using the following parameters: 94 °C for 3 min, 30 cycles of 94 °C for 45 s, 54 °C for 45 s, 72 °C for 3 min, once 72 °C for 12 min, and hold at 4 °C. Analyze PCR products on a 1 % agarose gel. *rnc* and *Δrnc-38* results in PCR products of 0.9 and 2.2 kbp, respectively.

17. Select a positive colony of ET505(DE3) *Δrnc-38* and inoculate it in 5 mL of LB + kanamycin. Incubate with aeration overnight. Prepare frozen stocks as described in **step 7**. In Fredens et al. [7] we isolated several positive clones and named one of these NJF01, which were used in subsequent studies.
18. To render *E. coli* NJF01 competent for transformation, grow NJF01 in LB + kanamycin with aeration overnight.
19. Inoculate 1 mL of the overnight culture into 100 mL LB + tetracycline. Incubate with aeration to $OD_{600} = 0.5$.
20. Incubate on ice for 15 min.
21. Pellet cells at 3,300 × *g* for 5 min and discard supernatant.
22. Resuspend cells in 40 mL TfbI and incubate on ice for 15 min.
23. Repeat **step 21**.
24. Resuspend cells in 4 mL TfbII and incubate on ice for 1 h.
25. Use for transformation within hours or make aliquots of 250 μL cells in 1.5 mL-tubes for long term storage at −80 °C.
26. For transformation, thaw competent *E. coli* NJF01 on ice. For each reaction transfer 50 μL cells to a 15-mL glass tube on ice and add 1 μL empty vector L4440 or vectors for RNAi [from either A. Fire (Stanford University) or from Ahringer RNAi library collection]. Mix carefully and incubate on ice for 30 min.
27. Heat shock in a water bath at 42 °C for 1 min.
28. Quickly place tubes on ice for 2 min.
29. Add 250 μL LB at RT and incubate with aeration for 1 h.
30. On respective LA plates with ampicillin spread 20 and 100 μL transformation mix using a spatula.
31. Allow the plates to dry and incubate overnight. Pick an isolated colony from one of the plates and restreak on a new LA plate with ampicillin and incubate overnight.
32. Pick a restreaked colony and grow in LB + ampicillin with aeration overnight. Prepare frozen stocks as described in **step 7**.

3.2 Labeling of C. elegans

Complete labeling of *C. elegans* is achieved at adulthood of first generation worms. However, the L4 larva is usually used for analysis and the proteome of adult worms is complicated by the presence of embryos and small larvae. Therefore, L4 larvae are preferred for analysis and labeling over two generations is necessary. This protocol describes a phosphoproteomic strategy with anticipated results using approximately 1 mg of protein in total, however, the presented phosphoproteomics strategy is applicable to lower amounts of material [17]. For proteomic studies without phosphopeptide enrichment, one plate of worms of each condition is sufficient.

1. Prepare minimal plates: Mix 12 g of agarose, 3 g of NaCl, 1 mL of 1 M $CaCl_2$, 1 mL of 1 M $MgCl_2$, 27 mL of 1 M K_2HPO_4/KH_2PO_4 (pH 6.0), and fill up to 1 L with H_2O [18]. Autoclave for 15 min and cool to 65 °C. Add 1 mL of 5 mg/mL cholesterol, 1 mL of 1 M IPTG, and 500 μL of 50 mg/mL carbenicillin (*see* **Note 4**). Mix by whirling without introducing air bubbles. Pour approximately 20 mL into at least 46 petri dishes.
2. Prepare lysine-free EZ medium: Mix 50 mL of 10× MOPS, 50 mL of 10× ACGU, 5 mL of 0.132 M K_2HPO_4, 5 mL of 20 % glucose, and 100 mL of 5× Supplement EZ without lysine. Fill up to 500 mL with H_2O and autoclave for 15 min. Store at 4 °C.
3. Restreak NJF01 transformed with either L4440 control plasmid or specific RNAi plasmids on LA plates with tetracycline and ampicillin. Incubate overnight.
4. Pre-label NJF01: in three 250-mL flasks, mix 50 mL of lysine-free EZ medium with ampicillin. To each flask add either unlabeled lysine (Lys0), Lys4, or Lys8 to a final concentration of 0.4 mM. Add transformed NJF01 strains to each flask and incubate overnight.
5. Transfer each labeled culture to a 50-mL tube and centrifuge at 3,300 × *g* for 10 min to pellet the cells. Aspirate the supernatant to a final volume of 5 mL and resuspend the cells by pipetting. Spread 1 mL of culture on each of 3 × 5 minimal plates and leave at RT overnight to dry (*see* **Note 5**).
6. Before labeling of the first generation, synchronize *C. elegans*: in a 15-mL tube, collect adult animals containing eggs in 3.5 mL of H_2O. Mix 0.5 mL of 5 N NaOH and 1 mL of 5 % NaOCl and add to the 15-mL tube. Incubate at RT for 7 min (*see* **Note 6**) while shaking the tube vertically every few minutes. Pellet eggs by centrifugation at 1,300 × *g* for 30 s. Aspirate the solution and add 10 mL of H_2O. Spin again and repeat the wash. Spin again and aspirate the solution. Add 10 mL of S-basal and 10 μL of cholesterol solution. Incubate on a vertically rotating wheel for 28 h. Transfer 10 μL to a cover slide and count the number of living larvae. Pellet L1 larvae by centrifugation at 2,800 × *g* for 5 min and aspirate the supernatant to obtain a concentration of 10–50 larvae/μL.
7. Place up to 1,750 L1 larvae onto each minimal plate seeded with bacteria from **step 5**. Incubate for 3 days.

8. Pre-label NJF01 transformed with vectors for RNAi: in three 500-mL flasks, prepare 100 mL of EZ medium with ampicillin and Lys0, Lys4, or Lys8 as described in **step 4**. Add NJF01 transformed with L4440 or vectors for RNAi and incubate overnight. Repeat **step 5**.
9. For labeling of the second generation, synchronize the three populations of worms from **step** 7 (*see* **Note** 7). Place 6,500 Lys0-labeled L1 larvae onto the plate with Lys0-labeled NJF01 L4440. Accordingly, transfer the other populations on their respective plates. Incubate for approximately 48 h until the nematodes reach L4 state.
10. Harvest the worms at L4 state by washing the plates with 0.9 % NaCl using glass Pasteur pipettes. Collect the three populations in separate 15-mL tubes, centrifuge at 180 × *g* for 1 min, aspirate the supernatant, and add 0.9 % NaCl to a final volume of 10 mL. Repeat the wash once and add 0.9 % NaCl to a final volume of 10 mL. Wait 20 min to allow the worms to empty their intestine. Spin again and aspirate the supernatant. Adjust the volume to 150 μL with H_2O and keep cold.

3.3 Protein Extraction and Digestion

In the following sections, use low-binding tubes to minimize loss of proteins and peptides on plastic surfaces. For proteomic studies without phosphopeptide enrichment, reduce all volumes by 50 % and do not split the sample in **step 9**.

1. Prepare 5× protease inhibitor stock: dissolve one tablet of Complete protease inhibitor in 10 mL of H_2O. Store at −20 °C.
2. Prepare SDS buffer: in a 15-mL tube, mix 250 μL of 2 M Tris–HCl (pH 7.5) with 2.5 mL of H_2O. Adjust to pH 6.9 with 2 M HCl. Add 2.5 mL of 10 % SDS, 2 mL of 50 % glycerol, 100 μL of 1 M DTT, 2 mL of 5× Complete protease inhibitor stock, and one tablet of PhosSTOP phosphatase inhibitor cocktail (Roche) (*see* **Note 1**). Fill up to 10 mL with H_2O and mix gently by pipetting. Store at −20 °C.
3. Add 150 μL of SDS buffer with protease and phosphatase inhibitor to the worms (*see* **Note 8**).
4. Keep the samples in an ice bath and sonicate five times for 30 s at 5–6 W.
5. Add 1 μL of 1 M $MgCl_2$ and 2 μL of benzonase. Keep on ice for 15 min.
6. Determine the protein concentration of each sample (*see* **Note 9**) by for example BCA protein assay kit. Combine the three samples 1:1:1 based on the protein concentrations (*see* **Note 10**). Adjust the volume to 1 mL with SDS buffer.
7. Reduce disulfide bonds by adding DTT to a final concentration of 10 mM and incubate at 56 °C for 30 min.

8. Cool to RT and add IAA in TEAB to a final IAA concentration of 25 mM. Incubate at RT in the dark for 20 min to alkylate cysteines.
9. Split the sample into two 1.5-mL tubes and purify proteins by methanol-chloroform precipitation [19]. Use ice-cold solutions and keep samples on ice. Adjust the volume in each tube to 600 μL with H_2O. Add 450 μL of methanol and vortex briefly. Add 150 μL of chloroform. Vortex and centrifuge at 14,000 × *g* for 1 min.
10. Aspirate the upper phase without disturbing the protein layer. Add 500 μL of methanol, vortex, and spin for 2 min (*see* **Note 11**).
11. Aspirate the supernatant and resuspend the pellet in one of the tubes in 100 μL of 6 M urea, 2 M thiourea by sonicating five times for 30 s at 2–3 W in an ice bath (*see* **Note 12**). Transfer the solution to the other tube and resuspend the pellet by sonication as described.
12. Prepare Lys-C stock solution: On ice, dilute Lys-C in H_2O to a final activity of 0.05 amidase units (AU)/μL. Store aliquots at −20 °C.
13. Dilute 5 μL Lys-C stock solution in 95 μL of 100 mM TEAB, add the 100 μL to the protein solution, and incubate at RT (*see* **Note 12**) for 3 h with gentle rotation.
14. Repeat **step 13** but incubate overnight with gentle rotation.
15. Subsequently, add 10 % TFA to a final concentration of 1 % followed by centrifugation at 14,000 × *g* for 10 min to precipitate insoluble material and transfer the supernatant to a new 1.5-mL tube.
16. Determine the peptide content of the sample (*see* **Note 9**) to allow for the determination of the optimal amount of TiO_2 beads to be used for the phosphopeptide enrichment [17] (*see* **Note 13** and Subheading 3.4).

3.4 Phosphopeptide Pre-enrichment Using TiO_2

1. Adjust the peptide solution to loading buffer conditions by adding 1.2 mL of 100 % ACN, 60 μL of 100 % TFA, and 114 mg of glycolic acid. Vortex to dissolve glycolic acid.
2. Add 8 mg of TiO_2 beads (*see* **Note 13**) and incubate in a thermomixer at RT at 1,400 rpm for 10 min. Pellet beads by brief centrifugation (2,000 × *g* for 2 min at RT) in a benchtop centrifuge and transfer the supernatant to a new tube.
3. Add 4 mg of TiO_2 beads to the supernatant and repeat **step 2**. Transfer the supernatant of non-phosphorylated peptides to a new tube (may be stored at −80 °C).
4. Add 300–500 μL of loading buffer to the tube with 4 mg TiO_2 beads. Mix by vortexing and transfer the solution and the beads to the tube with 8 mg TiO_2 beads. Mix by vortexing and

transfer solution and beads to a new tube. Pellet beads by brief centrifugation and transfer the supernatant to the tube with non-phosphorylated peptides from **step 3**.

5. Wash beads in 300 μL of buffer 1, mix, centrifuge (2,000 × *g* for 2 min at RT), and transfer the supernatant to the tube with non-phosphorylated peptides.
6. Wash beads in 300 μL of buffer 2 and repeat **step 5**. Dry beads by vacuum centrifugation for 10 min.
7. Add 200 μL of elution buffer, mix by vortexing, incubate for 10 min at RT in a thermomixer under continuous shaking at 1,400 for 10 min, and centrifuge (2,000 × *g* for 2 min at RT). Avoid disturbing the bead pellet while transferring the supernatant with phosphopeptides to a new tube. Add 50 μL of elution buffer to the beads, mix, and centrifuge (2,000 × *g* for 2 min at RT). Transfer supernatant without any beads to the tube with the first eluate.
8. Dry the phosphopeptide sample to completeness in a vacuum centrifuge.

3.5 Sequential Elution from IMAC Separation of Multiply and Mono-phosphorylated Peptides

1. Redissolve the phosphopeptide sample in 200 μL of 50 % ACN, 0.1 % TFA and adjust the pH to 1.8 using 10 % TFA. Wash 120 μL of IMAC slurry with 500 μL of 50 % ACN, 0.1 % TFA, spin the beads down, discard the solvent and repeat the washing step. Add the phosphopeptide sample to the beads.
2. Allow the phosphopeptides to bind to the IMAC beads by incubation for 30 min at RT under continuous shaking at 1,400 rpm. After incubation, apply the sample–bead slurry to two constricted 200 μL gel loader tips and capture the beads in the tip by applying air pressure with a syringe while collecting the flow-through sample containing mainly mono-phosphorylated and traces of non-phosphorylated peptides in a 1.5 mL tube (*see* **Note 14**).
3. Wash the resulting IMAC column with 70 μL of 50 % ACN, 0.1 % TFA and collect the wash along with the flow-through from **step 2**. Elute the remaining mono-phosphorylated peptides slowly off the IMAC column using 70 μL of 20 % ACN, 1 % TFA and pool it with the flow-through, thereby resulting in a mono-phosphorylated peptide sample (*see* **Note 14**).
4. Elute the multi-phosphorylated peptides from the IMAC material into a tube using 80 μL of 1 % NH_4OH, pH 11.3 and acidify the sample with 8 μL of 100 % FA (*see* **Note 14**).
5. Adjust the mono-phosphorylated peptide sample to 80 % ACN and 1 % TFA and incubate it for 10 min under shaking at

1,400 rpm with 8 mg of TiO_2 material as used in the TiO_2 pre-enrichment step.

6. After incubation, pellet the beads by centrifugation (2,000 × *g* for 2 min at RT) in a benchtop centrifuge and remove the supernatant, which is incubated with 4 mg of TiO_2 beads for further 10 min under shaking at 1,400 rpm.
7. After the second incubation, pellet the beads by centrifugation and remove the supernatant. Add 100 μL of 50 % ACN, 0.1 % TFA to both tubes with beads, vortex them shortly, and mix the two pools of bead slurries. After a quick centrifugation step to pellet the beads, remove the supernatant.
8. After drying the TiO_2 beads for 5 min in a vacuum centrifuge, elute the phosphopeptides using 100 μL of 1 % NH_4OH, pH 11.3 for 10 min at RT under continuous shaking at 1,400 rpm. Spin the beads down in a benchtop centrifuge and remove the eluate into a new low-binding microcentrifuge tube without disturbing the beads. Add 50 μL of 1 % NH_4OH, pH 11.3 to the beads, vortex, and spin the beads down. Transfer the supernatant to the tube containing the first eluate without removing any beads. Acidify the sample using 15 μL of 100 % FA.
9. Desalt mono- and multi-phosphorylated peptide samples on a 1-cm R3 microcolumn packed in a P200 pipette tip with a 3 M C18 disk in the tip by loading it onto the column by applying air pressure with a plastic syringe. Wash the column with 100 μL of 0.1 % TFA and elute the peptides into a tube with 50 % ACN, 0.1 % TFA. Dry the phosphopeptide samples in a vacuum centrifuge.

3.6 Preparation of Non-phosphorylated Peptides

1. Dry the non-phosphorylated peptide-containing solution (Subheading 3.4, **step 4**) by vacuum-centrifugation before desalting. Dissolve the peptides in 5 mL 0.1 % TFA.
2. Wash Sep-Pak cartridge with 5 mL 70 % ACN, 0.1 % TFA.
3. Equilibrate with 10 mL 0.1 % TFA.
4. Load solution of non-phosphorylated peptides onto the Sep-Pak cartridge. Collect flow-through and load it again onto the cartridge. Wash the cartridge with 5 mL 0.1 % TFA. Elute the peptides with 1.5 mL of 70 % ACN, 0.1 % TFA into a new 1.5-mL tube. Dry by vacuum-centrifugation.

3.7 Peptide Fractionation by HILIC

1. Redissolve the non-phosphorylated peptides in 1.2 μL of 10 % TFA, add 10.8 μL of H_2O, and slowly add 108 μL of ACN. Redissolve the mono-phosphorylated peptides in 0.4 μL of 10 % TFA, add 3.6 μL of H_2O, and slowly add 36 μL of ACN. Centrifuge the sample at 14,000 × *g* for 2 min to pellet insoluble material.

2. Load 40 µL sample onto the HILIC column of the capillary HPLC system at a flow rate of 12 µL/min. Separate the peptides at a flow-rate of 6 µL/min with an increasing aqueous gradient ranging from 100 to 60 % solvent B in 35 min. Collect fractions every minute.
3. Combine the fractions to approximately 16 samples in total (depending on the time available for LC-MS/MS analysis) based on intensity measured by UV detection. Dry fractions by vacuum-centrifugation. Redissolve in 0.5 µL of 100 % FA and add 10 µL of H_2O.
4. Analyze fractions by reversed-phase nanoLC-MS/MS as described previously [17].
5. Process the raw data and search the peak lists against the *C. elegans* Uniprot database, e.g., using Mascot Distiller, Proteome Discoverer, TransProteomic Pipeline, or, as in our case, MaxQuant. Perform the database search with carbamidomethyl (C) as fixed modification and acetylation (N-terminus) and oxidation (M) as variable modifications. For non-phosphorylated peptides, select enzyme specificity as Lys-C with up to 1 missed cleavage allowing 7 ppm peptide ion tolerance and 0.6 Da MS/MS tolerance. We allow charge states up to 7, a maximum of two labeled amino acids per peptide, and up to three modifications per peptide. Search phosphorylated peptides with phosphorylation (STY) as variable modification, up to two missed cleavages, a maximum of three labeled amino acids per peptide, and up to seven modifications per peptide.
6. The lists of proteins and phosphosites as well as their corresponding quantitative ratios can be imported into Microsoft Excel and phosphosite ratios can be normalized to protein ratios. Furthermore, Perseus, which is part of the MaxQuant software suite, can calculate an intensity-weighted significance of the ratios (significance B). We have employed the following thresholds to our data: posterior error probability (PEP) ≤ 0.01 for reliable identifications, ratio counts ≥ 2 for reliable quantifications, a significance $B \leq 0.05$ for significant regulation, and a localization probability ≥ 0.75 for reliable localization of phosphorylations. Since the functional annotation of *C. elegans* proteins is low compared to other model organisms, it might be useful to obtain additional information about protein functions as well as orthologs from better characterized organisms through for example gProfiler (http://biit.cs.ut.ee/gprofiler/). Motif enrichment and kinase prediction of the phosphosites can be performed using the respective online services motif-x (http://motif-x.med.harvard.edu/) and NetworKIN (http://networkin.info/version_2_0/search.php).

4 Notes

1. For proteomic studies without phosphopeptide enrichment, leave out phosphatase inhibitors.
2. Traditionally, SILAC is performed with labeled lysine and arginine, resulting in one labeled amino acid per peptide from a tryptic digest. However, in *C. elegans* and other organisms, arginine-to-proline conversion results in satellite peaks originating from heavy proline containing peptides, which reduces the overall quantitative accuracy as well as identification rates. In the described procedure, this is solved by incorporation of labeled lysine followed by Lys-C digestion. Alternatively, the arginine-to-proline conversion problem can be eliminated through experimental correction [20], genetic manipulation [8, 21], computational correction [22], or a frequent amino acid like leucine can be incorporated instead of arginine and combined with tryptic digestion to maximize the quantitative coverage of the proteome [23].
3. Allow the plate to dry at RT for several hours before the experiment to ease absorption of the high volume of liquid.
4. Carbenicillin selects for plasmids for RNAi and IPTG induces T7-driven expression of dsRNA. If RNAi is not a part of the experimental setup, leave out carbenicillin and IPTG.
5. As positive control for RNAi includes NJF01 transformed with a *dpy-13* RNAi vector that will result in a shortened phenotype in *C. elegans.* Grow the bacterium in LB + ampicillin and spread 1 mL on a minimal plate. Add 100 L1 larvae and incubate for 2 days.
6. If the worms containing eggs are exposed to the bleach mix for more than 10 min, the eggs may be damaged and fewer larvae can be obtained from the synchronization.
7. At second generation, the three populations should reach the L4 stage at the same time. Since most treatments and mutations affect the time of development, this must be accounted for at the time of synchronization.
8. For proteomic studies without phosphopeptide enrichment, it is sufficient to use a single plate of each *C. elegans* population. Sonicate the animals in half of the described volumes of H_2O and SDS buffer without phosphatase inhibitors. Determine protein concentration and mix the populations 1:1:1 in a 1.5-mL low-binding tube. Adjust the volume to 500 μL with SDS buffer.
9. Protein concentration can be determined using a BCA Protein Assay Kit or, more accurately, using protein acid hydrolysis

followed by amino acid analysis on for example a Biochrom 30 amino acid analyzer.

10. The degree of incorporation of the labeled lysines can be measured by analysis of unmixed populations of Lys4- and Lys8-labeled animals. This analysis does not require enrichment or fractionation.
11. In a fixed-angle centrifuge, the protein pellet may stick to the side of the tube. To avoid this, collect the protein pellet in the bottom of the tube by mechanical force or spin the pellet down in a centrifuge with swing out buckets prior to centrifugation at 14,000 × *g*.
12. Be aware that urea can carbamylate lysine residues at elevated temperatures. Hence, sonicate gently having occasional breaks between each round of sonication, and keep the solution in an ice bath. Furthermore, perform disulfide bond reduction as well as proteolytic digestion at RT.
13. Approximately 0.6 mg of TiO_2/100 μg of peptides is appropriate [17].
14. Make sure that there are no IMAC beads in the flow-through and eluates. Centrifuge the solutions at 14,000 × *g* for 2 min and transfer the supernatant if a pellet appears.

Acknowledgements

This work was supported by The Lundbeck Foundation and The Danish Council for Independent Research, Natural Sciences.

References

1. Jiang H, English AM (2002) Quantitative analysis of the yeast proteome by incorporation of isotopically labeled leucine. J Proteome Res 1:345–350
2. Ong SE, Blagoev B, Kratchmarova I et al (2002) Stable isotope labeling by amino acids in cell culture, SILAC, as a simple and accurate approach to expression proteomics. Mol Cell Proteomics 1:376–386
3. Gruhler A, Olsen JV, Mohammed S et al (2005) Quantitative phosphoproteomics applied to the yeast pheromone signaling pathway. Mol Cell Proteomics 4:310–327
4. Sury MD, Chen JX, Selbach M (2010) The SILAC fly allows for accurate protein quantification in vivo. Mol Cell Proteomics 9:2173–2183
5. Gruhler A, Schulze WX, Matthiesen R et al (2005) Stable isotope labeling of *Arabidopsis thaliana* cells and quantitative proteomics by mass spectrometry. Mol Cell Proteomics 4:1697–1709
6. Kruger M, Moser M, Ussar S et al (2008) SILAC mouse for quantitative proteomics uncovers kindlin-3 as an essential factor for red blood cell function. Cell 134:353–364
7. Fredens J, Engholm-Keller K, Giessing A et al (2011) Quantitative proteomics by amino acid labeling in *C. elegans*. Nat Methods 8: 845–847
8. Larance M, Bailly AP, Pourkarimi E et al (2011) Stable-isotope labeling with amino acids in nematodes. Nat Methods 8:849–851
9. Van Gilst MR, Hadjivassiliou H, Jolly A et al (2005) Nuclear hormone receptor NHR-49 controls fat consumption and fatty acid composition in *C. elegans*. PLoS Biol 3:e53

10. Larsen MR, Thingholm TE, Jensen ON et al (2005) Highly selective enrichment of phosphorylated peptides from peptide mixtures using titanium dioxide microcolumns. Mol Cell Proteomics 4:873–886
11. Pinkse MW, Uitto PM, Hilhorst MJ et al (2004) Selective isolation at the femtomole level of phosphopeptides from proteolytic digests using 2D-NanoLC-ESI-MS/MS and titanium oxide precolumns. Anal Chem 76:3935–3943
12. Kuroda I, Shintani Y, Motokawa M et al (2004) Phosphopeptide-selective column-switching RP-HPLC with a titania precolumn. Anal Sci 20:1313–1319
13. Thingholm TE, Jensen ON, Robinson PJ et al (2008) SIMAC (sequential elution from IMAC), a phosphoproteomics strategy for the rapid separation of monophosphorylated from multiply phosphorylated peptides. Mol Cell Proteomics 7:661–671
14. Babitzke P, Grange L, Olszewski J et al (1993) Analysis of mRNA decay and rRNA processing in *Escherichia coli* multiple mutants carrying a deletion in RNase III. J Bacteriol 175:229–239
15. Deutsch EW, Mendoza L, Shteynberg D et al (2010) A guided tour of the Trans-Proteomic Pipeline. Proteomics 10:1150–1159
16. Cox J, Mann M (2008) MaxQuant enables high peptide identification rates, individualized p.p.b.-range mass accuracies and proteome-wide protein quantification. Nat Biotechnol 26:1367–1372
17. Engholm-Keller K, Birck P, Storling J et al (2012) TiSH—a robust and sensitive global phosphoproteomics strategy employing a combination of TiO_2, SIMAC, and HILIC. J Proteomics 75:5749–5761
18. Krijgsveld J, Ketting RF, Mahmoudi T et al (2003) Metabolic labeling of *C. elegans* and *D. melanogaster* for quantitative proteomics. Nat Biotechnol 21:927–931
19. Wessel D, Flugge UI (1984) A method for the quantitative recovery of protein in dilute solution in the presence of detergents and lipids. Anal Biochem 138:141–143
20. Van Hoof D, Pinkse MW, Oostwaard DW et al (2007) An experimental correction for arginine-to-proline conversion artifacts in SILAC-based quantitative proteomics. Nat Methods 4:677–678
21. Bicho CC, de Lima Alves F, Chen ZA et al (2010) A genetic engineering solution to the "arginine conversion problem" in stable isotope labeling by amino acids in cell culture (SILAC). Mol Cell Proteomics 9:1567–1577
22. Park SK, Liao L, Kim JY et al (2009) A computational approach to correct arginine-to-proline conversion in quantitative proteomics. Nat Methods 6:184–185
23. Coute Y, Hernandez C, Appel RD et al (2007) Labeling of *Bifidobacterium longum* cells with 13C-substituted leucine for quantitative proteomic analyses. Appl Environ Microbiol 73: 5653–5656

Chapter 10

SILAC-Based Temporal Phosphoproteomics

Chiara Francavilla, Omid Hekmat, Blagoy Blagoev, and Jesper V. Olsen

Abstract

In recent years, thanks to advances in Mass Spectrometry (MS)-based quantitative proteomics, studies on signaling pathways have moved from a detailed description of individual components to system-wide analysis of entire signaling cascades, also providing spatio-temporal views of intracellular pathways. Quantitative proteomics that combines stable isotope labeling by amino acid in cell culture (SILAC) with enrichment strategies for post-translational modification-bearing peptides and high-performance tandem mass spectrometry represents a powerful and unbiased approach to monitor dynamic signaling events. Here we provide an optimized SILAC-based proteomic workflow to analyze temporal changes in phosphoproteomes, which involve a generic three step enrichment protocol for phosphopeptides. SILAC-labeled peptides from digested whole cell lysates are as a first step enriched for phosphorylated tyrosines by immunoaffinity and then further enriched for phosphorylated serine/threonine peptides by strong cation exchange in combination with titanium dioxide-beads chromatography. Analysis of enriched peptides on Orbitrap-based MS results in comprehensive and accurate reconstruction of temporal changes of signaling networks.

Key words Phosphorylation, Peptide enrichment, Phosphotyrosine, SCX, TiO_2, Orbitrap, Q-Exactive, In-solution digestion

1 Introduction

All living cells receive signals from the extracellular space and intracellular compartments and they need to process this information in a fast and efficient manner to generate the correct output with the right timing. Intracellular signal transduction pathways are constituted by protein complexes organized in networks. A delicate balance between stimulatory and inhibitory signals transmitted through the protein networks determines the strength and duration of responses, ultimately resulting in a specific biological outcome [1]. Many diseases, including cancers, are considered to be driven by aberrant alterations of signaling networks [2]. Therefore, qualitative and quantitative studies of the dynamics of signaling events are essential to understand how cells respond specifically to a multitude of perturbations.

Bettina Warscheid (ed.), *Stable Isotope Labeling by Amino Acids in Cell Culture (SILAC): Methods and Protocols*, Methods in Molecular Biology, vol. 1188, DOI 10.1007/978-1-4939-1142-4_10, © Springer Science+Business Media New York 2014

Many critical events involved in the tight regulation of cellular responses are mediated by protein post-translational modifications (PTMs) whose dynamic changes are essential for determining cellular behaviors. In particular phosphorylation, which in eukaryotes mainly modifies Ser, Thr, and Tyr residues, is believed to control all intracellular signaling cascades. In mammalian cells, phosphorylation affects at least one-third of all proteins and is among the most widely studied PTMs [3].

Mass spectrometry (MS)-based proteomics is the method of choice to study how changes in PTMs affect cellular signaling networks at a whole cell level and in an unbiased manner [4–6]. Proteomics typically involves whole cell lysate protein digestion by trypsin and analysis of the resulting peptide mixtures by nanoflow liquid chromatography-tandem mass spectrometry (LC-MS/MS). However, PTMs are in general sub-stoichiometric and thereby the PTM-modified peptides are often present in very low amounts. Therefore at least three main factors need to be considered before initiating a large-scale proteomics study of temporal changes of PTM-modified peptides derived from perturbed signaling networks (and in particular of entire phosphoproteomes).

The first important aspect to consider is how to quantify the relative PTM changes between different cell populations. To achieve this, two or more phosphoproteomes of interest must be distinguishable in the mass spectrometer. Despite recent advances in label-free approaches, in which the MS signal intensity of each peptide is compared between LC-MS runs [7], these are still not well suited for accurate temporal phosphoproteomic studies. Quantitative methods in proteomics are in general based on stable isotope labeling with ^{13}C, ^{15}N, ^{2}H, or ^{18}O that introduces a known mass difference between peptides with same amino acid sequence and modification state, thus allowing for direct relative quantitation of abundance changes. Incorporation of stable isotopes can be achieved either by metabolic labeling of entire proteomes [8] or by in-vitro reactions with stable isotope-containing chemicals [9, 10]. In the described protocol we will focus only on the popular metabolic labeling strategy termed Stable Isotope Labeling by Amino acid in Cell culture (SILAC) [11], which is one of the simplest, most powerful, and widely used approaches in MS-based quantitative proteomics and phosphoproteomics [4, 5]. SILAC is also the most accurate quantitative strategy when using cell culture systems to study signaling dynamics. SILAC encodes cellular proteomes through normal metabolic processes, incorporating nonradioactive, heavy stable isotope-enriched amino acids in newly synthesized proteins [8]. It is most often based on the use of both Arg and Lys amino acids containing the stable ^{13}C and ^{15}N isotopes. The specific mass difference between the heavy-labeled and unlabeled amino acids used during cell growth allows the phosphoproteomes to be distinguished in the mass spectrometer. Tryptic peptides will appear as distinct pairs separated by a defined mass in the mass spectrometer, thus enabling the relative quantification of even small changes between samples [8]. In combination with affinity enrichment of phosphopeptides, SILAC has been successfully used to quantify temporal changes in phosphoproteomes, for example in response to external stimuli like growth factors [12] or differentiation stimuli [13].

Another major difficulty in large-scale phosphoproteomics studies is the detection of phosphopeptides in complex samples due to their generally low abundance, which is caused by the sub-stoichiometric nature of this modification as well as rapid dephosphorylation by protein phosphatases. To solve these dynamic range issues, it is essential to perform an enrichment step at the level of phosphopeptides. Several robust workflows for phosphopeptide enrichment have already been described, either based on the use of specific antibodies [14, 15] or using several stages of chromatography (e.g., strong cation exchange (SCX)) in combination with metal-ion-based immobilized metal affinity (IMAC) [16, 17] or titanium dioxide (TiO_2)-beads chromatography [18, 19]. Antibodies are widely used for detection of PTMs on proteins including phosphorylated proteins by western blot analysis. However, high-quality site-specific antibodies are not always available for the PTM of interest. For example, it is not easy to generate specific high-affinity antibodies against phosphorylated Ser and/or Thr residues. More recently, approaches based on the isolation of modified peptides from tryptic digest by immunoaffinity purification have been successfully used for global analysis of lysine acetylation [20] or tyrosine phosphorylation [14]. The majority of phosphorylation events in mammalian cells occur on serine (90 %) and threonine (9–10 %) residues, whereas tyrosine phosphorylation only represents about 0.5–1 % of all human phosphorylation events. However, the importance of tyrosine phosphorylation is highlighted by its key role as signaling mediator resulting in the regulation of most major cellular processes, including cell growth, cell motility, and gene transcription [21]. Therefore, specific and high-affinity antibodies against phosphorylated tyrosines have been used in combination with tandem mass spectrometric analysis to specifically enrich and analyze tyrosine-phosphorylated proteins [22] or peptides [10, 14]. The most successful generic analytical strategy for enrichment of phosphopeptides takes advantage of the unique chemical properties of the phosphate group that has negative charge at low pH and is able to interact with ion exchange beads and to coordinate to immobilized metal ions or metal oxides. For instance, a TiO_2-based solid matrix has proven an efficient and specific enrichment tool for phosphopeptides from complex peptide mixtures in several large-scale phosphoproteomics studies [4, 5, 12, 13]. Peptide separation to reduce sample complexity prior to TiO_2-based enrichment has the advantage of improving enrichment specificity and dynamic range. In this respect, SCX chromatography, which separates peptides based on their in-solution charge-state, effectively fractionates and enriches tryptic phosphopeptides from their unmodified counterparts at low pH. SCX chromatography combined with IMAC/TiO_2 chromatography has so far been the most successful strategy used in global phosphoproteomics studies [4, 5, 12, 23].

The last point to consider in generating data of high-quality is the great advantage of performing the MS analysis on instruments with high resolving power, high mass accuracy, high dynamic range, and high sequencing speed. Most laboratories make use of the Orbitrap-based mass spectrometers like the linear ion-trap-Orbitrap (LTQ-Orbitrap Velos) [24] or quadrupole-Orbitrap (Q-Exactive) [25, 26] instruments for large-scale phosphoproteomics analyses. In combination with optimized sample preparation and data analysis, this results in an in-depth coverage of protein and PTM identifications. For computational proteomics, the MaxQuant software suite [27, 28] is a powerful program that automatically processes and analyzes raw LC-MS files by detecting peptide peaks, isotope clusters and SILAC pairs/triplets as three-dimensional peaks in the *m/z*, elution time and intensity space. Mass accuracy in the p.p.b. range is achieved by multiple measurements of peptide masses along the elution time profile resulting in high-confidence peptide and protein identification through the integrated Andromeda peptide search engine [29].

Here we present an optimized SILAC-based phosphoproteomics workflow to analyze dynamic changes in phosphoproteomes, which involves a generic three-stage enrichment protocol of phosphopeptides. SILAC-labeled whole cell lysates are digested into peptides by specific proteases like trypsin, the resulting peptides are first enriched for tyrosine-phosphorylated peptides and then further enriched for phosphorylated serine/threonine-containing peptides by SCX followed by TiO_2-based chromatography. Finally, nanoscale LC-MS/MS analysis is performed using the Q-Exactive mass spectrometer. The optimal experimental design of a SILAC experiment varies depending on the complexity of the biological system of interest. In the classical SILAC approach described here, it is possible to choose 2–5 cellular conditions to be compared. For analysis of dynamic changes in phosphoproteomes with more than two time points, the samples have to be split into more than one experiment and combined computationally based on a common experimental point. We start by describing the preparation of SILAC media and a protocol for evaluating the degree of incorporation of the heavy or medium labeled SILAC amino acids into the proteins. We then suggest a highly efficient strategy to digest proteins and prepare samples for phosphopeptide enrichment. The first step is to enrich for tyrosine-phosphorylated peptides by immunoaffinity purification using immobilized anti-pan-phosphotyrosine antibodies. After reducing sample complexity by SCX fractionation, a further phosphopeptide enrichment step will be performed using TiO_2-beads chromatography. Phosphopeptides are finally desalted and concentrated before LC-MS/MS analysis. The described protocol relies on the use of reversed-phase C_{18} nanoflow LC-MS/MS system for peptide identification. Peptides are separated on packed C_{18} porous bead columns by reversed-phase HPLC using a linear gradient of increasing acetonitrile in acidified water and directly electrosprayed for introduction of ionized phosphopeptides into the mass spectrometer for analysis. Fully tryptic phosphopeptides ionized in the positive ion mode generally appear with charge state of +2, which makes collisional-activated dissociation the preferred sequencing method. LC-MS/MS analysis is performed in the data dependent acquisition mode by which a full scan of the precursor ions is performed first for precise measurements of peptide masses and relative quantitation of SILAC pairs. This is followed by isolation and fragmentation of the top-N most abundant precursors by Higher-energy Collisional Dissociation (HCD) [30]. MS and MS/MS are both analyzed in the last generation mass analyzer, the Orbitrap [31]. The peptides' sequences and their modifications can then be identified from the resulting MS/MS spectra by *in-silico* matching against a protein sequence database. Quantification and statistical evaluation of identified peptides are performed in MaxQuant software [27–29].

The workflow discussed here can easily be adapted to several global, qualitative and quantitative studies of dynamic changes in protein phosphorylation, as already described [12, 13].

2 Materials

All solvents in this protocol are prepared with ultrapure MilliQ water of 18.2 MΩ cm resistivity.

2.1 Commonly Used Buffers

All buffers should be prepared and stored in glassware.

1. Phosphate-buffered saline (PBS) from Invitrogen.
2. Denaturation buffer: 6 M urea, 2 M thiourea, 10 mM Hepes, pH 8.0 (10 M NaOH). It can be frozen as stock solution at −20 °C.
3. Reduction buffer: 1 M dithiothreitol (DTT) in 25 mM ammonium bicarbonate (NH_4HCO_3). It can be frozen as stock solution at −20 °C.
4. Alkylation buffer: 550 mM chloroacetamide in 25 mM NH_4HCO_3. It can be frozen as stock solution at −20 °C.
5. Buffer A: 0.5 % (v/v) acetic acid (AA) in water. It can be stored as stock solution at room temperature for several weeks.

6. Buffer B: 80 % (v/v) acetonitrile (ACN) and 0.5 % (v/v) AA in water. It should be prepared on a weekly basis.
7. Buffer A*: 2 % (v/v) ACN and 1 % (v/v) trifluoroacetic acid (TFA) in water. It can be stored as stock solution at room temperature for several weeks.
8. Buffer A′: 3 % (v/v) ACN and 1 % (v/v) TFA in water. It can be stored as stock solution at room temperature for several weeks.
9. Buffer A″: 8 % (v/v) ACN and 0.5 % (v/v) AA in water. It can be stored as stock solution at room temperature for several weeks.

2.2 SILAC Labeling

1. Amino acids: L-arginine (Arg0), L-lysine (Lys0), L-arginine-U-$^{13}C6$ (Arg6), L-lysine-$^{2}H_4$ (Lys4), L-arginine-U-$^{13}C_6$-$^{15}N_4$ (Arg10), and L-lysine-U-$^{13}C_6$-$^{15}N_2$ (Lys8).
2. All the media (DMEM and RPMI) should be SILAC media without arginine and lysine.
3. Serum has to be "dialyzed serum" in order to avoid free amino acids from the serum.

2.3 Sample Preparation for Temporal Phosphoproteomics

1. Modified RIPA buffer: 50 mM Tris–HCl, pH 7.5 (1 M HCl), 150 mM NaCl, 1 % (v/v) NP-40, 0.1 % (w/v) sodium deoxycholate, 1 mM EDTA.
2. Phosphatase inhibitors: 500 mM β-glycerophosphate, 500 mM sodium fluoride (NaF), 100 mM sodium-orthovanadate (Na_3VO_4). They can be frozen as stock solution at −20 °C. Kinase inhibitor: divalent metal chelator EDTA is included in the buffer. Complete protease inhibitor cocktail from Roche Diagnostic, 1 tablet/10 mL of buffer.
3. Material for determination of protein concentration (Bradford assay).

2.4 In-Solution Protein Digestion

1. Ice-cold acetone.
2. Lysyl endopeptidase (LysC): prepare at a final concentration of 0.5 mg/ml in 50 mM NH_4HCO_3. It can be frozen as stock solution at −20 °C.
3. Trypsin, sequencing grade: resuspend the latter in 50 mM AA to a final concentration of 0.5 mg/ml. It can be frozen as stock solution at −20 °C.

2.5 Purification of Digested Peptides

1. SepPak C_{18}: Classic C_{18} (Waters).
2. 10 cc syringe with plunger removed.

2.6 Immunoaffinity Purification of Tyrosine-Phosphorylated Peptides

1. Immunoprecipitation buffer: 50 mM MOPS, pH 7.2 (1 M HCl), 10 mM sodium phosphate, 50 mM NaCl. It can be stored as stock at 4 °C.

2. Anti-phosphotyrosine antibody beads (for instance from Santa Cruz, Cell Signaling Technology or Millipore).
3. NaCl 50 mM.
4. 0.1 % TFA for peptide elution.

2.7 Strong Cation Exchange Chromatography

1. Semi-preparative SCX columns: Resource™ S column, 1-ml bed volume, 6.4 mm inner diameter × 30 mm length (GE Healthcare, Sweden) or polySULFOETHYL A, 17-ml bed volume, 9.4 mm inner diameter × 250 mm length (PolyLC).
2. HPLC or FPLC, e.g., ÄKTA FPLC system (GE Healthcare, Sweden).
3. SCX solvent A: 5 mM potassium dihydrogen phosphate, 30 % (v/v) ACN. The pH should be 2.7 (to adjust with 20 % TFA).
4. SCX solvent B: 5 mM potassium dihydrogen phosphate, 30 % (v/v) ACN, 350 mM potassium chloride. The pH should be 2.7 (20 % TFA).

2.8 Titanium Oxide Chromatography

1. Titansphere TiO_2 beads 10 μm (GL Science Inc, Japan).
2. Loading solution: 0.02 g/ml 2,5-dihydroxybenzoic acid (DHB) in buffer B.
3. Reversed-phase C_8 microcolumn: High Performance Extraction Disk C_8 (3 M, Empore).
4. Elution buffer 1: 5 % ammonia (NH_4OH) in water.
5. Elution buffer 2: 10 % NH_4OH, 25 % ACN.
6. Elution buffer 3: 5 % piperidine, pH 11.8 (10 M NaOH).
7. Acidification buffer: 1 % TFA, 5 % ACN.

2.9 Desalting and Concentrating Peptides on StageTips

1. Solid Phase Extraction Disk C_{18} (3 M, Empore).
2. C_{18} Elution buffer: 40 % (v/v) ACN and 0.5 % (v/v) AA in water. It can be stored as stock solution at room temperature for several weeks.
3. Methanol.

2.10 Analysis of Peptides on Q-Exactive

1. Reversed-phase C_{18} material for nano-HPLC columns: Reprosil AQUA-Pur 1.9 μm particles (Dr. Maisch, Germany).
2. 15 cm fused silica emitter (Proxeon Biosystem), 75 μm inner diameter, 5 μm laser-pulled tip.
3. Bomb-loader device (Proxeon Biosystem).
4. EASY-nanoflow LC system connected to a Q-Exactive (both from Thermo Fisher Scientific).
5. Buffer A and buffer B.

3 Methods

All steps in the protocol take place at room temperature unless stated otherwise.

3.1 SILAC Labeling

1. To start SILAC labeling, prepare the amino acid stock solutions (*see* **Notes 1** and **2**) and the different types of SILAC media (*see* **Notes 3–5**).
2. Trypsinize the cell lines of interest (required only if cells are adherent) (*see* **Note 6**).
3. Split the cells in equal portions into two or three tubes (depending on double or triple SILAC experiment) and wash cell pellets with PBS. After centrifugation, discard PBS and resuspend cells from each tube in the respective SILAC medium (*see* **Note 7**) (Fig. 1).

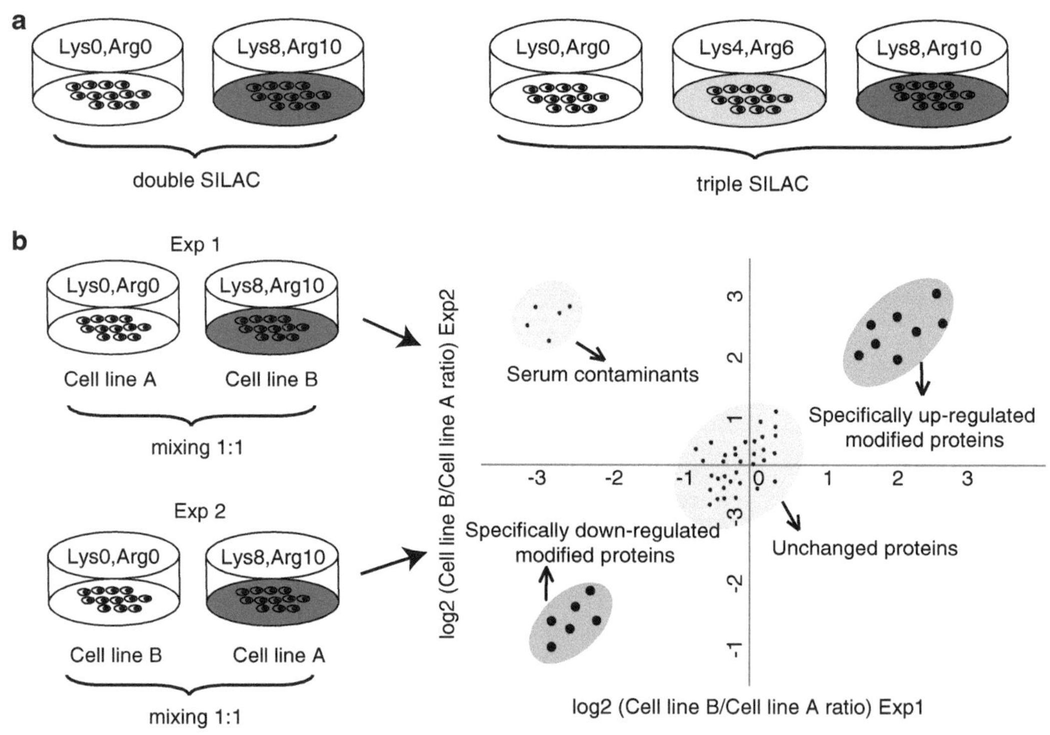

Fig. 1 Quantitation method based on SILAC. (**a**) Double (*left*) and triple (*right*) SILAC media are used to compare two or three different conditions, respectively. (**b**) Quantitation from replica experiments with inversed SILAC labeling. In the first biological replicate (Exp 1), cell line A is grown in light SILAC medium (Lys0, Arg0), while cell line B is grown in heavy SILAC medium (Lys8, Arg10) (*top*). In the second replicate (Exp 2), cell line B is grown in light SILAC medium (Lys0, Arg0), while cell line A is grown in heavy SILAC medium (Lys8, Arg10) (*bottom*). In both the experiments, proteins from cell lines A and B are then mixed 1:1. In the scatter plot, proteins enriched specifically in cell line B are shown as larger dots in the *dark grey circle*. Background proteins with a SILAC ratio around 1 are shown as small dots in the *light grey circle*. Log2-transformed SILAC ratios are used

4. Grow cells in SILAC media (*see* **Note 8**) for 5–10 doublings (*see* **Note 9**), changing the medium every 2 days or splitting them according to the standard procedure (*see* **Note 10**).
5. After ten doublings, freeze some cells (in order to have them ready for the experiment later) and process a small aliquot of the culture (10^6 cells) for labeling efficiency check (*see* **Note 11**).

3.2 Labeling Check

1. Wash the cell pellets three times with PBS, add 100 μl of Denaturation buffer for 1 h, and digest 100 μg of the sample according to the recommended workflow (*see* Subheading 3.4).
2. Desalt 5 μg of the digested peptides on reversed-phase C_{18} StageTips (*see* **Note 12**), elute peptides with C_{18} Elution buffer, remove ACN in a vacuum concentrator to reach a volume of 5 μl, dilute with buffer A* to a volume of 8–10 μl and inject 2–5 μl onto the column for nano-LC-MS/MS analysis (*see* Subheading 3.10).
3. Analyze the samples with the recommended software MaxQuant (*see* **Note 13**).

3.3 Sample Preparation for Temporal Phosphoproteomics

1. Grow a sufficient number of SILAC-labeled cells (from **step 5** in Subheading 3.1).
2. Harvest the cells at different time points (for instance every 6–24 h) or stimulate cells for specific shorter time intervals before harvesting (*see* **Note 14**).
3. After washing cells in cold PBS, lyse them on ice—if possible in the cold room—with modified RIPA buffer (*see* **Note 15**) and wait 20–30 min before scraping them off. Centrifuge at maximum speed in a microcentrifuge for 40 min at 4 °C (*see* **Note 16**).
4. Estimate protein concentration of the supernatant using established methods (for instance, Bradford) (*see* **Note 17**).
5. Acetone precipitate proteins over night at −20 °C in a 50-ml tube containing 80 % (v/v) ice-cold acetone (*see* **Note 18**).
6. Take out the tubes and centrifuge at 1,000 × *g* for 5 min. Discard acetone completely without drying the pellets.
7. Add Denaturation buffer and either resuspend the pellets by pipetting or leave them under rotation for several hours (*see* **Note 19**).
8. Estimate protein concentration of the supernatant again using established methods (for instance, Bradford) (*see* **Note 20**).
9. Mix SILAC-labeled samples in new tubes according to the experimental design (*see* **Note 21**) (Fig. 2).

3.4 In-Solution Protein Digestion

1. For protein reduction, add Reduction buffer to a final concentration of 1 mM DTT to the mixed samples (**step 9** of Subheading 3.3) for 1 h (*see* **Note 22**) (Fig. 3).

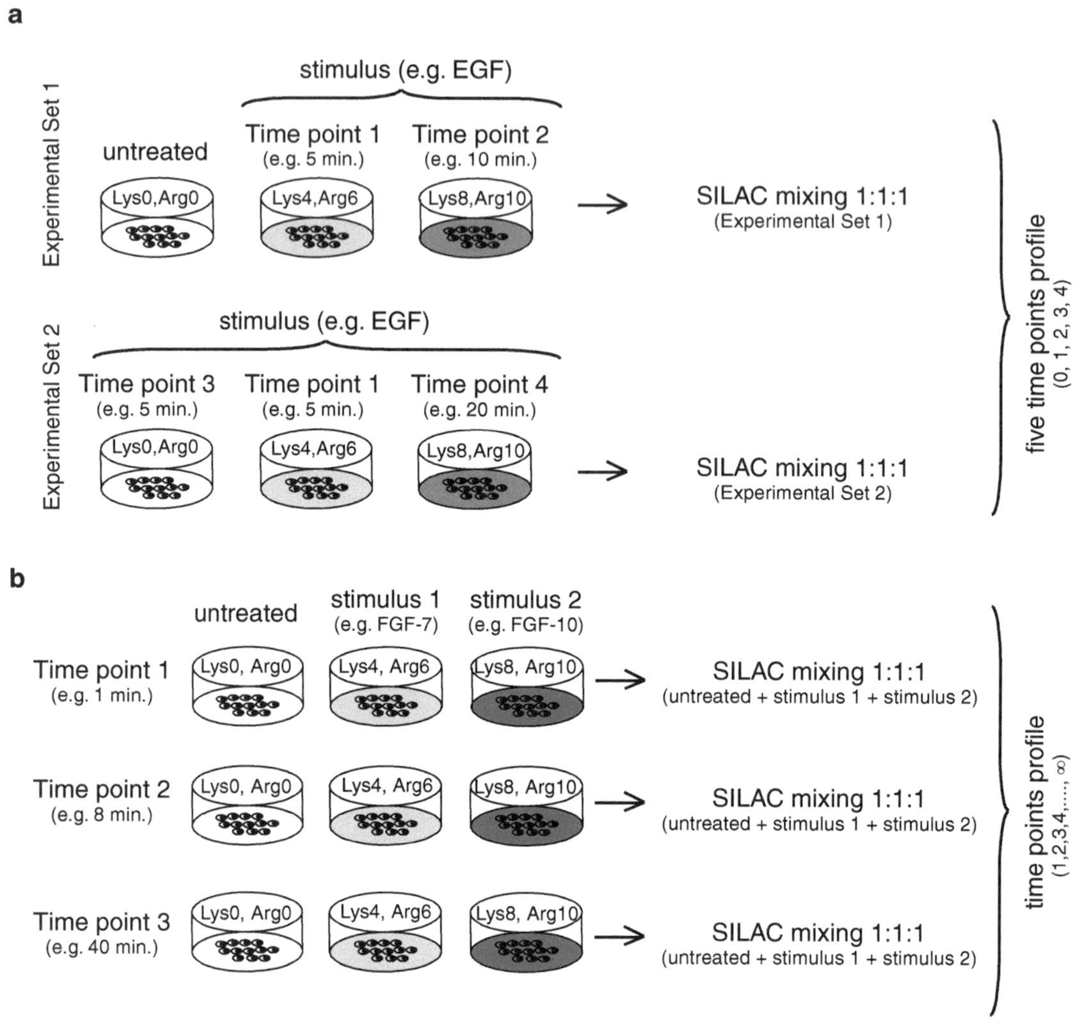

Fig. 2 Outline of two possible strategies to characterize temporal proteome/phosphoproteome changes during signaling. (**a**) Generation of multi-time point temporal profiles in response to specific stimulus. In the Experimental Set 1, triple SILAC-labeled cells are either left untreated or stimulated with a specific stimulus for two different time points. In the Experimental Set 2, triple SILAC-labeled cells are stimulated with the same stimulus for one of the two time points used in the Experimental Set 1 and for two additional time points. Proteins from each Experimental Set are then mixed 1:1:1, resulting in two sets of samples which cover five different time lengths. (**b**) Direct comparison of temporal profiles between two distinct stimuli. For each time point, the three SILAC cell populations are either left untreated or treated with two specific stimuli. Proteins are then mixed 1:1:1 resulting in a number of samples equal to the number of the time points of interest

2. For protein alkylation, add Alkylation buffer to a final concentration of 5.5 mM of chloroacetamide, mix well, and incubate for 45 min in the dark (*see* **Note 23**).
3. Check the pH (should be 8.0); adjust if necessary (*see* **Note 24**).
4. For protein digestion, add 1 μg of LysC per 100 μg of protein and incubate for 3–5 h (*see* **Note 25**).

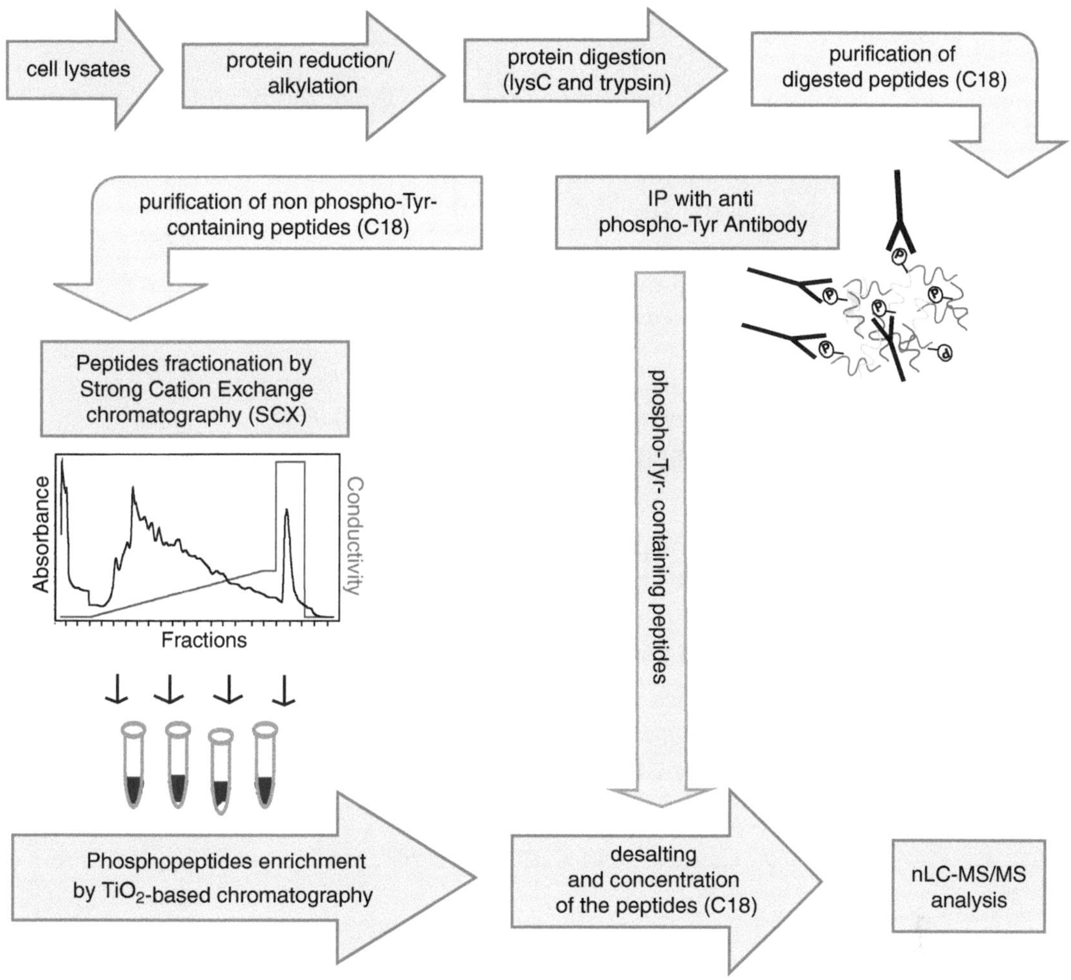

Fig. 3 Outline of the phosphoproteomics workflow

5. Dilute samples with 4 volumes of water (*see* **Note 26**).
6. Check the pH (should be 8.0); adjust if necessary (*see* **Note 24**).
7. Add 1 μg of trypsin per 100 μg of protein and incubate overnight (*see* **Note 27**).
8. Acidify the protein digest by adding TFA to a final concentration of 2 %. The pH should be ~2.5.
9. Centrifuge the samples in at maximum 1,000×*g* for 5 min to remove any precipitate that may form.

3.5 Purification of Digested Peptides

1. Connect a 10 cc reservoir (10 cc syringe with plunger removed) to the shorter end of a SepPak C_{18} column (*see* **Note 28**) (Fig. 3).
2. Pre-wet the column with 5 ml of 100 % ACN (*see* **Note 29**).

3. Wash the column twice with 4 ml of 0.1 % AA in MilliQ water (*see* **Note 30**).
4. Load the acidified solution of digested peptides from **step 10** of Subheading 3.4 onto the column (*see* **Note 31**).
5. Wash the column three times with 6 ml of 0.1 % AA in MilliQ water (*see* **Notes 32** and **33**).
6. Elute peptides into 50-ml tubes with 3 ml of 40 % ACN followed by 3 ml of 60 % ACN. At the very end, use the plunger (*see* **Note 34**).

3.6 Immunoaffinity Purification of Tyrosine-Phosphorylated Peptides

1. Remove ACN by vacuum concentrator at 60 °C to reach a volume of 20–50 μl (*see* **Note 35**).
2. Add 500 μl of the Immunoprecipitation buffer (*see* **Note 36**).
3. Check the pH (should be 7.4); adjust if necessary (*see* **Note 37**).
4. Let the peptides stand for 10 min and further dissolve peptides under gentle shaking overnight (*see* **Note 38**).
5. Clear solution by centrifugation in a microcentrifuge at maximum speed for 10 min.
6. Cool down on ice for 60 min.
7. Transfer the peptide solution into a new tube containing phosphotyrosine antibody beads (*see* **Note 39**) (Fig. 3).
8. Incubate for 2 h on a rotor at 4 °C.
9. Centrifuge in a cooled microcentrifuge at 500 × *g* for 2 min at 4 °C.
10. Accurately transfer the supernatant into a clean 15-ml tube (*see* **Note 40**).
11. Add 1 ml of Immunoprecipitation buffer to the beads, mix by inverting tubes five times, centrifuge in a cooled microcentrifuge at 500 × *g* for 1 min at 4 °C.
12. Repeat this step five times with the Immunoprecipitation buffer and twice with 50 mM NaCl.
13. At each step, the supernatant can be added to the new 15-ml tube mentioned in **step 10**.
14. Elute the peptides by adding 50 μl of 0.1 % TFA and incubating for 10 min on a rotor under gentle shaking (*see* **Note 41**).
15. Centrifuge in a cooled microcentrifuge at 500 × *g* for 1 min and transfer the supernatant into a clean Eppendorf tube.
16. Repeat **steps 14** and **15** at least three times.
17. Peptides can now be concentrated and purified on C_{18} StageTips (*see* Subheading 3.9 for a detailed protocol) before MS analysis (Fig. 3).
18. Add TFA to a final concentration of 0.1 % to the supernatant from **steps 10** and **13**.

19. Centrifuge the samples for 5 min at maximum 1,000×*g* to remove any precipitate that may form.
20. Repeat the steps described in Subheading 3.5 using 0.1 % TFA instead of AA in both the washing and the elution steps as specified in **Notes 30** and **34** (Fig. 3).

3.7 Strong Cation Exchange Chromatography

1. After elution from SepPak C_{18}, it is recommendable to save about 5 μg of peptides to run on the MS instrument in order to get an estimation of the total peptides present in the sample (*see* **Note 42**).
2. Equilibrate the SCX column on the FPLC/HPLC system by running a blank run (*see* **Note 43**).
3. Load the peptides in SCX solvent A at a flow rate of 1 ml/min (*see* **Notes 44** and **45**).
4. Elute the bound peptides with a linear gradient of 0–30 % SCX solvent B in 30 min at a flow rate of 1 ml/min and collect 2 ml fractions (*see* **Notes 46** and **47**) (Fig. 3).
5. Wash the SCX column with 5–10 column volumes of 100 % SCX Solvent B.
6. Run a blank run again (*see* **Note 48**).

3.8 Titanium Oxide Chromatography

1. Each SCX fraction is incubated with TiO_2 beads separately and is referred to as sample in this section (Fig. 3).
2. Weigh the TiO_2 beads into a dedicated tube; for each sample, weigh about 1.5–2 mg of beads.
3. Add Loading solution (*see* **Note 49**) to the beads (6 μl per 1.5 mg beads) and mix on a rotor under gentle agitation for 15 min.
4. Add 5 μl of loading solution containing DHB-coated TiO_2 beads to each sample and incubate on a rotor under gentle agitation for 30 min.
5. Centrifuge in a microcentrifuge at maximum 1,000×*g* for 5 min and transfer the supernatant—still rich in phosphopeptides—to a clean tube.
6. Repeat **steps 4** and **5** (*see* **Note 50**).
7. Wash sample-bound TiO_2 beads with 100 μl of SCX solvent B; centrifuge and discard the supernatant.
8. Wash sample-bound TiO_2 beads with 100 μl of freshly made 50 % buffer B + 50 % buffer A*; centrifuge and discard the supernatant.
9. Resuspend sample-bound TiO_2 beads in 50 μl of buffer B.
10. Prepare one C_8 microcolumn for each sample by placing a 1 mm^2 piece of Empore C_8 material into a 200 μl pipette tip as previously described [32].

11. Load sample-bound TiO_2 beads onto a dedicated microcolumn by centrifuging in a microcentrifuge at maximum 1,000 ×*g* for 2 min (*see* **Note 51**).
12. Slowly elute each sample into a 96 well microtiter plate: first with 2× 20 μl of Elution buffer 1, followed by 2× 20 μl of Elution buffer 2 and, eventually, by 20 μl of Elution buffer 3 (*see* **Note 52**).
13. Concentrate the eluates under vacuum to 5–10 μl.
14. Add 20 μl of Acidification buffer to acidify the samples.
15. Vortex the samples and centrifuge for 2 min.

3.9 Desalting and Concentrating Peptides on StageTips

1. Prepare one C_{18} StageTip per sample by adding two C_{18} disks to a 200 μl pipette tip (*see* **Note 53**).
2. Add 20 μl of methanol to activate the StageTip and wash the filter by centrifuging in a microcentrifuge at maximum 1,000 ×*g* for 1 min. All the liquid should go through.
3. Wash the tips by adding 20 μl of buffer B and centrifuge for 1–2 min.
4. Re-equilibrate the tips by adding 2× 20 μl of buffer A′ and centrifuge for 1–2 min (*see* **Note 54**).
5. Load the peptides (for instance from **step 15** of Subheading 3.8, from **step 17** of Subheading 3.6 or from **step 2** of Subheading 3.2) onto the StageTips and centrifuge until the solution has passed through the filters.
6. Wash the StageTips twice with 20 μl of buffer A″ and centrifuge for 1–2 min (*see* **Note 55**).
7. Elute the peptides into a 96-well microtiter plate by adding 2× 10 μl of C_{18} Elution buffer (*see* **Note 56**).
8. Concentrate under vacuum until the volume of sample in each well is around 5 μl.
9. Dilute with 3–5 μl of buffer A* and centrifuge for 1 min.

3.10 Analysis of Peptides on Q-Exactive

3.10.1 Online Liquid Chromatography

1. The LC part of the analytical LC-MS system described here consists of an EASY-nanoflow LC system connected to a Q-Exactive through a nanoelectrospray ion source [33].
2. Pack an analytical column in a 15 cm fused silica emitter, 75 μm inner diameter, 5 μm laser-pulled tip with a methanol slurry of reverse phase C_{18} particles at a constant helium pressure using a bomb-loader device, as described [33].
3. Connect the column directly to the HPLC auto-sampler through a 25 μm inner diameter fused silica transfer line and a micro Tee-connector (*see* **Note 57**).
4. Load 5 μl of the purified phosphopeptide mixture at a flow rate of 500 nl/min (back pressure of 130–300 mbar) for about 20 min using 3 % of buffer B.

5. After loading, reduce the flow rate to 250 nl/min and separate and elute the peptides with a 120 min linear gradient from 8 to 25 % buffer B followed by 10 min at 80 % buffer B.
6. At the end of the run, wash the analytical column by increasing the flow to 500 nl/min and reducing the percentage of buffer B to 5 %.

3.10.2 Mass Spectrometry

1. Operate the Q-Exactive quadrupole orbitrap mass spectrometer in data-dependent mode to automatically switch between full scan MS and MS/MS acquisition as described [25, 26].
2. Acquire survey full scan MS spectra (from m/z 300 to 1,750) in the orbirap with a resolution of70,000 (defined at $m/z=200$) after accumulation of ions at 1e6 target value based on predictive automatic gain control from the previous full scan.
3. Isolate and fragment the 8–12 most intense multiple-charged ions ($z \geq 2$) in the octopole collision cell by HCD with a fixed fill time of 250 ms and a resolution of 70,000.
4. Typical mass spectrometric parameters are:
   - Spray voltage: 2.1 kV
   - Sheath and auxillary gas flow: 0
   - Heated capillary temperature: 275 °C
   - Normalized collision energy: 25 %
   - Dynamic exclusion time: 30 ms
   - MS/MS ion selection threshold: 1e5 counts and 2.0 Da isolation width
   - "Lock Mass" option disabled

4 Notes

1. For SILAC labeling, proteins in one sample are labeled with heavy stable isotope-coded forms of essential amino acids while proteins in another sample contain the light forms of the corresponding amino acids. This creates a known mass difference between the same proteins/peptides in the mass spectra when the two samples are combined, making data quantification accurate and robust [8]. Users can choose the specific amino acids. The use of both lysine and arginine is recommended since, after protein digestion with trypsin, all peptides (except the protein's C-terminal peptide) are labeled and can be used for quantification.
2. To prepare amino acid stock solutions, dissolve arginine and lysine in PBS to a concentration of 84 and 146 mg/ml, respectively. The stock can be stored at 4 °C for one month.

Sterilize the stocks before adding them to the media using 0.22-μm syringe filters.

3. Depending on the type of SILAC experiment (double or triple SILAC are used to compare two or three different conditions, respectively), two or three media need to be prepared (Fig. 1a). Add serum (*see* **Note 4**), antibiotics, L-glutamine, sodium pyruvate and any other supplement that cells may need at the same concentration used for growing cells in the regular medium to SILAC media. Add 166 μl of L-arginine (Arg0) and L-lysine (Lys0), L-arginine-U-$^{13}C_6$ (Arg6) and L-lysine-$^2H_4$ (Lys4) or L-arginine-U-$^{13}C_6$-$^{15}N_4$ (Arg10) and L-lysine-U-$^{13}C_6$-$^{15}N_2$ (Lys8) to 500 ml of SILAC medium deprived of these amino acids to get light, medium and heavy SILAC media, respectively. This results in final concentrations of 28 mg/l arginine and 49 mg/l lysine, which works well for many cell lines. However, the final concentrations of amino acids may need to be adjusted or further titrated for some cell lines. Prepare the light and heavy SILAC media to perform a double SILAC experiment; prepare the light, medium and heavy SILAC media to perform a triple SILAC experiment.
4. Use dialyzed serum for SILAC media. It is recommendable to titrate the amount of serum and to compare the growth of the cell line of interest in regular and SILAC media performing, for example, viability tests or a growth assay.
5. SILAC media can be kept at 4 °C for a couple of weeks.
6. Trypsin must be removed every time cells are splitted.
7. In order to perform biological replicates it is possible to use the "labeling swap" strategy. For instance, to compare two different cell lines, they can be grown both in light and heavy media and then mixed accordingly (Fig. 1b). The crossover experiment represents a stringent criterion for specificity and removes background proteins in all cases.
8. Although SILAC appears to be one of the most accurate methods for quantification, it has been restricted to systems in which the complete proteome can be labeled by turnover excluding, for instance, human tissues samples. Recently, a method has been described in which a SILAC-labeled sample (usually in the heavy state) is used as reference to spike into each of the experimental samples grown in normal non-labeled (light) medium [34]. The choice between a classical SILAC strategy and a spike-in SILAC approach depends on the experimental question and on the system used. It is worth noting that quantification is more precise using the classical strategy [34].
9. The minimum number of cell doublings necessary to get full incorporation of amino acids depends on the cell lines of interest and needs to be tested. A minimum of 5 cell doubling is

required; however, it is recommendable to wait for 8–10 cell doublings and subsequently to scale up the cultures according to the number of cells needed at the end of the experiment.

10. For splitting the cells, follow standard procedures, always taking care to use the respective SILAC medium and to not mix SILAC media.
11. It is recommendable to perform an incorporation test on medium and heavy samples by analyzing those separately by MS.
12. C_{18} StageTips [32] can be substituted with equivalents and can be stored at 4 °C for several months (*see* also Subheading 3.9).
13. MaxQuant software is suitable for proteomics data analysis [27–29]. For labeling efficiency check, it is possible to follow the previously published protocol [28] with the following modifications. Analyze the files (corresponding to the medium and the heavy sample) as "double labeled" choosing the appropriate amino acids in the labeling options; add the variable modifications "methionine oxidation," "N-term acetylation," "heavy (or medium) proline" (to check arginine-to-proline conversion). In the Identify module, uncheck the "second peptide" and "re-quantify" options. Analyze the "evidence.txt" output file to calculate the incorporation rate. Eliminate CON and REV that represent contaminants and reverse hits, respectively. Distinguish between Lys- and Arg-containing peptides. For each of these subsets, determine the incorporation rate as 1 − (1/median ratio), using the not-normalized ratios. For both Lys and Arg, the rate should be above 0.95. In cases where Arg incorporation is significantly lower than Lys incorporation, there is probably proline-to-arginine conversion. Use a higher concentration of Arg for cell labeling. The number of cases in the "evidence.txt" table that contain heavy or medium proline indicates the extent of arginine-to-proline conversion. This number should not be higher than 1–2 %. Proline-containing peptides can be observed as extra-peaks in the spectrum. In this case, it is recommendable to lower the concentration of Arg in the SILAC media [8] or, alternatively, small amounts of unlabeled proline can be supplemented to the SILAC media [35].
14. At this step, SILAC-labeled samples are still treated as independent samples. Mix the samples at a later stage (*see* **Note 21**).
15. Add fresh inhibitors: 5 mM β-glycerophosphate, 5 mM NaF, 1 mM Na-orthovanadate, and protease inhibitor cocktail (for instance from Roche Diagnostic, 1 tablet/10 ml of lysis buffer).
16. If cells grow in suspension, wash cell pellets with PBS, add the modified RIPA buffer to the cell pellets on ice for 20–30 min and collect the supernatant after centrifugation.

17. Keep small aliquots at −20 °C for future validation studies.
18. Samples can be stored for several weeks in acetone at −20 °C.
19. It is fundamental to resuspend the samples at room temperature. Heating up the samples will result in carbamylation of free amino groups by urea during protein digestion. Start with an amount of Denaturation buffer equal to the volume of the pellets and increase progressively without exceeding 5 ml in order to avoid diluting the samples too much.
20. Compare the new concentration with the previous one (*see* **step 4** of Subheading 3.3 to estimate the percentage lost) in order to adjust the initial amount of cells to use for the next experiment. The protocol has been optimized for 2–30 mg of protein.
21. There are different strategies for preparing (and mixing) SILAC-labeled samples in order to study signaling cascade changes in a dynamic manner. Here, we describe two of these strategies that have been successfully applied to characterize temporal changes of Epidermal Growth Factor Receptor (EGFR) [12] and Fibroblast Growth Factor Receptor 2b (FGFR2b) [40] signaling. Both strategies make use of triple SILAC experiments. In the first strategy, the three SILAC-labeled populations of the first Experimental Set are stimulated with one specific ligand (EGF in this example) for different time length (for instance 0, 5, and 10 min) [12]. A second Experimental Set of triple SILAC-labeled cells is stimulated with the same ligand for three additional time points, one of which being in common to the first set (for instance 1, 5 and 20 min) [12]. The two time course experiments are then combined using the common time point (5 min EGF stimulation in the example), thus providing a five time points profile (Fig. 2a). With this approach it is possible to search for similar patterns in the time profiles of the regulated phosphoproteins, defining different temporal clusters. Using the second strategy, it is instead possible to compare two different stimuli for each time length of interest. After choosing the points of the extended time course (1, 8, and 40 min in this example), the three SILAC-labeled populations are either left untreated or treated with two different stimuli (in this example, FGF-7 and FGF-10 are two specific ligands for FGFR2b) [40]. All the time course experiments are then combined using the untreated condition as common point (Fig. 2b). The advantage of this approach is the combination of a "quality" component (the comparison between two stimuli) with the "temporal" dimension (evaluation of potentially infinitive time points). Of course, the quantitation among time points is not as strong as in the first strategy due to biological and technical variability. The choice among these or other strategies depends on the experimental question.

22. DTT reduces disulfide bonds.
23. Cysteines need to be alkylated to prevent reformation of disulfide bonds. In many digestion protocols, iodoacetamide instead of chloroacetamide is used as alkylating agent. However, iodoacetamide can make a covalent adduct to lysine residues by attachment of acetamide molecules that are isobaric to a lysine diglycine tag. To avoid introduction of such an artifact, it is fundamental to use chloroacetamide [36].
24. If necessary, adjust the pH with a very low volume of 1 M Tris–HCl, pH 8.0; however, the total salt concentration should not exceed 10 mM since it may interfere with SCX chromatography (*see* **Note 28**).
25. LysC works in high urea concentrations (8 M) and cleaves on the carboxyl side of lysine residues. This is a pre-digestion step. For larger amounts of protein (>10 mg), add 1 μg per 200 μg of protein. It is also possible to incubate the samples overnight.
26. To keep the salt concentration low, use pure MilliQ water rather than 50 mM NH_4HCO_3 to dilute samples before incubation with trypsin.
27. Trypsin cleaves at the carboxyl sites of both lysine and arginine, resulting in peptides of 8–9 residues on average. For larger amounts of protein (>10 mg), add 1 μg per 200 μg of protein.
28. SepPak C_{18} columns entail silica-based C_{18} hydrocarbon chains for reverse-phase (hydrophobic) extraction of peptides from aqueous solutions. Peptides bind to the chromatographic material and are then eluted by increasing ACN concentrations. Peptides up to 20 mg can be purified on one column. Digested peptides should be purified as soon as possible after digestion.
29. Each time a solution is applied to the column, air bubbles form on the narrow inlet of the column. These can be removed with a gel-loader tip placed onto a 200 μl micropipette.
30. Use 0.1 % TFA-containing washing buffer instead of 0.1 % AA-containing buffer if you do not intend to enrich for phosphorylated tyrosine-containing peptides.
31. If the volume of the acidified cleared solution is larger than the 10 ml reservoir, the sample must be applied several times. The column might turn yellow when the amount of peptides is higher than 5–10 mg.
32. In case the flow rate dramatically slows down after loading of the samples, it is possible to accelerate the washing step by applying gentle pressure to the column using the plunger that was originally removed. Do not apply vacuum.
33. Columns can be kept at 4 °C for several weeks before peptide elution.

34. If not interested in enrichment of phosphorylated tyrosine-containing peptides, peptides can be eluted two times with 2 ml of 40 % ACN, 0.1 % TFA followed by 2 ml of a buffer containing 60 % ACN and 0.1 % TFA. In this case, proceed directly to Subheading 3.7.
35. Lyophilization is not recommended to avoid an excessive loss of peptides. The removal of ACN in the vacuum concentrator will take a few hours and a white precipitate will be clearly observable.
36. Antibody-antigen binding is more efficient in aqueous buffers at physiological pH and ionic strength. The buffer should be cold.
37. The pH should be closer to neutral or at least not lower than 5–6. If the pH is moderately acidic, titrate the solution with concentrated Tris solution that has not been adjusted for pH. Do not use strong base.
38. This is a crucial step in order to dissolve all the peptides again before adding the antibody of choice.
39. A variety of anti-phosphotyrosine antibodies are commercially available. The three most commonly used are PY99 (Santa Cruz biotechnology, CA, USA), 4G10 (Millipore, USA), and PY100 (Cell Signaling Technology, USA). They can also be used in combination. We found PY100 both sensitive and specific. Most of these antibodies are supplied covalently linked to agarose beads, making them especially convenient for MS analysis. The amount of beads should be optimized for each application. The protocol described here is performed with 50 μl of PY100 beads (100 μl slurry) per 30 mg of digested peptides.
40. The supernatant is rich in other modified (including phosphorylated) peptides and can be conserved for further analysis (Fig. 3 and Subheadings 3.7 and 3.8).
41. The elution of the bound phosphopeptides is performed using strong acid such as TFA that disrupts antigen-antibody interactions.
42. After estimating the peptide concentration using established methods, add 100 μl of buffer A* and purify the peptides on C_{18} StageTips before MS analysis.
43. SCX chromatography is the method of choice (for instance over SDS-PAGE gels) to group tryptic peptides based on total charge and local charge distribution.
44. During loading, it is important to monitor conductivity: it should not be higher than 4 mS/cm to allow binding of all peptides. It is also possible to re-load the flow-through onto the SCX column after diluting it with water.

45. For high-resolution chromatography, it is important to keep the amount of peptides loaded below the maximum column capacity.
46. Conductivity—the best measure for gradient stability—should reach 13 mS/min at 30 % SCX solvent B.
47. Samples can be kept at 4 °C for a couple of days before further processing. It is possible either to analyze all the collected fractions or to combine some of them. For instance, all the fractions collected before the beginning of the gradient can be combined in a couple of tubes; the fractions collected at the very end of the gradient or during the wash with 100 % SCX solvent B can be also combined in few tubes. From each fraction, it is possible to save about 100 μl (or a volume roughly equivalent to 1–5 μg of peptides) to run on MS after adding buffer A* and after desalting on C_{18} StageTips.
48. Reproducible SCX chromatography is obtained with long and constant equilibration time.
49. The loading solution has to be fresh and is used to increase the binding specificity of TiO_2 beads for the phosphopeptides in the sample. It is possible to use different loading solutions such as 0.02 g/ml of DHB in buffer B [18, 37], 0.02 g/ml lactic acid [38], 0.01 g/ml DMSO, and 0.03 g/ml EDTA. DHB used at 0.02 g/ml gives the best result in terms of identified phosphopeptides without increasing the possibility of a contamination of the LC/MS instruments. Recently, it has been shown that the use of 6 % (v/v) TFA can improve phosphopeptide identification [39] without excluding the use of DHB.
50. The double or even triple incubation with TiO_2 beads can increase the number of identified phosphopeptides as shown in Fig. 4a. Some of the SCX fractions can be combined during the second incubation step.
51. Make sure that all the liquid passes through.
52. Increasing the pH of the solutions used for elution better disrupts the coordination between the phosphate group and the coordination group present in the beads. The step with piperidine is optional (*see* Fig. 4b for a comparison of different methods of elution; some of the SCX fractions can be combined and analyzed separately). We suggest using the described double-step ammonia elution protocol, which is highly reproducible (*see* Fig. 4c for a comparison between technical replicates of the entire protocol).
53. As already described [32], the principle behind it is that C_{18} hydrocarbon chains bind peptides through hydrophobic interactions. Peptides can be eluted by applying an organic solvent such as ACN or methanol that is less polar than water.

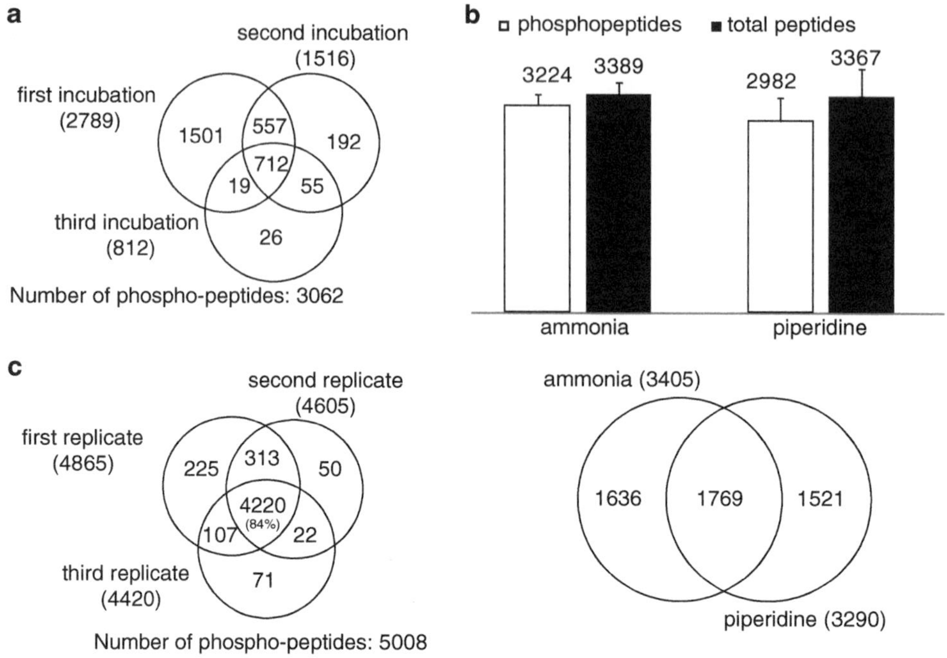

Fig. 4 The enrichment methodology for phosphopeptides. HEK293 tryptic lysate was pre-fractionated on SCX, early fractions were pooled and divided into equal aliquots (0.3–0.5 mg each) and used in the experiments. All experiments were done in triplicates and errors are standard deviations. (**a**) Venn diagram showing the effect of multiple sequential incubations of the sample with TiO_2 beads in improving the number of identified phosphopeptides. (**b**) Histograms (*top*) and Venn diagram (*bottom*) showing the effect of sequential elution of phosphopeptides from TiO_2 beads using solutions of increasing basicity (5 % ammonia followed by 5 % piperidine). The Venn diagram shows the overlap between one of the replicates shown in the histograms. (**c**) The Venn diagram shows the overlap between three technical replicates of the entire protocol

54. Make sure that the filter is still wet after the second wash before adding the samples. Otherwise repeat this step once more.
55. At this point, the StageTips can be stored at 4 °C for months before proceeding to the elution step.
56. To elute, force the solution through the filter into the chosen well of the plate by applying back pressure with a 10 ml syringe.
57. Note that there is no pre-column in this LC-MS setup.

Acknowledgements

The authors thank all members of the Department of Proteomics at the Novo Nordisk Foundation Center for Protein Research for fruitful discussion. The work was supported by the seventh framework program of the European Union (Contract no. 262067—PRIME-XS). The NNF Center for Protein Research was supported by a generous donation from the Novo Nordisk Foundation. Dr. Chiara Francavilla was supported by Marie Curie and EMBO Long-Term postdoctoral fellowships. Dr. Blagoev was supported by the Lundbeck Foundation.

References

1. Schlessinger J (2000) Cell signaling by receptor tyrosine kinases. Cell 103(2):211–225
2. Hanahan D, Weinberg RA (2011) Hallmarks of cancer: the next generation. Cell 144(5): 646–674
3. Cohen P (2001) The role of protein phosphorylation in human health and disease. The Sir Hans Krebs Medal Lecture. Eur J Biochem 268(19):5001–5010
4. Choudhary C, Mann M (2010) Decoding signalling networks by mass spectrometry-based proteomics. Nat Rev Mol Cell Biol 11(6): 427–439
5. Rigbolt KT, Blagoev B (2012) Quantitative phosphoproteomics to characterize signaling networks. Semin Cell Dev Biol 23(8):863–871
6. Dengjel J, Kratchmarova I, Blagoev B (2009) Receptor tyrosine kinase signaling: a view from quantitative proteomics. Mol Biosyst 5(10): 1112–1121
7. Luber CA, Cox J, Lauterbach H et al (2010) Quantitative proteomics reveals subset-specific viral recognition in dendritic cells. Immunity 32(2):279–289
8. Ong SE, Mann M (2006) A practical recipe for stable isotope labeling by amino acids in cell culture (SILAC). Nat Protoc 1(6):2650–2660
9. Ow SY, Salim M, Noirel J et al (2009) iTRAQ underestimation in simple and complex mixtures: "the good, the bad and the ugly". J Proteome Res 8(11):5347–5355
10. Boersema PJ, Foong LY, Ding VM et al (2010) In-depth qualitative and quantitative profiling of tyrosine phosphorylation using a combination of phosphopeptide immunoaffinity purification and stable isotope dimethyl labeling. Mol Cell Proteomics 9(1):84–99
11. Ong SE, Blagoev B, Kratchmarova I et al (2002) Stable isotope labeling by amino acids in cell culture, SILAC, as a simple and accurate approach to expression proteomics. Mol Cell Proteomics 1(5):376–386
12. Olsen JV, Blagoev B, Gnad F et al (2006) Global, in vivo, and site-specific phosphorylation dynamics in signaling networks. Cell 127(3):635–648
13. Rigbolt KT, Prokhorova TA, Akimov V et al (2011) System-wide temporal characterization of the proteome and phosphoproteome of human embryonic stem cell differentiation. Sci Signal 4(164):rs3
14. Rikova K, Guo A, Zeng Q et al (2007) Global survey of phosphotyrosine signaling identifies oncogenic kinases in lung cancer. Cell 131(6): 1190–1203
15. Matsuoka S, Ballif BA, Smogorzewska A et al (2007) ATM and ATR substrate analysis reveals extensive protein networks responsive to DNA damage. Science 316(5828):1160–1166
16. Nuhse TS, Stensballe A, Jensen ON et al (2003) Large-scale analysis of in vivo phosphorylated membrane proteins by immobilized metal ion affinity chromatography and mass spectrometry. Mol Cell Proteomics 2(11): 1234–1243
17. Andersson L, Porath J (1986) Isolation of phosphoproteins by immobilized metal (Fe3+) affinity chromatography. Anal Biochem 154(1): 250–254
18. Larsen MR, Thingholm TE, Jensen ON et al (2005) Highly selective enrichment of phosphorylated peptides from peptide mixtures using titanium dioxide microcolumns. Mol Cell Proteomics 4(7):873–886
19. Olsen JV, Macek B (2009) High accuracy mass spectrometry in large-scale analysis of protein phosphorylation. Methods Mol Biol 492: 131–142
20. Choudhary C, Kumar C, Gnad F et al (2009) Lysine acetylation targets protein complexes and co-regulates major cellular functions. Science 325(5942):834–840
21. Hunter T (2009) Tyrosine phosphorylation: thirty years and counting. Curr Opin Cell Biol 21(2):140–146
22. Blagoev B, Ong SE, Kratchmarova I et al (2004) Temporal analysis of phosphotyrosine-dependent signaling networks by quantitative proteomics. Nat Biotechnol 22(9):1139–1145
23. Zarei M, Sprenger A, Metzger F et al (2011) Comparison of ERLIC-TiO2, HILIC-TiO2, and SCX-TiO2 for global phosphoproteomics approaches. J Proteome Res 10(8): 3474–3483
24. Olsen JV, Schwartz JC, Griep-Raming J et al (2009) A dual pressure linear ion trap Orbitrap

instrument with very high sequencing speed. Mol Cell Proteomics 8(12):2759–2769

25. Kelstrup CD, Young C, Lavallee R et al (2012) Optimized fast and sensitive acquisition methods for shotgun proteomics on a Quadrupole Orbitrap mass spectrometer. J Proteome Res 11(6):3487–3497
26. Michalski A, Damoc E, Hauschild JP et al (2011) Mass spectrometry-based proteomics using Q Exactive, a high-performance benchtop quadrupole Orbitrap mass spectrometer. Mol Cell Proteomics 10(9):M111.011015
27. Cox J, Mann M (2008) MaxQuant enables high peptide identification rates, individualized p.p.b.-range mass accuracies and proteome-wide protein quantification. Nat Biotechnol 26(12):1367–1372
28. Cox J, Matic I, Hilger M et al (2009) A practical guide to the MaxQuant computational platform for SILAC-based quantitative proteomics. Nat Protoc 4(5):698–705
29. Cox J, Neuhauser N, Michalski A et al (2011) Andromeda: a peptide search engine integrated into the MaxQuant environment. J Proteome Res 10(4):1794–1805
30. Olsen JV, Macek B, Lange O et al (2007) Higher-energy C-trap dissociation for peptide modification analysis. Nat Methods 4(9): 709–712
31. Nagaraj N, D'Souza RC, Cox J et al (2010) Feasibility of large-scale phosphoproteomics with higher energy collisional dissociation fragmentation. J Proteome Res 9(12):6786–6794
32. Rappsilber J, Mann M, Ishihama Y (2007) Protocol for micro-purification, enrichment, pre-fractionation and storage of peptides for proteomics using StageTips. Nat Protoc 2(8):1896–1906
33. Ishihama Y, Rappsilber J, Andersen JS et al (2002) Microcolumns with self-assembled particle frits for proteomics. J Chromatogr A 979(1–2):233–239
34. Geiger T, Cox J, Ostasiewicz P et al (2011) Super-SILAC mix for quantitative proteomics of human tumor tissue. Nat Methods 7(5):383–385
35. Rigbolt KT, Blagoev B (2010) Proteome-wide quantitation by SILAC. Methods Mol Biol 658:187–204
36. Nielsen ML, Vermeulen M, Bonaldi T et al (2008) Iodoacetamide-induced artifact mimics ubiquitination in mass spectrometry. Nat Methods 5(6):459–460
37. Kelstrup CD, Hekmat O, Francavilla C et al (2011) Pinpointing phosphorylation sites: quantitative filtering and a novel site-specific x-ion fragment. J Proteome Res 10(7):2937–2948
38. Sugiyama N, Masuda T, Shinoda K et al (2007) Phosphopeptide enrichment by aliphatic hydroxy acid-modified metal oxide chromatography for nano-LC-MS/MS in proteomics applications. Mol Cell Proteomics 6(6):1103–1109
39. Zhou H, Low TY, Hennrich ML et al (2011) Enhancing the identification of phosphopeptides from putative basophilic kinase substrates using Ti (IV) based IMAC enrichment. Mol Cell Proteomics 10(10):M110.006452
40. Francavilla C, Rigbolt KT, Emdal KB et al (2013) Functional proteomics defines the molecular switch underlying FGF receptor trafficking and cellular outputs. Mol Cell 51(6): 707–722

Chapter 11

Global Ubiquitination Analysis by SILAC in Mammalian Cells

Zhiping Wu, Chan Hyun Na, Haiyan Tan, and Junmin Peng

Abstract

Ubiquitination is a versatile and dynamic posttranslational modification in cells, regulating almost all cellular events. With rapid developments of affinity capture reagents and high-resolution mass spectrometry, it is now feasible to globally analyze the ubiquitinated proteome (ubiquitome) using quantitative strategies, such as stable isotope labeling with amino acids in cell culture (SILAC). Here we describe in detail a SILAC protocol to profile the ubiquitome in mammalian cells including protein labeling, antibody-based enrichment, and analysis by mass spectrometry.

Key words Ubiquitin, SILAC, Antibody, Quantitative proteomics, Mass spectrometry

1 Introduction

Protein ubiquitination regulates a wide range of cellular physiology, including protein degradation, protein trafficking, and DNA repair [1], and its dysfunction contributes to the development of numerous human diseases such as cancer and neurodegenerative disorders [2]. Ubiquitin (Ub), a polypeptide of 76 amino acids, can modify thousands of protein substrates in eukaryotic cells [3–5]. Ubiquitination occurs through the catalytic activities of Ub-activating enzymes (E1), Ub-conjugating enzymes (E2), and Ub ligases (E3), forming an isopeptide bond between the Ub C-terminal carboxyl group and an amine group in protein substrates (i.e., the epsilon amine of lysine residues or the alpha amine at the N-terminus). Moreover, Ub itself can be further conjugated by other Ub molecules through seven lysine residues (K6, K11, K27, K29, K33, K48, and K63) as well as its N-terminus (M1), assembling a variety of polyUb chains [6]. The modification can be reversed by deubiquitinases (DUBs) that hydrolyze the isopeptide bond between Ub and the substrates. The complex Ub chains are recognized by Ub receptors harboring Ub-binding domains and, thus, modulate binding, activity, and localization of modified sub-

Bettina Warscheid (ed.), *Stable Isotope Labeling by Amino Acids in Cell Culture (SILAC): Methods and Protocols*, Methods in Molecular Biology, vol. 1188, DOI 10.1007/978-1-4939-1142-4_11,

strates. Whereas conventional K48-linked polyUb chains primarily mediate protein degradation by the 26S proteasome, monoubiquitination and K63-linked polyUb chains function in proteasome-independent pathways. Recently, K11 polyUb linkages have been discovered as an alternative signal for proteasomal degradation [6]. The physiological roles of other linkages remain unclear. As the scope of the ubiquitination network is enormous, evidenced by the finding of a large number of Ub enzymes (i.e., 38 E2s, >600 E3s, and ~95 DUBs) in the human proteome [1], it is an opportunity and a challenge to analyze the dynamics of all ubiquitinated species (ubiquitome) in cells.

In the past decade, significant progress has been made toward the global analysis of ubiquitination, largely owing to developments in mass spectrometry (MS)-based proteomics, as well as the invention of affinity capture reagents. Because the stoichiometry of ubiquitinated conjugates is typically low in cells, enriching the Ub conjugates prior to proteomics analysis is essential. The enrichment methods are based on epitope-tagged Ub (e.g., His, FLAG, biotin, HA, myc), Ub-binding domains, or Ub antibodies [3, 7–12]. The isolated Ub conjugates are then digested with trypsin and the resulting peptides are analyzed by MS. As trypsin cleaves Ub conjugates to generate a diglycine tag (GG, monoisotopic mass of 114.0429 Da) on modified residues, this unique tag enables the site determination. More recently, two monoclonal antibodies have been produced to recognize the GG tag on lysine residues (K-GG), allowing for the enrichment of GG peptides instead of ubiquitinated proteins [4, 13]. This new strategy dramatically improves the capacity of analyzing the GG peptides, identifying up to ~20,000 ubiquitinated sites in cells and tissues [4, 5, 14–18]. Furthermore, these Ub enrichment methods can be combined with quantitative MS strategies, such as stable isotope labeling with amino acids in cell culture (SILAC) [19], to study the dynamics of the ubiquitome under various conditions [4, 12, 15–17, 20].

In this chapter, we present a modified procedure to quantitatively analyze the ubiquitome in mammalian cells by the SILAC method. Experimental details of the enrichment of K-GG peptides, MS data acquisition, and bioinformatics data analysis are described (Fig. 1). The implementation of strong cation exchange (SCX) chromatography can further improve the throughput of the analysis [17, 18] and is also included in the procedure.

2 Materials

2.1 Metabolic Labeling of Mammalian Cell Lines

1. HEK293 cells or other cell lines of choice (*see* **Note 1**).
2. Dulbecco's modified Eagle's medium (DMEM) deficient in L-lysine.
3. Lysine: light L-lysine (Lys0) and heavy $^{13}C_6{}^{15}N_2$-lysine (Lys8).

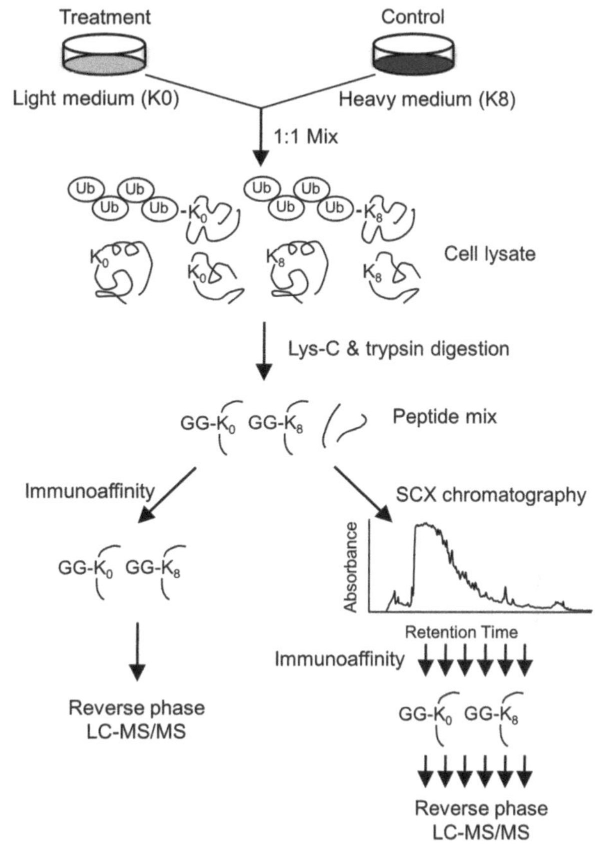

Fig. 1 Strategy for the SILAC based ubiquitome analysis. Mammalian cells (e.g., HEK293) are cultured in light and heavy SILAC medium, respectively (e.g., Lys0 *versus* Lys8). After complete labeling, the cell lysates are mixed at a 1:1 ratio followed by in-solution digestion with LysC and trypsin. Resulting K-GG peptides can be directly enriched by K-GG antibodies. Alternatively, the peptides can be fractionated by SCX chromatography and then purified by the immunoaffinity method. Finally, the enriched peptides are analyzed by LC-MS/MS for peptide/protein identification and quantification

4. SILAC media: lysine deficient medium supplemented with 10 % dialyzed fetal bovine serum (FBS), 0.26 mM light or heavy lysine, 1 U/ml penicillin, and 1 μg/ml streptomycin.
5. 100-mm cell culture dishes.

2.2 Lysis of SILAC-Labeled Cells and Protein Digestion

1. Phosphate Buffered Saline (PBS): 137 mM NaCl, 2.7 mM KCl, 10 mM Na_2HPO_4, 2 mM KH_2PO_4, pH 7.4.
2. BCA protein assay kit (Thermo Scientific).
3. Lysis buffer: 0.1 M Tris–HCl, pH 8.5, 8 M urea, 10 mM iodoacetamide, and 0.15 % sodium deoxycholate.
4. Glass beads (1 mm diameter, Sigma).
5. LysC enzyme (Sigma).

6. Sequencing grade trypsin (Promega).
7. Sample dilution buffer: 0.1 M Tris–HCl, pH 8.5 in 5 % Acetonitrile (ACN).
8. Sep-Pak C_{18} spin columns (Waters).
9. ZipTip C_{18} (Millipore).
10. SpeedVac (Thermo Scientific).
11. Trifluoroacetic acid (TFA).
12. 40 % ACN in 0.1 % TFA.

2.3 Enrichment of GG Peptides by SCX Chromatography (Optional)

1. SCX binding buffer: 5 mM KH_2PO_4, pH 3, 25 % ACN.
2. SCX elution buffer: 5 mM KH_2PO_4, pH 3, 1 M KCl, 25 % ACN.
3. HPLC (e.g., 1220 Infinity LC, Agilent Technologies).
4. SCX column (250 mm × 94 mm, PolySULFOETHYL A™, PolyLC).

2.4 Enrichment of GG Peptides by Affinity Chromatography

1. K-GG antibody coupled to protein A agarose (PTMScan Ubiquitin Remnant Motif (K-epsilon-GG) antibody beads conjugates, Cell Signaling Technology) (*see* **Note 2**)
2. Immunoaffinity purification buffer: 50 mM MOPS–NaOH, pH 7.2, 10 mM Na_2HPO_4, 50 mM NaCl.
3. Wash buffer: Immunoaffinity purification buffer with 0.15 % sodium deoxycholate.
4. 5 mM ammonium bicarbonate.
5. 0.15 % TFA.

2.5 Identification of GG Peptides by LC-MS/MS

1. Solvent A: 0.2 % formic acid.
2. Solvent B: 70 % ACN in 0.2 % formic acid.
3. C_{18} reversed-phase column (150 mm × 75 μm, 2.7 μm HALO™ beads, Michrom Biosources).
4. LTQ-Velos or Elite Orbitrap mass spectrometer (Thermo Scientific).
5. nanoAcquity UPLC (Waters) or Easy-nLC 1000 (Thermo Scientific).
6. Software package including Sequest algorithm [21].

3 Methods

A large amount of starting material is required for global ubiquitome profiling because of the low level of Ub conjugates as well as the large dynamic range of these modified proteins in cells. It is recommended to start with at least 20 mg of protein which can be harvested from ~10 cell culture dishes (150 mm). If SCX fractionation of the peptides is preferred (Fig. 1), more starting material may be used.

3.1 Metabolic Labeling of Mammalian Cell Lines

1. Seed two 100-mm cell culture dishes with HEK293 cells and grow the cells in SILAC medium containing light lysine and heavy lysine, respectively (*see* **Note 3**).
2. Culture the cells for at least seven generations and use a small aliquot to examine labeling efficiency. Validate that the labeling efficiency is >95 % (*see* **Note 4**).
3. During the labeling, it is important to confirm that the change to SILAC medium does not affect biological readouts in the proposed study. This analysis is critical to verify the experimental condition for the subsequent large-scale analysis.
4. Treat the labeled cells under two different conditions as designed (e.g., with or without a proteasome inhibitor).

3.2 Lysis of SILAC-Labeled Cells and Protein Digestion

1. Harvest the cells by washing them with cold PBS thrice and then adding lysis buffer (buffer/cell volume ratio ~8). Scrape and collect the cells. Add glass beads equal to ~30 % of the volume and vortex for 10 s with 20 s cooling in between for 10 cycles (*see* **Notes 5–7**).
2. Centrifuge at 20,000 ×g at 4 °C for 5 min to remove cell debris and glass beads.
3. Collect the supernatant and make at least three aliquots (two small aliquots of ~10 μl for the determination of protein concentration, and a pilot SILAC mixing experiment, respectively, as well as the remaining large aliquot for the large-scale analysis).
4. Take one aliquot for protein BCA assay (*see* **Note 8**).
5. Based on the measured protein concentration, perform a pilot SILAC analysis by mixing the samples at a 1:1 ratio (*see* **Note 9**). Use the results of the pilot SILAC experiment to adjust the mixing volumes of the two samples to ensure a ratio of 1:1.
6. Add LysC to digest the sample at room temperature (RT) for 1 h (enzyme–substrate ratio: 1:200, *see* **Note 10**).
7. Dilute the sample with 0.1 M Tris–HCl, pH 8.5 and 5 % ACN to reduce the urea concentration to 2 M.
8. Add trypsin to digest the sample at 37 °C overnight (enzyme–substrate ratio: 1:200).
9. Terminate the digestion reaction by adding TFA to a final concentration of 1 %.
10. Centrifuge at 20,000 ×g at RT for 10 min and collect the supernatant.

11. Desalt the peptides with a reverse phase SPE column (Sep-Pak C_{18}, *see* **Note 11**). Load the samples onto a preconditioned column, wash with 0.1 % TFA, and elute the peptides with 40 % ACN in 0.1 % TFA.
12. Dry the desalted peptides by SpeedVac and store them at −80 °C until further analysis.

3.3 Enrichment of GG Peptides by SCX Chromatography (Optional, See Notes 12 and 13)

1. Dissolve the desalted peptides in SCX binding buffer at ~20 mg/ml (as to the amount of peptides, refer to the concentration of protein in starting material).
2. Inject the peptides onto a pre-equilibrated SCX column, separate peptides with a gradient from 18 to 38 % of SCX elution buffer over 10 min at a flow rate of 1.5 ml/min, and collect fractions every minute.
3. Analyze a small aliquot of each fraction to determine the charge state distribution along the elution profile [18].
4. Design a pooling strategy for reducing the sample number for antibody-based affinity enrichment (*see* **Note 14**).
5. Acidify the SCX fractions by adding TFA to a final concentration of 1 % and desalt the peptides with a reverse phase C_{18} column.
6. Dry the desalted peptides by SpeedVac and store at −80 °C until further analysis (*see* **Note 15**).

3.4 Enrichment of GG Peptides by Affinity Chromatography

1. Dissolve desalted peptides in immunoaffinity purification buffer at ~40 mg/ml (*see* **Note 16**).
2. Examine the pH of the samples to be neutral (~pH 7.2).
3. Centrifuge at 20,000 × *g* for 10 min at RT and collect the supernatant.
4. Incubate the peptides with K-GG antibody beads at 4 °C with gentle rotation for 1 h (~1 μg antibody per μl beads, use 2 μl of beads per mg of starting protein).
5. Spin down the beads at 1,500 × *g* at 4 °C for 5 s (*see* **Note 17**).
6. Wash the beads with wash buffer (at least 30 bed volumes) thrice.
7. Wash the beads with ice-cold 5 mM ammonium bicarbonate once.
8. After removing the supernatant, elute GG peptides with 0.15 % TFA (3 bed volumes, 5 min incubation at RT).
9. Spin down the beads at 1,500 × *g* at RT for 5 s and collect the supernatant.
10. Repeat the elution step and combine the collected supernatants.
11. Desalt the eluted peptides (*see* **Note 18**).

3.5 Identification of GG Peptides by LC-MS/MS

1. Analyze the samples on an optimized LC-MS/MS system [22] including HPLC (e.g., NanoAcquity UPLC), a sensitive MS (e.g., LTQ Velos or Elite Orbitrap), and a C_{18} reversed phase column (e.g., 150 mm × 75 μm ID, 2.7 μm HALO™ beads).
2. Elute the peptides with a 2-h gradient of 10–30 % ACN using solvent A and solvent B at a flow rate of 250 nl/min.
3. Detect the peptides by orbitrap analysis (350–1,500 *m/z*; resolution of 60,000; automatic gain control value of 1,000,000; and 100 ms maximal ion filling time) followed by 20 low resolution data-dependent MS/MS scans in the LTQ (automatic gain control value of 5,000; 200 ms maximal ion filling time; isolation window of 3 *m/z*; 35 % collision energy; and a dynamic exclusion time of 30 s).
4. Search the MS/MS spectra against a composite target/decoy database by computer software, e.g., Sequest [21]. The database is composed of a selected protein database (www.uniprot.org) and common contaminants. The decoy components contain reversed sequences of all target proteins [3, 23]. Search parameters include fully tryptic restriction, mass tolerance of precursor ions (20 ppm) and product ions (0.5 Da), fixed modification of Cys (+57.0215 Da for alkylation by iodoacetamide), dynamic modifications of Met (+15.9949 Da for oxidation) and Lys (+8.0142 Da for SILAC heavy amino acid labeling, +114.0429 Da for GG modification), and maximal 5 modifications in a single peptide.
5. Filter the peptide matches by mass accuracy and matching scores to reduce the peptide false discovery rate under 1 % (*see* **Note 19**).
6. Analyze the peptides with C-terminal GG modification. As GG modification of lysine residues leads to trypsin miscleavage [3, 24], C-terminal GG peptides derived from the database search are typically considered as false positives [4]. However, we have found that some peptides are derived from proteins that end with lysine at the C-terminus, which should be considered to be true positives after manual examination. In addition, in some cases, GG modification sites are misassigned to C-terminal lysine residues instead of the adjacent lysine at the -1, -2 or -3 position. These peptides may be rescued by repositioning the GG modification sites in the peptides [18].
7. When interpreting the data, one caveat for this analysis is that some GG peptides are generated from proteins modified by Nedd8 or ISG15 instead of Ub (*see* **Note 20**).
8. The SILAC quantification is performed as previously described [20, 25] (*see* **Note 21**).

4 Notes

1. Although SILAC labeling was initially developed to label cells in culture, the method is now extended to fully label nematodes [26], Drosophila [27, 28], and rodents [29] by feeding with SILAC-labeled food.
2. Among the two K-GG monoclonal antibodies commercially available, the antibody from Lucerna was generated against GG-modified histone [13], the one from Cell Signaling Technology was produced against the sequence CXX XXXXKGGXXXXXX (X = any amino acid except Cys, Trp) [4]. Both antibodies gave similar and partially overlapping results in a recent comparative study, but the antibody from Lucerna led to ~30 % less coverage of the ubiquitome [5]. Thus, we describe the use of the antibody from Cell Signaling Technology in this article. In addition, both antibodies cannot bind to M1-GG peptides (linear peptide modified on the N-terminal amine group) or iodoacetamide-modified pseudo-GG peptides [13, 18].
3. The mass difference between the light and heavy stable isotope-labeled lysine should be at least 4 Da to minimize peak overlap during quantification. We use Lys0 and Lys8 for double SILAC and Lys0, Lys8, and Lys17 ($^{13}C_6{}^{15}N_2{}^2H_9$-lysine) for triple SILAC experiments, although other variants of heavy lysine (e.g., $^2H_4$-lysine (Lys4), $^{13}C_6$-lysine (Lys6), and 15N22H9-lysine (Lys11)) may also be used. The incorporation of deuterium may affect the retention time of peptides during LC-MS/MS. With only lysine heavy isotopes, all the Ub-derived, lysine-modified GG peptides will be labeled for the SILAC quantification. To minimize experimental errors, it is recommended to perform at least one biological replicate by swapping the SILAC label.
4. Cell growth needs to be closely monitored to ensure that the properties of the cells in SILAC media are the same as in regular media. SILAC media should be changed every 2–3 days [19]. The dialyzed FBS in the SILAC media may be the source of light amino acids, which prevents full labeling in some cases. The FBS concentration can be decreased to 5 % to limit this factor. The protocol for analyzing the labeling efficiency in a small aliquot of cells includes cell lysis, protein digestion, peptide desalting, and LC-MS/MS, which are described in the following sections.
5. Because of the cell dimensions and residual PBS buffer in the plates, extra solid urea may be needed to maintain the final urea concentration at 8 M. As some DUBs are still active under the buffer condition, it is critical to minimize sample

handing time during cell harvest and keep samples at 4 °C. If not used, freeze the samples on dry ice immediately and store them at −80 °C.

6. Light-sensitive iodoacetamide is freshly prepared and used as a cysteine alkylation reagent to inhibit DUBs, most of which are cysteine proteases. However, at high temperature (>37 °C), iodoacetamide can also modify lysine residues to form a pseudo-GG tag with exact the same mass as the Ub-derived GG tag [30]. To reduce this side reaction to a negligible level, it is important to use a low level of iodoacetamide (e.g., 10 mM) and keep samples at low temperature [20]. A less reactive chemical (i.e., chloroacetamide) may be used as an alternative alkylation reagent [30]. In addition, the pseudo GG peptides can be eliminated during the K-GG antibody-based affinity enrichment [18].
7. Protease inhibitor cocktail is not added to the buffer to avoid inhibition of the enzyme used for protein digestion.
8. The total cell lysate contains various non-protein components, many of which (e.g., reducing species which are capable of reducing Cu^{2+} to Cu^{1+}) may interfere with the BCA analysis, leading to the overestimation of protein concentration. But the calculated concentrations can be used for relative comparison among harvested samples as these samples contain similar levels of the non-protein components.
9. The protocol for analyzing a small aliquot of the SILAC mixing experiment includes protein digestion, peptide desalting, and LC-MS/MS, which are described in the following sections. The analysis reveals systematic bias during sample mixing, which allows for the correction of sample mixing volumes to guarantee a 1:1 ratio for the large-scale analysis.
10. As LysC is active to cleave proteins in the 8 M urea-containing buffer, it is often used in combination with trypsin to improve the efficiency of in-solution digestion. But this step is optional.
11. To ensure efficient recovery of peptides, Sep-Pak resin (~1 ml) is used for peptides digested from 20 mg of protein.
12. As SCX fractionation can reduce the complexity of peptide samples, it can be implemented prior to the antibody-based enrichment step (Fig. 1). The charge state of a peptide in SCX fractions (at pH 3.0) can be calculated from the total number of lysine, arginine, and histidine residues and the N-terminal amine in the peptide. Most tryptic peptides are doubly charged, but typical GG peptides are at least triply charged in SCX buffer, so that the SCX chromatography allows for enrichment of GG peptides. However, additional modifications (e.g., phosphorylation and N-terminal acetylation) can decrease the charge state of GG peptides [18].

13. If the amount of starting material is less than 5 mg of protein, SCX fractionation may be skipped since the loss of GG peptides during SCX outweighs the gain of GG peptide separation for the identification.
14. While the fractions that mainly contain triply charged peptides are analyzed individually, fractions with either earlier or later retention time are pooled into two samples, respectively, as the detectable GG peptides in non-triply charged fractions are low [18].
15. The TFA buffer should be completely dried. Otherwise, the remaining acid will affect the antibody-based affinity purification.
16. A high concentration of GG peptides is critical for efficient binding to the K-GG beads. A concentration of 40 mg/ml or more is recommended.
17. High-speed spins may destroy the agarose beads.
18. The K-GG antibody is non-covalently bound to protein A beads. Thus, the antibody is co-eluted with GG peptides and may affect the LC-MS/MS analysis. Depending on the summed amount of peptides and antibody eluted from the beads, Sep-Pak C_{18} (>10 μg) or ZipTip C_{18} (<10 μg) may be selected for desalting the sample and removing the antibody.
19. To reduce false discoveries during data processing, it is recommended to perform a null experiment to examine the software package used for database searching and filtering [18]: (a) take an LC-MS/MS run from non-GG peptide-containing samples (e.g., *E. coli* cell lysate); (b) analyze the data using the bioinformatics pipeline; (c) confirm the results as anticipated (i.e., no GG peptides in the *E. coli* sample).
20. The GG peptides are generated not only from Ub-conjugates but also from Nedd8- and ISG15-modified proteins. However, only a small portion of the GG peptides (<6 %) consists of Nedd8- and ISG15-derived GG peptides [4, 18]. The level of these Ub-like protein modifications may vary in different starting materials.
21. The comprehensive ubiquitination dynamics should be normalized to the total protein levels in the cells [20]. Therefore, the peptide mix prior to K-GG antibody enrichment should also be analyzed in parallel for this purpose.

Acknowledgements

This work was partially supported by National Institutes of Health grant NS081571 and American Cancer Society grant RSG-09-181, and ALSAC (American Lebanese Syrian Associated Charities).

References

1. Komander D, Rape M (2012) The ubiquitin code. Annu Rev Biochem 81:203–229
2. Schwartz AL, Ciechanover A (2009) Targeting proteins for destruction by the ubiquitin system: implications for human pathobiology. Annu Rev Pharmacol Toxicol 49:73–96
3. Peng J, Schwartz D, Elias JE et al (2003) A proteomics approach to understanding protein ubiquitination. Nat Biotechnol 21:921–926
4. Kim W, Bennett EJ, Huttlin EL et al (2011) Systematic and quantitative assessment of the ubiquitin-modified proteome. Mol Cell 44:325–340
5. Wagner SA, Beli P, Weinert BT et al (2012) Proteomic analyses reveal divergent ubiquitylation site patterns in murine tissues. Mol Cell Proteomics 11:1578–1585
6. Kulathu Y, Komander D (2012) Atypical ubiquitylation – the unexplored world of polyubiquitin beyond Lys48 and Lys63 linkages. Nat Rev Mol Cell Biol 13:508–523
7. Meierhofer D, Wang X, Huang L et al (2008) Quantitative analysis of global ubiquitination in HeLa cells by mass spectrometry. J Proteome Res 7:4566–4576
8. Kirkpatrick DS, Denison C, Gygi SP (2005) Weighing in on ubiquitin: the expanding role of mass-spectrometry-based proteomics. Nat Cell Biol 7:750–757
9. Xu P, Peng J (2006) Dissecting the ubiquitin pathway by mass spectrometry. Biochim Biophys Acta 1764:1940–1947
10. Franco M, Seyfried NT, Brand AH et al (2011) A novel strategy to isolate ubiquitin conjugates reveals wide role for ubiquitination during neural development. Mol Cell Proteomics 10(M110):002188
11. Chen PC, Na CH, Peng J (2012) Quantitative proteomics to decipher ubiquitin signaling. Amino Acids 43:1049–1060
12. Bustos D, Bakalarski CE, Yang Y et al (2012) Characterizing ubiquitination sites by peptide-based immunoaffinity enrichment. Mol Cell Proteomics 11:1529–1540
13. Xu G, Paige JS, Jaffrey SR (2010) Global analysis of lysine ubiquitination by ubiquitin remnant immunoaffinity profiling. Nat Biotechnol 28:868–873
14. Lee KA, Hammerle LP, Andrews PS et al (2011) Ubiquitin ligase substrate identification through quantitative proteomics at both the protein and peptide levels. J Biol Chem 286:41530–41538
15. Emanuele MJ, Elia AE, Xu Q et al (2011) Global identification of modular cullin-RING ligase substrates. Cell 147:459–474
16. Wagner SA, Beli P, Weinert BT et al (2011) A proteome-wide, quantitative survey of in vivo ubiquitylation sites reveals widespread regulatory roles. Mol Cell Proteomics 10(M111): 013284
17. Udeshi ND, Mani DR, Eisenhaure T et al (2012) Methods for quantification of in vivo changes in protein ubiquitination following proteasome and deubiquitinase inhibition. Mol Cell Proteomics 11:148–159
18. Na CH, Jones DR, Yang Y et al (2012) Synaptic protein ubiquitination in rat brain revealed by antibody-based ubiquitome analysis. J Proteome Res 11:4722–4732
19. Ong SE, Mann M (2006) A practical recipe for stable isotope labeling by amino acids in cell culture (SILAC). Nat Protoc 1:2650–2660
20. Xu P, Duong DM, Seyfried NT et al (2009) Quantitative proteomics reveals the function of unconventional ubiquitin chains in proteasomal degradation. Cell 137:133–145
21. Eng J, McCormack AL, Yates JR (1994) An approach to correlate tandem mass spectral data of peptides with amino acid sequences in a protein database. J Am Soc Mass Spectrom 5:976–989
22. Xu P, Duong DM, Peng J (2009) Systematical optimization of reverse-phase chromatography for shotgun proteomics. J Proteome Res 8: 3944–3950
23. Elias JE, Gygi SP (2007) Target-decoy search strategy for increased confidence in large-scale protein identifications by mass spectrometry. Nat Methods 4:207–214
24. Seyfried NT, Xu P, Duong DM et al (2008) Systematic approach for validating the ubiquitinated proteome. Anal Chem 80: 4161–4169
25. Seyfried NT, Gozal YM, Dammer EB et al (2010) Multiplex SILAC analysis of a cellular TDP-43 proteinopathy model reveals protein inclusions associated with SUMOylation and diverse polyubiquitin chains. Mol Cell Proteomics 9:705–718
26. Larance M, Bailly AP, Pourkarimi E et al (2011) Stable-isotope labeling with amino acids in nematodes. Nat Methods 8:849–851
27. Sury MD, Chen JX, Selbach M (2010) The SILAC fly allows for accurate protein quantification in vivo. Mol Cell Proteomics 9:2173–2183
28. Xu P, Tan H, Duong DM et al (2012) Stable isotope labeling with amino acids in Drosophila for quantifying proteins and modifications. J Proteome Res 11:4403–4412

29. Kruger M, Moser M, Ussar S et al (2008) SILAC mouse for quantitative proteomics uncovers kindlin-3 as an essential factor for red blood cell function. Cell 134:353–364

30. Nielsen ML, Vermeulen M, Bonaldi T et al (2008) Iodoacetamide-induced artifact mimics ubiquitination in mass spectrometry. Nat Methods 5:459–460

Chapter 12

Quantifying In Vivo, Site-Specific Changes in Protein Methylation with SILAC

Ho-Tak Lau, Karen A. Lewis, and Shao-En Ong

Abstract

Interest in protein methylation has grown rapidly in recent years. Mass spectrometry-based proteomics is ideally suited to characterize protein modifications, but the multiplicity of methylated residues and the lack of efficient methods to enrich methylated proteins have limited the proteomic identification of protein methylation sites. In this protocol, we compare two metabolic labeling approaches, stable isotope labeling by amino acids in cell culture (SILAC) and its variant heavy methyl SILAC, for studying protein methylation. Instead of heavy lysine and arginine in the typical SILAC experiment, heavy methyl SILAC uses ^{13}C, ^{2}H methionine as the labeling amino acid. As cells convert methionine to *S*-adenosylmethionine, heavy methyl SILAC encodes a 4 Da mass tag for each methyl group, distinguishing between degrees of methylation is possible from mass difference alone. We provide a protocol for SILAC-based analyses of protein methylation and highlight the strengths and weaknesses of each method for targeted and proteomic analyses.

Key words Protein methylation, SILAC, Quantification, Histones, Proteomics, Mass spectrometry, Posttranslational modification, Epigenetics, Proteomics

1 Introduction

Discovered over a half century ago [1], the functional roles of protein methylation in the cell are only just being uncovered [2–4]. For decades, protein methylation was thought to be a "permanent" posttranslational modification (PTM) until the recent discovery of a lysine demethylase, LSD1, by Shi and colleagues [5]. The tantalizing possibility that methylation may be regulated in a manner similar to other reversible, "on-off" switch-like PTMs like phosphorylation and ubiquitination is leading to new avenues of research in the field. Cellular methyltransferases use the activated methyl donor, *S*-adenosylmethionine (AdoMet), to catalyze the covalent substitution of $-CH_3$ groups at the epsilon amino groups of lysine and arginine residues. Methylations of lysine and arginine are catalyzed by lysine methyltransferases (KMTs) [6] and protein arginine methyltransferases (PRMTs) [7, 8], respectively. Lysine

Bettina Warscheid (ed.), *Stable Isotope Labeling by Amino Acids in Cell Culture (SILAC): Methods and Protocols*, Methods in Molecular Biology, vol. 1188, DOI 10.1007/978-1-4939-1142-4_12, © Springer Science+Business Media New York 2014

can be monomethylated, dimethylated, and trimethylated, while arginine can exist in monomethylated, and symmetric and asymmetric dimethylated forms.

Protein methylation plays a key role in transcriptional regulation and its deregulation can lead to severe outcomes during development as well as cancer. For instance, disruption of the histone methyltransferase GLP/G9a complex causes defects in learning and motivation in mice [9], possibly altering neuronal functions by affecting gene regulation through histone modification. PRMT1 is overexpressed in various types of cancers [10]; the KMT SET7/9 was shown to methylate the tumor suppressor p53, which in turn allows it to be acetylated by TIP60 to activate downstream DNA damage responses [11]. Despite the importance of protein methylation, relatively few sites of protein methylation are known [12–14]. The development of a robust approach for identification and quantification of protein methylation sites would accelerate the study of this important PTM.

To date, proteome-wide analyses of protein methylation have been limited and most methylated proteins have been identified by targeted approaches [15, 16]. This may in part be due to the lack of efficient enrichment strategies for methylated proteins. The development of specific antibodies is challenging due to the relatively low antigenicity of methylated amino acid residues and the multiplicity of methylation states. Most of the antibodies developed thus far target histone tails [17] and typically recognize additional amino acids flanking the methylation site. Researchers have also used cocktails of antibodies [16, 18] or antibodies against a particular methylated residue, like the pan-dimethylated lysine antibody, Abcam #ab7315, for affinity purification [19]. Issues with specificity and enrichment of multiple methylation states remain, limiting the overall utility of these reagents for global methylation analysis. Exacerbating the problem, proteomic identification of sites of protein methylation by mass spectrometry is challenging because the mass modification to a peptide is indistinguishable from certain amino acid substitutions. For instance, a peptide containing an unmethylated lysine/arginine and an alanine is identical in mass to its monomethylated form with an alanine-to-glycine substitution; an MS/MS spectrum containing the requisite discriminating ions would be needed to correctly identify the peptide.

Since its introduction in 2002 [20], stable isotope labeling by amino acids in cell culture (SILAC) has gained popularity as a quantitative proteomics approach due to its simplicity, robustness, and applicability to most experimental workflows. In SILAC, two populations of cells are grown in medium containing distinct forms of amino acids with carbons and nitrogens of either natural isotopic abundance or highly enriched, "heavy" nonradioactive stable isotopes of ^{13}C and ^{15}N. As cells grow and divide, the light and heavy amino acids are incorporated in de novo synthesized proteins to generate two proteomes, distinguishable by the specific mass difference of the encoded amino acids. SILAC allows for complete in vivo labeling of a whole proteome. Differential changes in protein expression are measured by the relative abundance of light and heavy peptides in mass spectrometry. The choice of SILAC amino acid is determined by the experimental goal, but because trypsin cleaves specifically after arginine and lysine residues [21], they are commonly used in SILAC as labeling amino acids to generate essentially fully quantifiable proteomes. With additional isotopologues of arginine and lysine, SILAC can quantify three [22] and up to five cell states [23] in a single experiment. SILAC is being applied in a variety of proteomics workflows [for a recent review, see [24]], including proteomic surveys of PTMs like phosphorylation [25], SUMOylation [26], ubiquitination [27], and methylation [28, 29].

We further developed a variant of the SILAC approach to study the protein methylome [16]. In heavy methyl SILAC, cells are grown in the presence of either normal (light) or [$^{13}C^2H_3$] (heavy) methionine. In the cell, methionine is metabolically converted to light/heavy AdoMet, which acts as the primary methyl donor for KMTs and PRMTs and results in a methylation-state specific mass difference encoded for all methylation sites (Fig. 1).

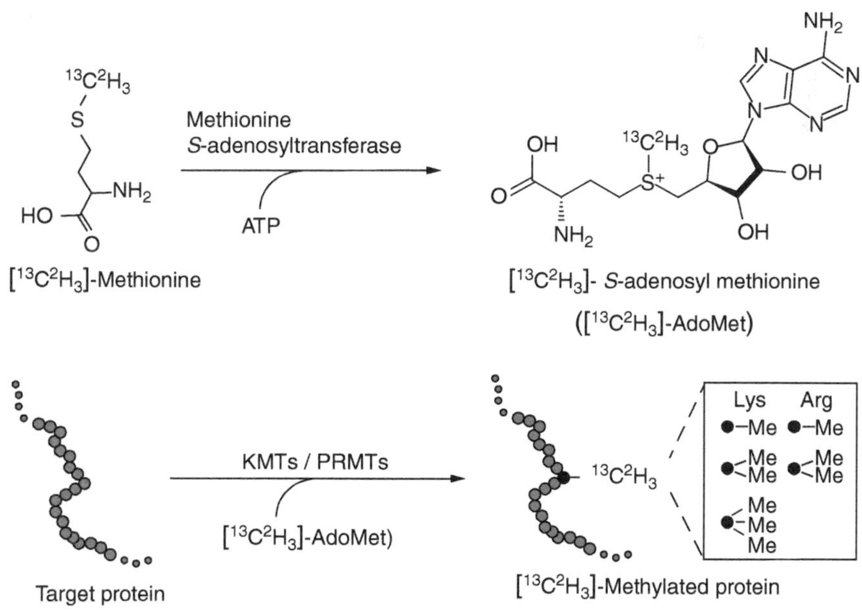

Fig. 1 Metabolic conversion and incorporation of [$^{13}C^2H_3$] in heavy methyl SILAC. [$^{13}C^2H_3$]-methionine is activated in vivo to the primary methyl donor [$^{13}C^2H_3$]-*S*-adenosylmethionine, which is used by lysine and arginine methyltransferases (KMT and PRMTs, respectively) to methylate proteins. The methylated residue (*black ball*) represents either lysine or arginine residues that can be substituted to multiple methylated forms as shown

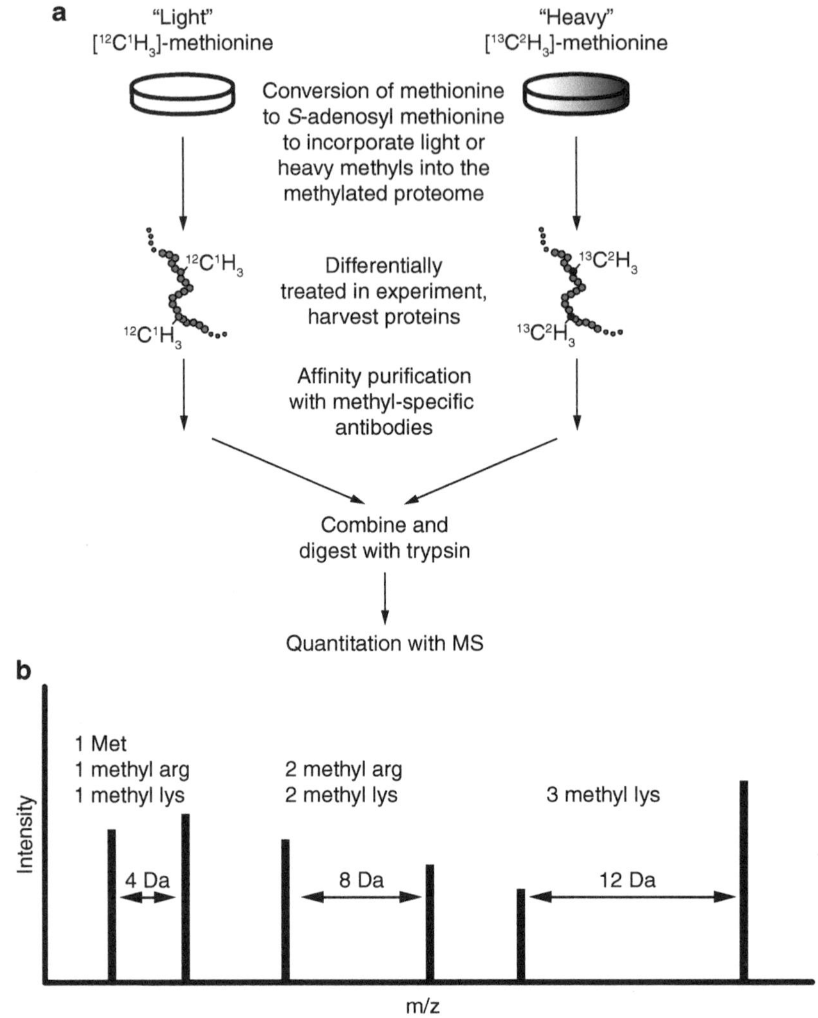

Fig. 2 Overview of a heavy methyl SILAC experiment. (**a**) Cells are labeled in medium containing either light [$^{12}C^1H_3$]-methionine (the control) or heavy [$^{13}C^2H_3$]-methionine (the treated/perturbed condition). Methylated proteins are enriched with specific antibodies and combined for a single MS analysis. (**b**) Quantification of light and heavy methylated peptide pairs is performed in the same way as SILAC and other MS1-based isotopic labeling methods. The encoding of individual methyl groups in heavy methyl SILAC allows for different methylation states to be identified simply by the mass separation between light and heavy methylated peptides

The experimental setup for heavy methyl SILAC is identical to the conventional SILAC experiment, allowing for identification and quantification of methylated peptides from cellular proteomes (Fig. 2a). For each methyl group added, heavy methyl SILAC increases the mass of a modified peptide by 4 Da when compared to the light peptide (Fig. 2b).

Heavy methyl SILAC has a distinct advantage over other quantitative proteomics approaches for the study of protein methylation because the isotopic tag is encoded in the methyl group itself,

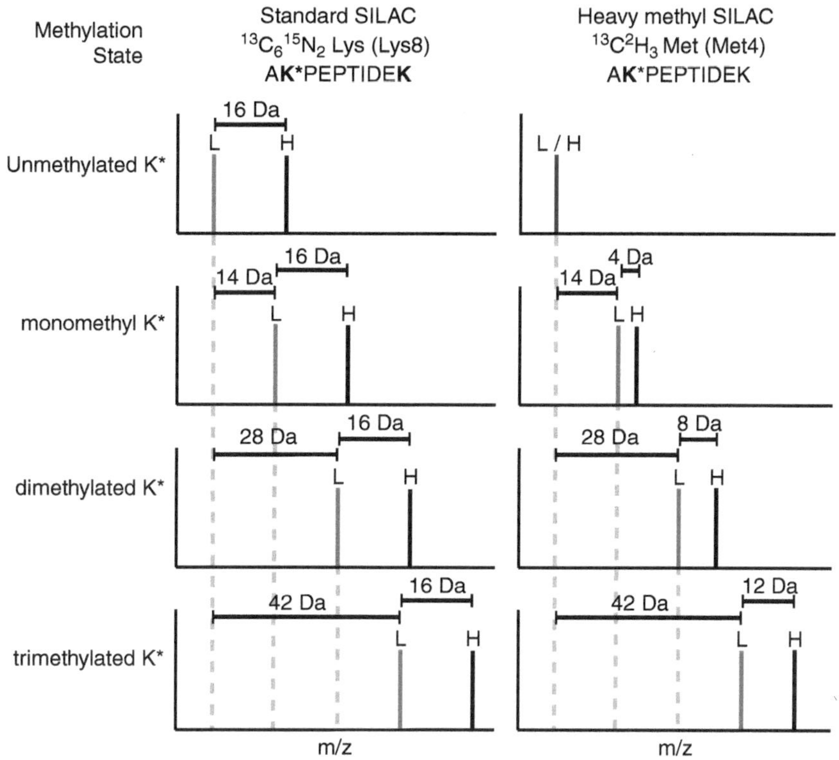

Fig. 3 Mass differences of light and heavy methylated peptides in SILAC and heavy methyl SILAC. Theoretical light and heavy peptide pairs are shown for a peptide bearing two lysines, with a site of possible modification (marked *). (*Left panel*) With each differential methylation state, the SILAC light (L) and heavy (H) peptides are separated by the 16 Da introduced by the two Lys8 residues; the SILAC L and H pair shift +14 Da with each additional methylation. (*Right panel*) In heavy methyl SILAC, the unmethylated peptide is present as a single form. With each additional methylation, the mass separation increases by 4 Da and the difference between monomethylated, dimethylated, and trimethylated lysine is identifiable by the mass separation alone

producing a modification state-specific mass separation between light and heavy peptides. Therefore, the methylation state of a peptide can easily be determined from the mass difference between light and heavy peptides in heavy methyl SILAC; this is not the case in normal SILAC since the isotopic labels are borne within the SILAC amino acid rather than within the methyl group (Fig. 3). By mixing the light and heavy labeled samples in equal proportions, the characteristic doublets allow for ready assignment of methylation states from MS spectra, greatly simplifying validation of identified hits [16].

This distinctive feature of heavy methyl SILAC is especially useful where the methylation state of the protein of interest is unknown, or if multiple methylation states of target protein(s) are expected. Exemplifying this approach, Garcia and colleagues used heavy methyl SILAC to monitor the dynamics of histone turnover

Table 1
Distinctive features of SILAC-based approaches for analyzing protein methylation

Feature	SILAC	Heavy methyl SILAC
Labeling amino acids	Lys, Arg	Met
Labeled peptides	All peptides containing Lys, Arg	All Met peptides, all methylated peptides
Encoded mass difference	All peptides are separated by a fixed mass difference for each SILAC amino acid	Methylated peptides are separated by a methylation state-specific mass difference: +4 Da for each methyl group; +4 Da for each Met
Number of mass modifications to consider at peptide search time[a]	Nine (four SILAC, five variable) SILAC L/H Lys, Arg Monomethyl, dimethyl, trimethyl Lys Monomethyl, dimethyl Arg	Twelve (two SILAC, five SILAC variable) SILAC L/H Met L/H monomethyl, dimethyl, trimethyl Lys L/H monomethyl, dimethyl Arg
Data analysis tools	Peptide identification and quantification software package for SILAC data analysis	No specialized tools available

[a]MS database search algorithms may treat mass modifications either as modification-residue combinations or grouped by mass (e.g., +14 Da for both methyl-Arg and methyl-Lys), resulting in fewer variable modifications in the latter case.

and site-specific methylation from a monomethylated, dimethylated state to trimethylated state [30, 31].

Although heavy methyl SILAC allows for facile identification of distinct degrees of methylation, it may not be the immediate choice for global analyses of the methylome. We highlight the current strengths and limitations of heavy methyl SILAC and a global labeling approach like SILAC or iTRAQ since the choice of method depends heavily on experimental goals (Table 1). We also list a number of key improvements in data analysis, biochemical enrichment, and reagents that would greatly expand the use of heavy methyl SILAC for proteome-wide analyses of methylation.

A practical challenge of methylation analysis by MS, particularly in heavy methyl SILAC, is the large number of variable modifications introduced (Table 1).

The search space for peptide identification expands greatly with the addition of each variable mass modification considered [32–34], and the multiplicity of methylated forms on several amino acid residues exceeds the specifications of most peptide identification software packages. Most current peptide identification and quantification software impose restrictions on mixed labeling states, i.e., disallowing peptides with both light and heavy SILAC amino acids, to limit false positive matches. Performing two searches, therefore, separating light and heavy labeled modifications, and combining the results post-search is a practical and necessary solution, but is far from optimal since additional scripts would be required to integrate database search results. The MaxQuant/Andromeda [35, 36] software package from the Mann lab detects isotopically labeled pairs prior to search time, identifying SILAC precursor–MS/MS combinations before searching them either as heavy or light SILAC peptides to significantly improve identification rates. Using the experimental design to specify MS peak detection effectively constrains database search space and shortens search times while increasing confidence in peptide identification. The multiplicity of methylation states complicates a similar implementation for heavy methyl SILAC data; but even an automated software tool that takes heavy methyl peptide mass differences into account when validating methylation site identifications would be very valuable.

There is an urgent need for a highly specific and efficient enrichment method to enable proteomic analyses of protein methylation. Because current approaches are non-selective, the majority of enriched peptides in an experiment are unmethylated. We and others have used antibodies and protein A/G beads to enrich methylated peptides but their performance varies considerably. Therefore, performing pilot experiments to evaluate a panel of antibodies from different sources is highly recommended. There is growing commercial interest in this market; Cell Signaling Technology now offers a MethylScan® service for identifying monomethylated arginine, based on a similar platform as their antibody-based reagents for enriching phosphorylated peptides. In a recent study, Acuto and colleagues utilized hydrophilic interaction liquid chromatography (HILIC) to enrich methylated tryptic peptides from Jurkat cells [29]. Their approach shows promise, identifying 215 methylated arginine sites on 115 proteins from 1 mg of heavy methyl SILAC labeled cells but comparisons of the human and mice protein methylome suggest that we are far from a full catalog of methylated sites [37]. An ability to enrich methylated peptides to near homogeneity would take full advantage of the methylation state-specific mass differences encoded by heavy methyl SILAC. Notably, the number of phosphorylation sites known prior to 2004 numbered in the hundreds, but with recent advances in phosphoproteomics methods, it is not uncommon for a published dataset to identify tens of thousands of phosphopeptides. We fully expect that state-of-the-art proteomic analyses with improved MS sensitivity and throughput coupled with robust methods to specifically enrich any PTM would have a similar transformative effect.

Heavy methyl SILAC requires methionine-bearing stable isotopes of ^{13}C and ^{2}H within its terminal methyl group (Fig. 1). Labeling cells with [$^{13}C^{2}H_3$]-methionine produces a 4 Da increase in mass in heavy peptides containing either a single methionine or a monomethylated residue (Fig. 2b), thus limiting the ability to identify a monomethylated peptide by mass separation alone. If additional isotopologues of methionine become commercially available, for example, methionine fully substituted with ^{13}C and ^{15}N in addition to the terminal $^{13}C^{2}H_3$, their application in heavy methyl SILAC would easily distinguish methionine-containing peptides (+9 Da) from the monomethylated (+4 Da) species by mass alone. This would provide additional specificity for peptide search algorithms and quantification software alike, and improve the specificity of novel real-time MS data acquisition strategies [38] for methylome analyses.

We compare heavy methyl SILAC to a conventional SILAC approach for methylation analysis (Table 1). The complexity of data analysis of heavy methyl SILAC data using current MS peptide identification workflows may limit its adoption for global methylation studies, but its unique ability to directly label a PTM provides a simple way to monitor the dynamics of methylation at specific sites [30]. The choice of method depends largely on the experimental goals, quality of sample enrichment for methylated peptides, and access to data analysis workflows. We expect that the expanding field of protein methylation will bring exciting new developments, including improvements in methylated peptide enrichment, new antibodies, and labeling amino acids that will greatly expand the current toolbox for methylation analyses.

2 Materials

2.1 Cell Culture, Lysis, and Immunoprecipitation

1. Cell line of choice.
2. Any defined tissue culture medium can be used in SILAC experiments by leaving out the labeling amino acid(s) during formulation. Medium lacking methionine was custom-synthesized (Caisson Labs, North Logan UT) for heavy methyl SILAC.
3. Normal ("Light") amino acid: L-methionine.
4. Heavy isotopes labeled ("Heavy") amino acid: [$^{13}C^{2}H_3$]-L-METHIONINE (methyl-$^{13}CD_3$).
5. Phosphate buffered saline (PBS).
6. 0.22 μm tissue culture sterile filter units to filter SILAC media.
7. Antibiotics (penicillin and streptomycin) and L-glutamine, as sterile 100× supplements.
8. Dialyzed fetal bovine serum (FBS) (*see* **Note 1**).
9. 0.25 % trypsin with EDTA.
10. Cell scrapers.

11. Ice-cold modified RIPA buffer (50 mM Tris–HCl, pH 7.8, 150 mM NaCl, 1 % NP-40, 0.25 % sodium deoxycholate, protease inhibitors).
12. Protein quantification assay (e.g., Bradford assay).
13. Methylated amino acid-specific antibody (e.g., rabbit polyclonal pan-methyl lysine antibody, Abcam #ab7315).
14. Protein A/G-agarose beads.
15. End-over-end rotator.
16. Low pH buffer for elution, such as 50 mM glycine, pH 2.4 (optional).

2.2 Gel Electrophoresis and Trypsin Digestion of Proteins

1. Scalpel.
2. NuPAGE® Novex 4–12 % Bis–Tris gel system with MES buffer system (Life Technologies).
3. NuPAGE® LDS Sample Buffer (4×) (Life Technologies).
4. Colloidal Coomassie stain (Simply Blue, Life Technologies).
5. Gel destaining solution (50 % (v/v) absolute ethanol in 50 mM ammonium bicarbonate (ABC)).
6. Dehydrating solution (absolute ethanol).
7. 100 mM triethylammonium bicarbonate (TEAB).
8. Sequencing grade-modified trypsin at 12.5 ng/μl with 100 mM TEAB, prepare as needed, store on ice.
9. 1 M tris(2-carboxyethyl)phosphine (TCEP).
10. 600 mM chloroacetamide.
11. Trifluoroacetic acid (TFA).
12. Acetonitrile (ACN).
13. Extraction solution (5 % acetonitrile in 0.1 % trifluoroacetic acid).
14. Vacuum centrifuge (SpeedVac).

2.3 Liquid Chromatography (LC)-MS and Data Analysis

1. StageTips [39], prepare as needed.
2. Acetonitrile (ACN).
3. Trifluoroacetic acid (TFA).
4. StageTip solvent B for elution (80 % ACN/0.1 % TFA).
5. StageTip solvent A for washing (5 % ACN/0.1 % TFA).
6. HPLC grade water.
7. LC buffers, HPLC grade: Solvent A (0.1 % acetic acid in water), Solvent B (80 % ACN in 0.1 % acetic acid).
8. Software to generate search engine-compatible MS/MS peak list file supplied by MS machine vendor.

9. Search engine to analyze MS data (e.g., Mascot, Sequest, MaxQuant/Andromeda).
10. Peak quantification software (ProteomeDiscoverer, MSQuant [40], MaxQuant, etc.).

3 Methods

A schematic workflow of the experimental design is illustrated in Fig. 2.

3.1 Preparation of SILAC Medium

1. Dissolve SILAC labeling amino acids in PBS, noting that the concentration of the stock depends on amino acid solubility. For convenience, we prepare our amino acid stocks in reference to a commonly used medium formulation, like DMEM. For instance, we prepare methionine as a 1,000× stock at 30 g/l (DMEM Met concentration is 30 mg/l). Be sure to take account of the increased formula weight of heavy amino acids in preparing stock solutions.
2. Supplement heavy and light amino acids to separate bottles of culture medium and sterilize media by passing through the 0.22 μm filter units.
3. Use dialyzed FBS (10,000 Da molecular weight cutoff, MWCO) instead of normal FBS (*see* **Note 2**). Supplement media with additional antibiotics and glutamine if necessary.

3.2 Adaptation and Labeling of Cells

1. Subculture cells from normal medium to separate dishes containing light and heavy SILAC media; passage cells when needed in the respective SILAC medium.
2. Adapted cells should be kept in their respective medium for at least five cell doublings, allowing for at least ~97 % ($1-(½)^5$) of the proteins to be labeled (*see* **Note 3** and **4**).
3. Subculture and expansion should be performed as needed.

3.3 Harvest of Cells, Protein Extraction, and Immuno-precipitation

Any standard protocol can be used for cell harvesting and protein extraction but the lysis buffer should be compatible with subsequent steps. A buffer containing a mild detergent like NP-40 or Triton, such as modified RIPA buffer, is commonly used because it maintains protein–protein interactions and is suitable for immunoprecipitation of methylated proteins.

1. Wash cells twice with ice-cold PBS to remove abundant serum proteins. Aspirate as much PBS as possible. Harvest cells by scraping in ice-cold modified RIPA buffer.
2. Keep lysates on ice for 15 min and vortex intermittently. Centrifuge lysate at 10,000 × *g* at 4 °C and transfer supernatant to new tubes.

3. Determine the protein concentration in lysates with a standard protein assay like the Bradford assay.
4. If a heavy methyl SILAC experiment is designed to compare a perturbed/treated condition to a control (Fig. 2a), **steps 2–10** should be performed separately for each labeled population. Alternatively, if a 1-to-1 mix is performed to aid identification of methylated peptide pairs [16], this mixing step is best performed at this stage.
5. Incubate 2 mg of protein with protein-A/G agarose beads on an end-over-end rotator for 2 h at 4 °C. This step may reduce background binding to the agarose beads.
6. Spin down beads at 600 × *g* and transfer each supernatant to a new tube.
7. Immunoprecipitate methylated proteins with the desired antibody for 2 h at 4 °C using an end-over-end rotator. The amount of antibody used should be estimated or optimized in a separate experiment.
8. Recover protein–antibody conjugates by adding protein-A/G agarose beads and incubate on the end-over-end rotator at 4 °C for an additional 4 h.
9. Spin beads down at 600 × *g* for 1 min. Collect supernatants for western blotting and controls, if desired.
10. Wash the beads with 1 ml of lysis buffer, spin at 600 × *g* to collect beads.
11. Combine light and heavy immunoprecipitation samples, if applicable (*see* **step 4**).
12. Repeat the wash steps twice.
13. Captured proteins can be eluted in different ways. For small volumes, NuPAGE® LDS Sample Buffer can be added to the beads, the mixture boiled, and proteins separated by PAGE gel electrophoresis. For volumes larger than 100 μl, proteins can be eluted with a low pH buffer and concentrated. Eluted proteins can be stored at −80 °C for several months before continuing on to the next step.

3.4 Reduction of Disulfide Bonds, Alkylation, and Gel Electrophoresis

1. Reduce disulfide bonds between cysteine residues by adding TCEP to a final concentration of 1 mM.
2. Sulfhydryl groups are alkylated to prevent the disulfide bond from reforming by the addition of chloroacetamide to a final concentration of 2 mM.
3. Incubate at 37 °C for 30 min.
4. Add LDS sample buffer to final concentration of 1× (include volume of protein A/G beads in calculations). Boil for 15 min and separate proteins by 1D PAGE.

5. Stain the gel with colloidal Coomassie stain or another protein dye.
6. Using a clean scalpel, divide the stained gel into slices with approximately the same amount of protein. Cut each gel slice into 1 mm^3 pieces and transfer the gel pieces into a 1.7 ml microcentrifuge tube. Gel slices can be stored at room temperature at this point.
7. Destain gel slices in 1 ml of destaining solution for 10 min. Repeat. Gel slices may not be fully destained. Remove the solution and discard.
8. Dehydrate gel slices in ethanol until gel pieces are opaque.
9. Add sequencing-grade modified trypsin to digest proteins. Add enough trypsin to rehydrate the gel slices, then add 3-times the gel volume of TEAB buffer. Incubate at 37 °C overnight.
10. Stop digestion by acidifying the solutions by adding TFA to a final concentration of 1 %. Mix and incubate for 15 min at 37 °C.
11. Collect the digested peptide solutions in new tubes.
12. Add approximately 2-times the gel volume of extraction buffer to the gel pieces and incubate for 15 min.
13. Combine peptide extractions from **steps 10–12**. If the solutions contain any organic solvents, remove them by drying in a vacuum evaporator before proceeding to the C18 desalting step.

3.5 LC-MS and Data Analysis

1. Desalt peptide digests using C18 microcolumns [StageTip [39] or equivalent].
2. Elute peptides from StageTips with 10 column volumes of StageTip buffer B.
3. Remove organic solvent using vacuum centrifugation and resuspend peptides in StageTip buffer A.
4. Analyze peptides using a standard proteomics LC-MS analysis workflow. Process raw MS data files for peptide identification, including the appropriate SILAC and methylation mass modifications, as needed.
5. Semi-automatic quantification of heavy methyl SILAC data is supported by the Mascot-MSQuant analysis package [40]. In general, any software tool that detects and quantifies peak volumes from raw MS data can be adapted to analyze heavy methyl SILAC experiments.
6. SILAC experiments to study methylation can be processed with a typical SILAC analysis workflow.

4 Notes

1. Dialyzed FBS is commercially available. We recommend a test labeling with SILAC to evaluate individual lots of dialyzed FBS as certain lots may contain residual natural isotope abundance amino acids that affect the level of incorporation of heavy SILAC amino acids.
2. SILAC medium should be tested because the use of dialyzed serum (10,000 MWCO) in SILAC media may result in the loss of certain low-molecular-weight growth factors and nutrients. Residual amino acids at undefined levels in normal, undialyzed FBS may also cause differences between the two sera.
3. Cells should be cultured in SILAC medium for a sufficient period of time to allow for full incorporation of the stable isotope. Incomplete labeling will cause errors in quantification. It is important to perform a pilot experiment and evaluate with MS to determine the optimal culturing time and levels of SILAC amino acids used if you are using a new cell line.
4. Arginine can be metabolically converted to proline in certain cell lines [41]. Titration of arginine to lower levels or supplementing with additional light proline can prevent this [42, 43], but empirical verification of this should be done at the same time as **Note 3**.

Acknowledgements

We are grateful to Jimmy Eng for helpful discussions. This work was supported by NIH ARRA Grant P30DA028846 and the Department of Pharmacology, University of Washington School of Medicine.

References

1. Ambler RP, Rees MW (1959) Epsilon-N-methyl-lysine in bacterial flagellar protein. Nature 184:56–57
2. Passos DO, Bressan GC, Nery FC et al (2006) Ki-1/57 interacts with PRMT1 and is a substrate for arginine methylation. FEBS J 273:3946–3961
3. Shen EC, Henry MF, Weiss VH et al (1998) Arginine methylation facilitates the nuclear export of hnRNP proteins. Genes Dev 12:679–691
4. Lee YH, Coonrod SA, Kraus WL et al (2005) Regulation of coactivator complex assembly and function by protein arginine methylation and demethylimination. Proc Natl Acad Sci U S A 102:3611–3616
5. Shi Y, Lan F, Matson C et al (2004) Histone demethylation mediated by the nuclear amine oxidase homolog LSD1. Cell 119:941–953
6. Zhang X, Wen H, Shi X (2012) Lysine methylation: beyond histones. Acta Biochim Biophys Sin (Shanghai) 44:14–27
7. Bedford MT, Richard S (2005) Arginine methylation an emerging regulator of protein function. Mol Cell 18:263–272
8. Di Lorenzo A, Bedford MT (2011) Histone arginine methylation. FEBS Lett 585:2024–2031
9. Schaefer A, Sampath SC, Intrator A et al (2009) Control of cognition and adaptive behavior by the GLP/G9a epigenetic suppressor complex. Neuron 64:678–691

10. Yoshimatsu M, Toyokawa G, Hayami S et al (2011) Dysregulation of PRMT1 and PRMT6, Type I arginine methyltransferases, is involved in various types of human cancers. Int J Cancer 128:562–573
11. Kurash JK, Lei H, Shen Q et al (2008) Methylation of p53 by Set7/9 mediates p53 acetylation and activity in vivo. Mol Cell 29: 392–400
12. Tachibana M, Sugimoto K, Fukushima T et al (2001) Set domain-containing protein, G9a, is a novel lysine-preferring mammalian histone methyltransferase with hyperactivity and specific selectivity to lysines 9 and 27 of histone H3. J Biol Chem 276:25309–25317
13. Schultz DC, Ayyanathan K, Negorev D et al (2002) SETDB1: a novel KAP-1-associated histone H3, lysine 9-specific methyltransferase that contributes to HP1-mediated silencing of euchromatic genes by KRAB zinc-finger proteins. Genes Dev 16:919–932
14. Iberg AN, Espejo A, Cheng D et al (2008) Arginine methylation of the histone H3 tail impedes effector binding. J Biol Chem 283: 3006–3010
15. Pang CN, Gasteiger E, Wilkins MR (2010) Identification of arginine- and lysine-methylation in the proteome of Saccharomyces cerevisiae and its functional implications. BMC Genomics 11:92
16. Ong SE, Mittler G, Mann M (2004) Identifying and quantifying in vivo methylation sites by heavy methyl SILAC. Nat Methods 1:119–126
17. Egelhofer TA, Minoda A, Klugman S et al (2011) An assessment of histone-modification antibody quality. Nat Struct Mol Biol 18:91–93
18. Boisvert FM, Cote J, Boulanger MC et al (2003) A proteomic analysis of arginine-methylated protein complexes. Mol Cell Proteomics 2:1319–1330
19. Couttas TA, Raftery MJ, Padula MP et al (2012) Methylation of translation-associated proteins in Saccharomyces cerevisiae: identification of methylated lysines and their methyltransferases. Proteomics 12:960–972
20. Ong SE, Blagoev B, Kratchmarova I et al (2002) Stable isotope labeling by amino acids in cell culture, SILAC, as a simple and accurate approach to expression proteomics. Mol Cell Proteomics 1:376–386
21. Olsen JV, Ong SE, Mann M (2004) Trypsin cleaves exclusively C-terminal to arginine and lysine residues. Mol Cell Proteomics 3:608–614
22. Blagoev B, Ong SE, Kratchmarova I et al (2004) Temporal analysis of phosphotyrosine-dependent signaling networks by quantitative proteomics. Nat Biotechnol 22:1139–1145
23. Molina H, Yang Y, Ruch T et al (2009) Temporal profiling of the adipocyte proteome during differentiation using a five-plex SILAC based strategy. J Proteome Res 8:48–58
24. Ong SE (2012) The expanding field of SILAC. Anal Bioanal Chem 404:967–976
25. Ibarrola N, Kalume DE, Gronborg M et al (2003) A proteomic approach for quantitation of phosphorylation using stable isotope labeling in cell culture. Anal Chem 75: 6043–6049
26. Vertegaal AC, Andersen JS, Ogg SC et al (2006) Distinct and overlapping sets of SUMO-1 and SUMO-2 target proteins revealed by quantitative proteomics. Mol Cell Proteomics 5:2298–2310
27. Meierhofer D, Wang X, Huang L et al (2008) Quantitative analysis of global ubiquitination in HeLa cells by mass spectrometry. J Proteome Res 7:4566–4576
28. Cuomo A, Moretti S, Minucci S et al (2011) SILAC-based proteomic analysis to dissect the "histone modification signature" of human breast cancer cells. Amino Acids 41: 387–399
29. Uhlmann T, Geoghegan VL, Thomas B et al (2012) A method for large-scale identification of protein arginine methylation. Mol Cell Proteomics 11:1489–1499
30. Zee BM, Levin RS, Xu B et al (2010) In vivo residue-specific histone methylation dynamics. J Biol Chem 285:3341–3350
31. Zee BM, Britton LM, Wolle D et al (2012) Origins and formation of histone methylation across the human cell cycle. Mol Cell Biol 32:2503–2514
32. Mann M, Wilm M (1994) Error-tolerant identification of peptides in sequence databases by peptide sequence tags. Anal Chem 66: 4390–4399
33. Perkins DN, Pappin DJ, Creasy DM et al (1999) Probability-based protein identification by searching sequence databases using mass spectrometry data. Electrophoresis 20: 3551–3567
34. Yates JR 3rd, Eng JK, McCormack AL et al (1995) Method to correlate tandem mass spectra of modified peptides to amino acid sequences in the protein database. Anal Chem 67:1426–1436
35. Cox J, Mann M (2008) MaxQuant enables high peptide identification rates, individualized p.p.b.-range mass accuracies and proteome-wide protein quantification. Nat Biotechnol 26:1367–1372
36. Cox J, Neuhauser N, Michalski A et al (2011) Andromeda: a peptide search engine integrated

into the MaxQuant environment. J Proteome Res 10:1794–1805

37. Bulau P, Zakrzewicz D, Kitowska K et al (2006) Quantitative assessment of arginine methylation in free versus protein-incorporated amino acids in vitro and in vivo using protein hydrolysis and high-performance liquid chromatography. Biotechniques 40:305–310
38. Graumann J, Scheltema RA, Zhang Y et al (2012) A framework for intelligent data acquisition and real-time database searching for shotgun proteomics. Mol Cell Proteomics 11(M111):013185
39. Rappsilber J, Ishihama Y, Mann M (2003) Stop and go extraction tips for matrix-assisted laser desorption/ionization, nanoelectrospray, and LC/MS sample pretreatment in proteomics. Anal Chem 75:663–670
40. Mortensen P, Gouw JW, Olsen JV et al (2010) MSQuant, an open source platform for mass spectrometry-based quantitative proteomics. J Proteome Res 9:393–403
41. Ong SE, Kratchmarova I, Mann M (2003) Properties of 13C-substituted arginine in stable isotope labeling by amino acids in cell culture (SILAC). J Proteome Res 2:173–181
42. Bendall SC, Hughes C, Stewart MH et al (2008) Prevention of amino acid conversion in SILAC experiments with embryonic stem cells. Mol Cell Proteomics 7:1587–1597
43. Lossner C, Warnken U, Pscherer A et al (2011) Preventing arginine-to-proline conversion in a cell-line-independent manner during cell cultivation under stable isotope labeling by amino acids in cell culture (SILAC) conditions. Anal Biochem 412:123–125

Chapter 13

Applying SILAC for the Differential Analysis of Protein Complexes

Karsten Boldt, Christian J. Gloeckner, Yves Texier, Felix von Zweydorf, and Marius Ueffing

Abstract

Pull-downs based on tag fusion proteins as well as immunoprecipitations (IP) are widely used methods to analyze protein interactions. Selectivity and specificity of both methods are compromised by nonspecific binding to the capture agent or carrier beads thereby generating false positives. Here, we provide a method combining stable isotope labeling of amino acids in cell culture (SILAC) with affinity purification, coupled to quantitative tandem mass spectrometry. It permits the analysis of protein interactions with high sensitivity, while being able to discriminate contaminants and nonspecific binders. Besides pruning out contaminants, high-resolution MS data combined with quantitative proteomics software allow the comparative analysis of protein interaction patterns of different protein variants, for example mutated versus normal protein variant or of regulatory changes in a given protein complex due to different states of activity.

Key words Protein complexes, Quantitative complex analysis, SILAC, SF-TAP, Affinity purification, Mass spectrometry

1 Introduction

Over the last decades, multiple methods have been developed to study a specific protein's function via genetically manipulating its expression in cells and model animals [1–4]. In vivo analyses of protein function, especially when tied to transgenic expression or knockout in animals and plants, have turned out to be very laborious and time-consuming. Bioanalytical methods such as immunoprecipitation or yeast two-hybrid interaction studies have been prone to produce a high degree of error. Recently, mass spectrometry (MS)-based strategies to identify proteins coupled to affinity purification have proven to provide information on protein interactions of a given protein of interest (POI), yet false positive results due to nonspecific binding of proteins to the capture agent or tag fused to the POI may impair proper analysis.

Bettina Warscheid (ed.), *Stable Isotope Labeling by Amino Acids in Cell Culture (SILAC): Methods and Protocols*, Methods in Molecular Biology, vol. 1188, DOI 10.1007/978-1-4939-1142-4_13, © Springer Science+Business Media New York 2014

With the advent of quantitative MS, the sensitive and at the same time confident analysis of protein complexes has significantly advanced. Here, we describe a combination of affinity-based enrichment of a POI as a tag fusion protein with its corresponding protein interactors. Discrimination of true *versus* false positive binders can be achieved by quantitative MS-based comparison to a negative control (Tag alone). When compared via differential isotope labeling of both POI and its corresponding control, this allows to differentially compare both data sets and to subtract unspecific binders common to both sets. Thus, subtractive discrimination of specific interactors can be achieved and, after subsequent analysis and interpretation of the resulting MS data, it is possible to get insights into the protein's interactome. When further analyzed by *in silico* tools that visualize protein functional network or pathway connectivity, this approach can provide clues about higher order biological functions associated with a given protein [5].

The composition of protein complexes may be highly regulated. Single or multiple complex components or even complete subcomplexes can be either recruited to or retracted from a complex [6, 7]. These changes can happen within minutes or even seconds, depending on the stimulation or inhibition, which acts on a cell [8, 9]. These rapid mechanisms enable cells to react on changes in their environment within a very short period of time. The processes of recruitment or retraction can be regulated by various mechanisms, for instance by posttranslational modifications like phosphorylation [10, 11] or by recruiting a protein that replaces a component by competing for the same binding motif [12–14]. The regulation of protein complexes and especially their composition can strongly alter the function of the complex and thereby not only influence or modulate the activity state but also induce a switch between certain functions. The complexes can also be altered by mutations in the protein sequence, which affect the connectivity within a network [5]. In consequence, this may lead to a loss or gain of function, especially if the mutation increases the affinity to certain target proteins. The modular use of proteins within dynamically regulated protein complexes increases systemic flexibility by which the cell can quickly adapt to environmental changes and respond to extrinsic or intrinsic signals.

As discussed above, a rapid and straightforward way to get insights into the function of a protein is the analysis of its protein complexes, i.e., by affinity-based methods like tandem affinity purification (TAP) or one-step affinity purification in combination with qualitative MS [15] or quantitative MS. It can be done within a few weeks and can be the first step to analyze the function of a protein. To achieve knowledge on the functional consequences of a mutation, associated complexes formed by the wild type (WT) and the mutated protein can be quantitatively compared. This can be achieved by purifying and quantitatively comparing the complexes of both the WT and the mutant protein. Given that a certain protein or even a sub-complex appears weaker or stronger associated with the mutant, this discrepancy can suggest the gain or loss of function and, thus, provides valuable insights into a disease's mechanism. The same principle can be applied to detect any other effect like the ones induced by mitogenic activation or inhibition of certain targets like receptors by compounds.

This protocol describes the expression, purification, and quantitative comparison of protein complexes from HEK293 cells, applying SILAC-based quantification. A prerequisite for the application of the protocol is the availability of a vector allowing the expression of a protein of interest as Strep-FLAG (SF)-TAP tag fusion. As the generation of these vectors as well as its application in tandem affinity purification has previously been described [16–18], the following protocol focuses on the application of a SILAC-based quantitative MS approach following single-step affinity purification via the Strep-tag.

2 Materials

2.1 SILAC Cell Culture and Transfection

1. SILAC medium: DMEM without L-lysine and L-arginine. For triple labeling experiments, use the following amino acid combinations in the medium:

   Light: L-lysine (final concentration: 0.55 M), L-arginine (final concentration: 0.4 M), L-proline (final concentration: 0.5 M) (*see* **Note 1**).

   Medium: $^{2}H_4$-lysine, $^{13}C_6$-arginine, L-proline; same concentrations as used for the *light* medium (*see* **Note 1**).

   Heavy: $^{13}C_6{}^{15}N_2$-L-lysine, $^{13}C_6{}^{15}N_4$-L-arginine, L-proline; same concentrations as used for the *light* medium (*see* **Note 1**).

   Sterilize the solutions by filtering using a 0.22 μm sterile filter before adding 10 % (v/v) of dialyzed fetal bovine serum, 50 units/ml Penicillin, and 0.05 mg/ml Streptomycin.
2. Polyethylenimine (PEI) solution: dissolve 100 mg of PEI in 900 ml of sodium chloride (NaCl) solution (150 mM, adjusted to pH 5.5 with HCl) by heating the solution to 80 °C under constant stirring. Adjust the pH to 7.8 by adding 10 mM HCl. Add the NaCl solution to a final volume of 1,000 ml. Sterilize the solution by filtering using a 0.22 μm sterile filter. Store aliquots at 4 °C.
3. SF-TAP-tag expression construct for the protein of interest. The generation of this construct is described in detail elsewhere [18].
4. Cell culture plates 10 cm/14 cm.
5. Serological pipettes.
6. Steritop-GP filter unit, 0.22 μm.
7. Phosphate buffered saline (PBS).

2.2 Cell Lysis and Complex Purification

1. Cell scraper.
2. Microspin columns.
3. Microcon YM-10 (cutoff: 10 kDa) centrifugal filter units (Millipore).
4. 10× Tris-buffered saline (TBS): to prepare 1 l of 10× TBS, dissolve 36.3 g of tris(hydroxymethyl)aminomethane (Tris) and 87.7 g of NaCl in water, adjust the pH with HCl to 7.4, and fill up to 1 l with water.
5. 50× protease inhibitor cocktail: dissolve one tablet of the protease inhibitor cocktail (Roche) in 1 ml of water.
6. Lysis buffer: to prepare 10 ml of lysis buffer, add 1 ml of 10× TBS, 200 μl of the 50× protease inhibitor cocktail, 100 μl of phosphatase inhibitor cocktail II and III (Sigma-Aldrich), and 50 μl of NP-40 to 8.55 ml of water
7. Wash buffer: to prepare 10 ml of wash buffer, add 1 ml of 10× TBS, 100 μl of phosphatase inhibitor cocktail II and III, and 10 μl of NP-40 to 8.79 ml of water.
8. TBS buffer: to prepare 10 ml, dilute 1 ml of 10× TBS in 9 ml of water.
9. Desthiobiotin elution buffer: dilute 10× buffer E (IBA) 1:10 in water (final concentration: 2 mM desthiobiotin).
10. Strep-Tactin Superflow (IBA).
11. Bio-Rad Protein Assay Kit.
12. Bovine serum albumin (BSA).
13. Disposable cuvettes.
14. UV/VIS spectrometer.

2.3 Sample Preparation for MS

1. 96-multiwell plate with lid.
2. 10 %, 10-well NuPAGE Gels (Life Technologies).
3. MES or MOPS buffer (Life Technologies).
4. 5× Laemmli buffer: 100 mM Tris–HCl pH 6.8, 5 % (w/v) sodium dodecyl sulfate, 50 % (v/v) glycerol, 500 mM β-mercaptoethanol, 0.05 % (w/v) bromophenol blue.
5. Coomassie solution: dissolve 4 g of CBB-G250 (Coomassie Brilliant Blue G250) in 1 l deionized water and store in the dark at RT.
6. Fix and destaining solution: 50 % (v/v) methanol, 12 % (v/v) acetic acid in deionized water.
7. DTT solution (5 mM): dissolve 154 mg of 1,4-dithiothreitol (DTT) in 2 ml of HPLC grade water to gain a 500 mM DTT stock solution; dilute 1:100 in HPLC grade water.

8. IAA solution (25 mM): dissolve 23 mg of 2-iodoacetamide (IAA) in 5 ml of HPLC water; store at RT in darkness until use; always prepare fresh.
9. ABC solution (50 mM): dissolve 198 mg of ammonium bicarbonate (ABC) in 50 ml of HPLC water; always prepare a fresh solution.
10. 40 % (v/v) LC/MS grade acetonitrile (ACN) in HPLC grade water.
11. 2.5 % (v/v) trifluoroacetic acid (TFA), protein sequencing grade (Merck).
12. 0.5 % (v/v) TFA in 50 % (v/v) ACN.
13. 0.5 % (v/v) TFA in 99.5 % (v/v) ACN.
14. Trypsin solution: dissolve 20 μg of lyophilized trypsin in 20 μl of 1 mM HCl to obtain a 1 μg/μl trypsin stock solution and store at −20 °C. Dilute the trypsin stock solution 1:100 with ABC solution to obtain a final concentration of 0.01 μg/μl trypsin (*see* **Note 2**).

2.4 Mass Spectrometry

1. LTQ-OrbiTrap Velos mass spectrometer.
2. Ultimate 3000 Nano-RSLC liquid chromatography system.
3. 2 ml autosampler vials.
4. Inserts for 2 ml autosampler vials.
5. Caps for 2 ml autosampler vials.
6. Nano trap column 75 μm i.d.×2 cm, packed with Acclaim PepMap100 C18, 3 μm, 100 Å.
7. Analytical column 75 μm i.d.×25 cm, Acclaim PepMap RSLC C18, 2 μm, 100 Å.
8. Liquid chromatography solvents:
   (a) Buffer A: 2 % ACN and 0.1 % formic acid in HPLC-grade water.
   (b) Buffer B: 80 % ACN and 0.08 % formic acid in HPLC-grade water.
   (c) Buffer C: 0.1 % TFA in HPLC-grade water.

2.5 Data Analysis

1. MaxQuant (http://www.maxquant.org/) (*see* **Note 3**).
2. Perseus (http://www.maxquant.org/) (*see* **Note 3**).

3 Methods

This protocol is optimized for HEK293 cells. Adaptation of experimental conditions may be required for other cell lines than HEK293.

Table 1
Study design for the protein complex detection as well as quantitative comparison of protein complexes formed by WT and mutant forms of a protein

Experiment	Light labeled	Medium labeled	Heavy labeled
1	ST-TAP	WT	Mutant
2	Mutant	SF-TAP	WT
3	WT	Mutant	SF-TAP

A vector-only (SF-TAP) condition serves as specificity control

3.1 Study Design

For all experiments, at least three biological replicates should be performed. An example study design for a comparative WT/mutant/control triple labeling protein interactome screen is shown in Table 1.

We advise to consider label switching to exclude false positive results due to non-labeled contaminants as well as any effects that might be induced due to the use of different amino acids in the different labeling states. Although stable isotopes should not be biologically different from the native amino acids, the level of purity, for example, could be different and may thereby lead to slightly different behavior of the cells.

3.2 Cell Culture and Cell Lysis

1. Grow HEK293 cells in light, medium, and heavy SILAC DMEM media.
2. Split one confluent 10-cm dish to up to three 14-cm dishes the day before transfection. The cells should be 40–80 % confluent at the time of transfection.
3. Transfect the cells with the expression construct for the SF-TAP alone or an protein unrelated to the POI as control (light SILAC-labeled), for the POI (medium SILAC-labeled), and for a mutant of the POI (heavy SILAC-labeled). If alterations based on exogenous noxa are to be analyzed, the WT protein should be expressed in both medium and heavy medium (*see* **Note 4**).

   For 14-cm dishes, add 8 μg of each expression vector to 1 ml of PEI solution. Vortex briefly and incubate the mixture for 10 min at RT before adding it dropwise to the cells and grow them for 48 h. For HEK293 cells, it is not necessary to remove the medium containing the PEI reagent.
4. Optional: exchange the growth medium 16 h prior to lysis to a serum-free medium before application of a stimulus like a growth factor or inhibitor.
5. Remove the medium and wash the cells once with 10 ml of warm PBS. Transfer plates onto ice and add 1 ml of lysis

buffer. Scrape off the cells from the plates using a cell scraper and transfer the suspended cells into a 2-ml reaction tube.

6. Continue the lysis by incubation for 20 min at 4 °C under constant agitation.
7. Centrifuge the samples for 10 min at 10,000 × *g* and 4 °C to remove cell debris and nuclei.

3.3 Affinity Purification

We here describe the affinity purification based on Strep-Tactin resin (Fig. 1). Do not combine the samples before the final elution is achieved (*see* **Note 5**).

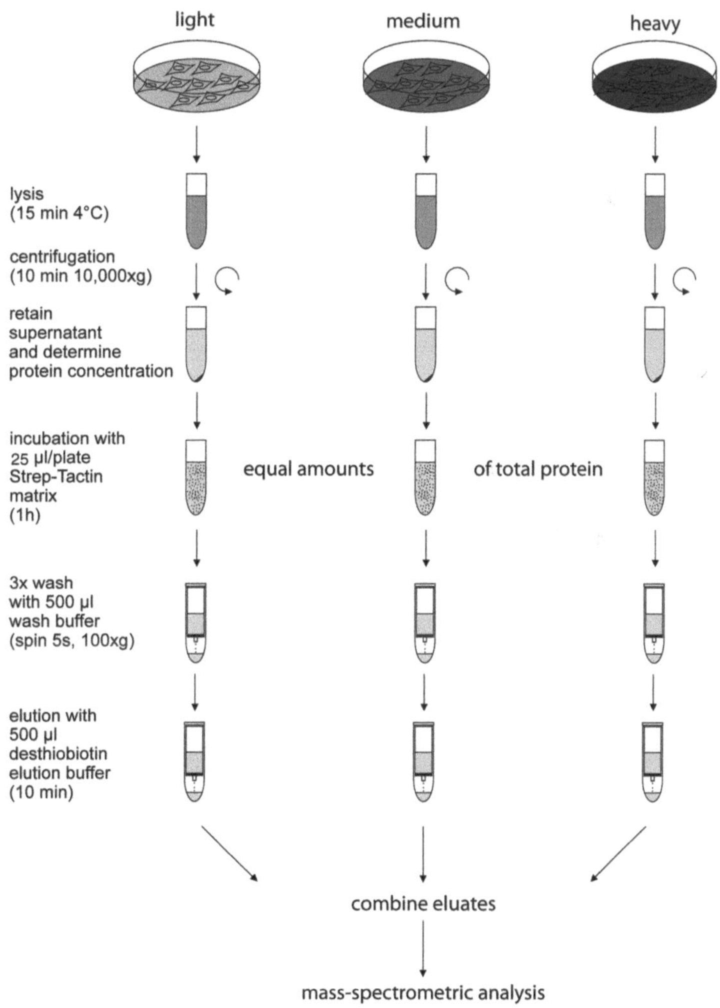

Fig. 1 General workflow for the protein complex analysis using a triple SILAC labeling strategy. The scheme shows the sample processing steps from cell culture until MS-based analysis and, most importantly, the separate processing of the samples until the final elution step to exclude interactor exchanges during sample processing and, thereby, failure to detect interactors by their significant enrichment compared to the control (*see* **Note 5**)

1. Preparation of the Strep-Tactin affinity matrix (*see* **Note 6**): wash beads 3× with 500 μl of lysis buffer. Carefully suspend the beads before preparing aliquots. 25 μl of packed beads should be used per 14-cm dish (a max. volume of 100 μl of settled beads per microspin column should not be exceeded).
2. Determine the protein concentration for each sample by using the Bio-Rad Protein Assay Kit according to the manufacturer's instructions.
3. Add equal protein amounts for all samples to the beads. Retain a small aliquot of each sample for optional characterization by western blot and store at −80 °C. Adjust the volume for all samples to the same final volume by adding lysis buffer.
4. Incubate the samples for 1 h under constant agitation at 4 °C to keep the resin suspended.
5. Centrifuge the samples for one min at 7,000 × *g* at 4 °C and carefully remove the supernatant to a residual transfer volume of maximal 500 μl per microspin column.
6. Remove top cap sealing plugs from the column outlet, mount the microspin columns into 2-ml reaction tubes and place them on ice to allow optimal cooling of the samples.
7. Transfer the beads to microspin columns.
8. Drain the columns either by gravity flow or by centrifugation for 5 s at 100 × *g*. Do not let the beads run dry (*see* **Note 7**).
9. Wash the beads 3× with 500 μl of wash buffer and drain the columns as described in **step 8**.
10. Replug the column outlets, add 500 μl of elution buffer, recap the columns, and incubate the samples for 10 min at 4 °C under constant agitation.
11. Open the bottom plug, place the column into a fresh 2-ml reaction tube and collect the eluate containing the complexes by centrifugation for 30 s at 1,000 × *g*.
12. Combine the corresponding light/medium/heavy labeled sample triplets and concentrate the samples to less than 30 μl using Microcons or comparable centrifugal units with a cutoff of 10 kDa or less.

3.4 Sample Preparation for MS

In many cases, pre-fractionation of the samples can be advantageous, especially when the complexity is rather high, as often seen for one-step purified samples, and if the abundance of the purified proteins differs strongly. In such cases, pre-fractionation does lead to a clearly increased depth of analysis. Because gel-based pre-fractionation also leads to a decreased sensitivity, it should only be considered if the total protein concentration is not the limiting factor (*see* **Note 8**).

1. Add 1/5 of the sample volume of Laemmli buffer to the sample, mix, and incubate for 5 min at 96 °C.
2. Separate the proteins on a NuPAGE 10 %, 10-well gel. Stop the electrophoretic separation after the bromophenol blue front has migrated approximately 1–1.5 cm from the bottom of the well.
3. Remove the gel from the cassette and incubate it twice for 15 min in fixation solution.
4. Stain the gel for a few minutes in Coomassie solution until the first signs of bands appear. Avoid overstraining as it is difficult to fully remove Coomassie from the samples in later steps (see **step 8**).
5. Immediately remove the staining solution and replace it by fixation solution to remove background stain. Incubate for further 15–30 min and exchange the solution twice.
6. Fractionate the gel lane with visible bands into 4–10 fractions, depending on the sample complexity and the sequencing capacity of the MS instrument. Use a clean (keratin-free) scalpel to cut the gel. Cut the gel fractions into pieces of approximately 1 mm^3 and transfer the pieces into a 96-well plate.
7. Wash the gel pieces by incubation in 100 μl HPLC-grade water.
8. De-stain the gel pieces using sequential incubation with 40 % ACN and 100 % ACN for 10 min each. If the de-staining is incomplete, repeat the procedure.
9. Remove the ACN solution and dry the gel pieces for 10 min before adding 100 μl of DTT solution and incubation for 15 min at 60 °C.
10. Remove the DTT solution, let the plate cool down to less than 37 °C and add 100 μl of IAA solution. Incubate for 45 min in the dark.
11. Remove the IAA solution and incubate the gel pieces twice in 40 % ACN for 5 min followed by incubation for 5 min in 100 % ACN. Remove the ACN and dry the gel pieces for 10 min at RT.
12. Add 50 μl of trypsin solution to the dried gel pieces. Let the gel pieces rehydrate for 5 min. If necessary, add additional 20 μl of trypsin solution. After complete rehydration, the gel pieces should still be covered by trypsin solution.
13. Incubate overnight at 37 °C.
14. Add 10 μl of 5 % TFA and incubate for 10 min under gentle agitation.
15. Collect the supernatant in a new reaction tube and add 100 μl 0.5 % TFA, 50 % ACN to the gel pieces. Incubate again for 10 min under gentle agitation.

16. Collect the supernatant and extract the peptides from the gel pieces by adding 50 µl 0.5 % TFA, 99.5 % ACN to the gel pieces and incubation for 10 min under gentle agitation. Combine this supernatant with supernatants resulting from **steps 15** and **16**.
17. Dry the combined supernatants in a SpeedVac to almost complete dryness and adjust the sample to 15 µl with 0.5 % TFA just prior to MS analysis.

3.5 Mass Spectrometry

As an example, the mass spectrometric analysis of the samples by an LTQ Orbitrap Velos coupled on-line to an Ultimate 3000 nano RSLC liquid chromatography (LC) system will be described here. In principle, the analysis can be done on various mass spectrometers but different software packages are in many cases necessary.

1. Transfer the sample into an autosampler vial.
2. Prepare the samples sequence in Xcalibur by selecting unknown sample as sample type, giving the sample a meaningful name and selecting the corresponding position for the autosampler vial.
3. As injection volume select 5 µl.
4. Prepare the LC method as follows: Use µlPickUp as sample injection method and load the samples for 5 min onto a nano-trap column at a flow rate of 6 µl/min in 98 % buffer C and 2 % buffer B.
5. Use a LC gradient with a length of 80 min from 2 % buffer in buffer A to 35 % buffer B in buffer A in 80 min at a flow rate of 300 nl/min followed by a short gradient from 35 % B to 95 % B in buffer A within 5 min.
6. After 5 min with 95 % buffer B, go back to 2 % buffer B in buffer A for at least 20 min (also *see* **Note 9**).
7. The eluting peptides can be analyzed by the LTQ Orbitrap Velos with a top tenth order double play method with CID (collision induced dissociation) fragmentation. Each full survey scan is performed in the Orbitrap analyzer (FTMS) with a resolution of 30,000 and 2 microscans in positive ion mode with profile as data type. The peptides are fragmented by CID with a normalized collision energy of 35. Singly charged ions and those with a minimal signal of less than 200 counts are rejected. The fragment spectra are acquired as centroid data in the ion trap mass analyzer (also *see* **Note 10**).
8. Enable the dynamic exclusion option and exclude every fragmented ion for 20 s.
9. The resulting raw data files can be used for MaxQuant analysis.

3.6 Data Analysis

Besides the settings mentioned below, we recommend using the standard settings of MaxQuant. The following protocol is described for the current MaxQuant version (1.3.X) and Perseus version (1.3.X). Future versions of the software may require different handling/settings. For installation and initial setup, please follow instructions on www.maxquant.org or the corresponding MaxQuant help on Google groups (*see* **Note 3**). Protein sequence databases can be obtained from the UniProtKB server (www.uniprot.org).

1. Load the raw files into the MaxQuant software.
2. Use a multiplicity of 3 to analyze triple labeling experiments.
3. Select Lys4 and Arg6 as medium, Lys8 and Arg10 as heavy label. Do not select any light label.
4. *Optional*: if phosphorylation or any other posttranslational modification is of interest, this should be additionally selected as variable modification.
5. Select a recent human subset of the UniProt database in the sequences field.
6. We recommend to set the threshold for peptides and quantification as follows: Unique: 1, Razor: 2, Ratio count: 3.
7. For any further information regarding MaxQuant, we would like to refer to the MaxQuant web page.
8. After the analysis is finished, load the normalized ratios as well as the corresponding intensities found in the proteingroups.txt file into the Perseus software package for statistical analysis.
9. Remove any contaminants, only identified by site and reverse hits by filtering.
10. Log2-transform all ratios and intensities.
11. Calculate the significance B-value. Use the left-sided test for the detection of specific interactors (Ratio WT/SF-TAP, M/L as well as mutant or stimulated/SF-TAP, H/L). For the comparison of complexes (H/M), use the two-sided test. As significance threshold we propose to use a *p*-value of 0.01.
12. Create scatter plots for all comparisons. You should have three scatter plots, two showing the enrichment of potential interactors compared to the control for the WT complex as well as for the mutant form of the complex or the stimulated situation and one showing the comparison of the complexes (Fig. 2) (*see* **Note 11**).
13. Remove all proteins from the list that were not detected as potential interactors either for the WT protein or the mutant/stimulated situation. Only the potential interactors should be considered for the WT/mutant or non-stimulated to stimulated comparison.

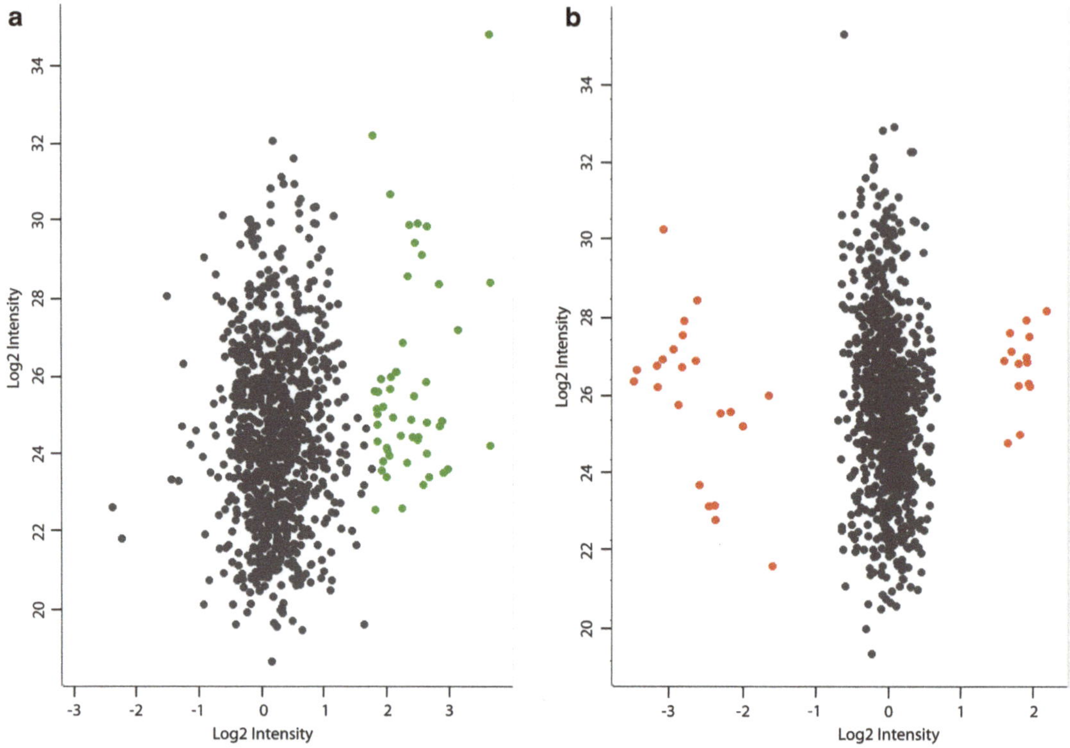

Fig. 2 Scatter plots exemplifying potential results gained from quantitative analysis of protein complex to detect protein complex components and alterations within protein complexes. The Perseus software package was used for the generation of these plots. In both scatter plots, each dot corresponds to a single protein. Plotted are the log 2 ratios (*X*-axes) and the log 2 intensities (*Y*-axes) from mean values of triplicate experiments. (**a**) Detection of specific protein complex components. Proteins specifically enriched compared to the negative control ($p < 0.01$) are highlighted in *green*. (**b**) Quantitative comparison of protein complexes. Proteins showing a significant alteration in association to the complex of interest ($p < 0.01$) depending on a mutation, stimulation or inhibition are highlighted in *red*

4 Notes

1. Additional L-proline is added to the medium to prevent arginine to proline conversion. This is a common problem that leads to severe errors in quantification, especially for peptides containing more than one proline residue. It can for many cell lines be prevented by simply adding additional L-proline to the growth medium [19].
2. The trypsin solution should always be prepared freshly.
3. The software tool used here (MaxQuant) relies on the use of Orbitrap-type instruments (Thermo Scientific). The method can in principle be used with many other MS instrumentation. Because MaxQuant is specifically designed for Orbitrap type instruments, different software solutions would be necessary

for other instrument types. The MaxQuant software is provided free of charge by the Proteomics and Signal Transduction research group at the Max Planck Institute of Biochemistry in Martinsried, Germany.

4. Before starting the comparison of the complexes, we strongly advise to check for equal expression of the WT and mutant forms of the protein. This should also be done in case of complex comparisons after stimulation or inhibition of a target. As short-term stimuli will most likely not lead to alterations in expression of the bait protein, this could well be the case for long-term ones.
5. Do not combine the samples before the final elution step. Although it might sound attractive to combine the samples earlier, this will lead to an exchange of interactors due to dissociation and association events. Depending on the interaction, this can be a fast process that will lead to an almost complete exchange between light, medium, and heavy labeled proteins within few minutes or even faster.
6. Alternatively to the Strep-Tactin-based purification system, one can also use the FLAG-based system. The SF-TAP-tag is comprised of both Strep and FLAG moieties. Especially for low abundant proteins, the FLAG-tag might increase the efficiency of the purification because it has a higher affinity to the FLAG resin as compared to the affinity of the Strep-tag to the Strep-Tactin resin. In principle, any other tag compatible with MS analysis and affinity purification of complexes can be used as well.
7. Do not let the beads run dry. Depending on the centrifuge used, this can happen quickly when using the microspin columns. This should be tested with beads only before starting the real experiment. If the beads run dry, this might dramatically affect the purification efficiency but also the stability of the complexes to be analyzed.
8. If pre-fractionation is applied or not, strongly depends on the type of sample and the power of the MS instrumentation. In principle, affinity-enriched samples can be analyzed without pre-fractionation due to the clearly reduced complexity compared to whole lysate or tissue samples. Nevertheless, pre-fractionation can still increase the analysis depth due to the possibility to separate high and low abundant proteins or peptides before the LC-MS/MS analysis. This can lead to a higher dynamic range and ultimately to the identification of lower abundant and weaker bound interactors.
9. Both the LC and the mass spectrometer need constant surveillance and maintenance. Before each analysis, the performance should be checked by the use of external standards like a digest of bovine serum albumin or cytochrome C. This is a critical step that needs to be evaluated for every system regularly.

10. To enable internal calibration the lockmass option [20] can be used with a background ion with a mass of 445.120020.

11. The quantification of the bait protein can be problematic, especially when the mutation leads to truncation. In such a case, a portion of the peptides cannot be quantified because of the presence in the WT protein only. Therefore, the peptides present in both WT and mutant should be quantified separately. The same is true if there are additional peptides in the mutant form.

References

1. Fire A, Xu S, Montgomery MK et al (1998) Potent and specific genetic interference by double-stranded RNA in Caenorhabditis elegans. Nature 391:806–811
2. Papaioannou VE, McBurney MW, Gardner RL et al (1975) Fate of teratocarcinoma cells injected into early mouse embryos. Nature 258:70–73
3. Smithies O, Gregg RG, Boggs SS et al (1985) Insertion of DNA sequences into the human chromosomal beta-globin locus by homologous recombination. Nature 317:230–234
4. Thomas KR, Folger KR, Capecchi MR (1986) High frequency targeting of genes to specific sites in the mammalian genome. Cell 44: 419–428
5. Boldt K, Mans DA, Won J et al (2011) Disruption of intraflagellar protein transport in photoreceptor cilia causes Leber congenital amaurosis in humans and mice. J Clin Invest 121:2169–2180
6. Sasagawa S, Ozaki Y, Fujita K et al (2005) Prediction and validation of the distinct dynamics of transient and sustained ERK activation. Nat Cell Biol 7:365–373
7. von Kriegsheim A, Pitt A, Grindlay GJ et al (2006) Regulation of the Raf-MEK-ERK pathway by protein phosphatase 5. Nat Cell Biol 8:1011–1016
8. Niethammer P, Kronja I, Kandels-Lewis S et al (2007) Discrete states of a protein interaction network govern interphase and mitotic microtubule dynamics. PLoS Biol 5:e29
9. Santos SD, Verveer PJ, Bastiaens PI (2007) Growth factor-induced MAPK network topology shapes Erk response determining PC-12 cell fate. Nat Cell Biol 9:324–330
10. Deng X, Gao F, May WS (2008) Protein phosphatase 2A inactivates Bcl2's antiapoptotic function by dephosphorylation and up-regulation of Bcl2-p53 binding. Blood 113(2): 422–428
11. Guillaud L, Wong R, Hirokawa N (2008) Disruption of KIF17-Mint1 interaction by CaMKII-dependent phosphorylation: a molecular model of kinesin-cargo release. Nat Cell Biol 10:19–29
12. Ke B, Oh E, Thurmond DC (2007) Doc2beta is a novel Munc18c-interacting partner and positive effector of syntaxin 4-mediated exocytosis. J Biol Chem 282:21786–21797
13. Miller SE, Collins BM, McCoy AJ et al (2007) A SNARE-adaptor interaction is a new mode of cargo recognition in clathrin-coated vesicles. Nature 450:570–574
14. Yeung K, Seitz T, Li S et al (1999) Suppression of Raf-1 kinase activity and MAP kinase signalling by RKIP. Nature 401:173–177
15. Gloeckner CJ, Boldt K, Schumacher A et al (2007) A novel tandem affinity purification strategy for the efficient isolation and characterisation of native protein complexes. Proteomics 7:4228–4234
16. Boldt K, van Reeuwijk J, Gloeckner CJ et al (2009) Tandem affinity purification of ciliopathy-associated protein complexes. Methods Cell Biol 91:143–160
17. Gloeckner CJ, Boldt K, Schumacher A et al (2009) Tandem affinity purification of protein complexes from mammalian cells by the Strep/FLAG (SF)-TAP tag. Methods Mol Biol 564:359–372
18. Gloeckner CJ, Boldt K, Ueffing M (2009) Strep/FLAG tandem affinity purification (SF-TAP) to study protein interactions. Curr Protoc Protein Sci. Chapter 19, Unit 19.20
19. Bendall SC, Hughes C, Stewart MH et al (2008) Prevention of amino acid conversion in SILAC experiments with embryonic stem cells. Mol Cell Proteomics 7:1587–1597
20. Olsen JV, de Godoy LM, Li G et al (2005) Parts per million mass accuracy on an Orbitrap mass spectrometer via lock mass injection into a C-trap. Mol Cell Proteomics 4:2010–2021

Chapter 14

Defining Dynamic Protein Interactions Using SILAC-Based Quantitative Mass Spectrometry

Xiaorong Wang and Lan Huang

Abstract

Protein–protein interactions are essential to various physiological processes in living cells. A full characterization of protein interactions is critical to our understanding of their roles in the regulation of protein functions. Affinity purification coupled with mass spectrometry (AP-MS) has become one of the most effective approaches to systematically study protein–protein interactions. In combination with quantitative mass spectrometry, specific interacting proteins can be efficiently distinguished from nonspecific background proteins. Based on interaction affinity and kinetics, protein interactions can be classified into different categories such as stable and dynamic interactions. Standard biochemical methods are effective in capturing and identifying stable protein interactions but are not sufficient enough to identify dynamic interactors. In this chapter, we describe integrated strategies to allow the identification of dynamic interactors of protein complexes by incorporating new sample preparation methods with SILAC-based quantitation.

Key words AP-MS, Protein–protein interaction, Quantitative mass spectrometry, HB-tag, Dynamic interactors, Stable interactors, Proteasome interacting proteins, PAM-SILAC, MAP-SILAC, Tc-PAM SILAC

1 Introduction

Protein complexes are dynamic and functional entities that are of critical importance for various biological processes in living cells. Protein–protein interactions play key regulatory roles in controlling the assembly, structure, and function of protein complexes in response to diverse cellular cues [1]. It has been well recognized that aberrant protein–protein interactions can lead to various human diseases including cancer [2, 3]. Therefore, a comprehensive characterization of interaction networks of protein complexes not only improves our understanding of cellular processes but also provides potential targets for future therapeutics. Due to technological advancement in recent years, affinity purification–mass spectrometry (AP-MS) has become the method of choice for globally mapping protein–protein interactions from various organisms

Bettina Warscheid (ed.), *Stable Isotope Labeling by Amino Acids in Cell Culture (SILAC): Methods and Protocols*, Methods in Molecular Biology, vol. 1188, DOI 10.1007/978-1-4939-1142-4_14, © Springer Science+Business Media New York 2014

with speed and sensitivity [4–6]. In combination with quantitative mass spectrometry, highly reliable interaction data can be obtained in which specific protein interactors can be effectively distinguished from nonspecific background proteins [6, 7]. This is important since nonspecific binding to the affinity matrix cannot be completely eliminated in resin-based affinity purification processes.

Although various quantitative mass spectrometry methods can be incorporated with AP-MS strategies, the stable isotope labeling of amino acids in cell culture (SILAC) strategy appears to be more attractive owing to global protein labeling during cell culture prior to any sample preparation procedures, thus minimizing sample loss during AP-MS experiments [7, 8]. When the standard SILAC approach is used, cells expressing the tagged bait protein are labeled metabolically with the light isotope and control cells (e.g., cells expressing the tag alone) are metabolically labeled with the heavy isotope or vice versa. For simplicity, in the following text, only the former situation will be described. After metabolic labeling and cell lysis, equal amounts of light and heavy labeled cell lysates are mixed before purification. We term this kind of standard SILAC strategy PAM (*p*urification *a*fter *m*ixing)-SILAC as shown in Fig. 1a [9]. After purification, the samples are subject to digestion and mass spectrometry analysis. When the purified proteins are present in both light and heavy labeled forms, their resulting peptides will be detected in MS as peptide pairs (light *vs.* heavy) with defined mass differences depending on the number of stable isotope-labeled amino acids in each peptide. By comparing mass spectral peak intensities of peptide pairs, their relative abundance ratios (i.e., SILAC ratios = light/heavy) can be calculated, which are the basis for distinguishing specific proteins from nonspecific background. Since the abundance of a specific interacting partner purified from the tagged bait sample should be significantly higher than the one from the control, its SILAC ratio would be much higher than 1. The higher the ratios, the more specific the interactions are. In contrast, the abundance of nonspecifically bound background proteins should be comparable from both the sample and the control, resulting in their SILAC ratios close to 1. Thus, specifically interacting proteins can be determined quantitatively using the PAM-SILAC method (the original SILAC approach).

In addition to specificity, proteins interact with each other with different affinity and kinetics. Only protein interactions with high enough affinity can be preserved during AP-MS experiments due to extensive washing steps. Among these interactions, proteins that interact with the bait at fast on and slow off rates are considered as stable interactors, whereas proteins that interact with the bait at fast on/off rates are known as dynamic interactors. With the PAM-SILAC method, protein purification is carried out after mixing the cell lysates from two types of cells (sample *vs.* control) that have been differentially labeled; all proteins are present in both the light

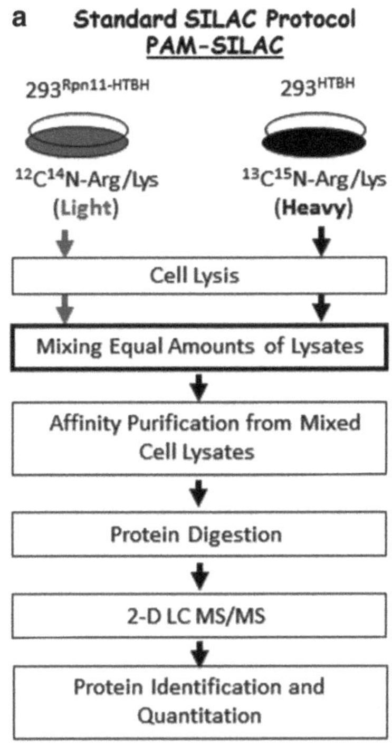

Fig. 1 SILAC-based AP-MS strategies to capture and identify dynamic and stable human proteasome-interacting proteins. $293^{Rpn11\text{-}HTBH}$ cells are grown in light SILAC medium containing $^{12}C_6{}^{14}N_4$-Arg/$^{12}C_6{}^{14}N_2$-Lys (*gray* color), whereas 293^{HTBH} cells are grown in heavy SILAC medium containing $^{13}C_6{}^{15}N_4$-Arg/$^{13}C_6{}^{15}N_2$-Lys (*black*). Two experimental schemes are depicted. (**a**) The standard SILAC method: PAM (purification after mixing)-SILAC; (**b**) the modified SILAC method: MAP (mixing after purification)-SILAC

(sample) and heavy (control) labeled forms during the purification. Although the presence of the two differentially labeled cell lysates does not affect stable interactions, it does interfere with the interactions between the dynamic interactors and the bait. As a result, some of the light labeled dynamic interactors initially bound to the bait can be replaced by their corresponding heavy labeled forms from the control cell lysate, thus leading to decreased SILAC ratios and hampering their identification as specific interactors. Depending on the interaction kinetics, an equilibrium can be achieved between the two differentially labeled forms of the dynamic interactors that are bound to the bait at a given incubation time, which will decrease SILAC ratios of these interactors close to those of background proteins. Thus, these specific but dynamic interactors cannot be effectively distinguished from background proteins based on their SILAC ratios determined by the original SILAC approach. Therefore, the PAM-SILAC method is not best suited for unambiguous identification of dynamic interactions. To circumvent this problem, we have

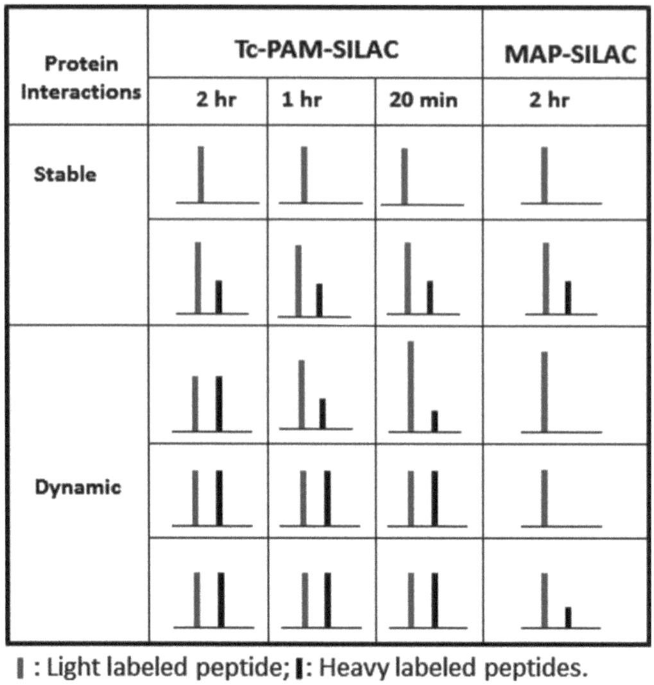

Fig. 2 Characteristic PAM-SILAC and MAP-SILAC ratio profiles of dynamic and stable interactors. The relative abundance of proteins is calculated based on the ratios of mass spectral peak intensities of the observed peptide pairs colored in *gray* (light form) and *black* (heavy form). Several typical examples are shown here. *Tc*: time controlled

developed the Tc (Time controlled)-PAM-SILAC method [9], in which different incubation times (e.g., 20 min, 1 h, 2 h) can be selected to facilitate the identification of dynamic interactors (Fig. 2). This is based on the observations that SILAC ratios for dynamic interactors are dependent on incubation time and they increase with less incubation time due to decreased interaction exchange between the light and heavy labeled proteins. If the on/off rates are not too fast, the dynamic interacting partners can be identified with shortened incubation times. Since stable interactors have SILAC ratios independent of incubation time, stable and dynamic interactors can be distinguished by the Tc-PAM-SILAC method. Although effective, the Tc-PAM-SILAC method may not be sufficient to identify dynamic interactors with very fast on/off rates. This is due to the fact that: (1) there is a limit on experimentally feasible incubation time; and (2) shortened incubation time often sacrifices binding efficiency and thus leads to compromised sensitivity.

In order to quantitatively identify all of the dynamic interacting proteins with different on/off rates, we have further developed a new sample preparation strategy, MAP (*m*ixing *a*fter *p*urification)-SILAC, which allows the complete elimination of interaction interferences from proteins in control cell lysates during purification (Fig. 1b). In the MAP-SILAC strategy, protein purification is carried out separately from equal amounts of the two cell lysates to be compared (sample *vs.* control) that have been differentially labeled. After the purification, the purified protein complexes are mixed for digestion and MS analysis. With this approach, there is no interaction exchange between differentially labeled forms and dynamic interactors can preserve their high SILAC ratios for unambiguous identification as specific interacting proteins. By comparing protein SILAC ratios obtained from MAP-SILAC and PAM-SILAC experiments, dynamic and stable interactors can be effectively distinguished [9, 10]. These new methods significantly expand the ability of AP-MS strategies to study protein interactions, allowing not only the identification of important but previously unidentifiable interacting proteins, but also the characterization of the nature of protein interactions.

These new integrated strategies have been successfully applied to characterize proteasome- [9] and COP9 signalosome-interacting proteins [10]. In this chapter, we use the study of dynamic interacting proteins of the human 26S proteasome complex as an example to illustrate the experimental workflow. The 26S proteasome is a multi-catalytic proteinase complex responsible for ubiquitin/ATP-dependent protein degradation [11]. His-Bio- (HB-) tag based affinity purification strategy is employed to isolate the human proteasome complex in a single step [12, 13]. In combination with MAP-SILAC and Tc-PAM-SILAC, a number of dynamic interactors of the 26S proteasome have been identified, most of which are key regulators in the ubiquitin–proteasome degradation system [9]. This further demonstrates the critical importance of identifying biologically significant dynamic interactors of protein complexes.

2 Materials

2.1 Cell Culture and Metabolic Labeling

1. A HEK293 cell line stably expressing C-terminal HTBH-tagged Rpn11: $293^{Rpn11\text{-}HTBH}$ (*see* **Note 1**).
2. A HEK293 cell line stably expressing the HTBH-tag alone: 293^{HTBH}.
3. Culture medium: EMEM (deficient in lysine and arginine) (e.g., from Sigma-Aldrich) (*see* **Note 2**).
4. Heavy isotope-labeled amino acids: $^{13}C_6{}^{15}N_4$-Arginine and $^{13}C_6{}^{15}N_2$-lysine. Make two 100× stock solutions: 2.8 mg/ml arginine and 7.3 mg/ml lysine in sterile water.
5. Light isotope-labeled amino acids: $^{12}C_6{}^{14}N_4$-arginine and $^{12}C_6{}^{14}N_2$-lysine. Make two 1,000× stock solutions: 28 mg/ml arginine and 73 mg/ml lysine in sterile water.

6. Heavy SILAC medium: EMEM supplemented with 28 μg/ml $^{13}C_6{}^{15}N_4$-arginine, 73 μg/ml $^{13}C_6{}^{15}N_2$-lysine, 10 % dialyzed fetal bovine serum, and 1 % penicillin/streptomycin (the 100 % solution contains 10,000 U penicillin and 10,000 U streptomycin).
7. Light SILAC medium: EMEM supplemented with 28 μg/ml $^{12}C_6{}^{14}N_4$-arginine, 73 μg/ml $^{12}C_6{}^{14}N_2$-lysine, 10 % dialyzed fetal bovine serum, and 1 % penicillin/streptomycin.

2.2 HB-tag Based Affinity Purification for MAP-SILAC and Tc-PAM-SILAC Experiments

1. 1× protease inhibitor cocktail: 1 μg/ml phenylmethylsulfonyl fluoride, leupeptin, aprotinin, pepstatin. Make a 100× stock solution for phenylmethylsulfonyl fluoride, store at 4 °C, and 1,000× stock solutions for leupeptin, aprotinin, and pepstatin, store at −20 °C.
2. 1× phosphatase inhibitor cocktail: 5 mM NaF, 0.1 mM Na_3VO_4, 2.5 mM $Na_4P_2O_7$, 1 mM EDTA. Make a 10× stock solution; store at −20 °C.
3. Trypsin–EDTA.
4. PBS.
5. Lysis buffer A:100 mM sodium chloride, 50 mM sodium phosphate, 10 % glycerol, 5 mM ATP, 1 mM DTT, 5 mM $MgCl_2$, 1× protease inhibiter cocktail, 1× phosphatase inhibitor cocktail, 0.5 % NP-40, pH 7.5. Make the buffer right before the experiment.
6. 20 gauge needles.
7. ImmunoPure Streptavidin (Thermo Scientific).
8. TEB buffer: 50 mM Tris–HCl, pH 7.5.
9. AcTEV protease (Life Technologies).
10. Microspin column (Bio-Rad).
11. Siliconized tubes (Axygen).

2.3 Protein Digestion and LC-MS/MS Analysis

1. Trichloroacetic acid (TCA), 100 % (w/v): dissolve 100 g TCA into 70 ml H_2O and keep at 4 °C.
2. Acetone, keep at −20 °C.
3. 50 mM NH_4HCO_3 with 8 M urea.
4. Trifluoroacetic acid (TFA).
5. 1 μg/μl sequencing-grade endopeptidase LysC stock solution.
6. 0.4 μg/μl sequencing-grade trypsin stock solution: dissolve 20 μg of trypsin in 50 μl of 1 mM TFA (*see* **Note 3**).
7. 10 % formic acid.
8. 2.1 mm × 10 cm PolySULFOETHYL A column (Nest Group).
9. Strong cation exchange (SCX) chromatography: AKTA Basic 10 (GE Healthcare).

10. AKTA buffer A: 5 mM KH_2PO_4, 0.1 % formic acid, 30 % acetonitrile, pH 2.7. For 1 l of the solution, add 0.68 g KH_2PO_4 and 1 ml of formic acid to 700 ml water, adjust the pH with formic acid, filter the solution using a 0.45 μm filter, and then add 300 ml of acetonitrile. The solution needs to be degassed for 20 min.
11. AKTA buffer B: 5 mM KH_2PO_4, 0.1 % formic acid, 30 % acetonitrile, 300 mM KCl, pH 2.7. For 1 l of the solution, add 0.68 g KH_2PO_4, 1 ml of formic acid, and 26.1 g KCl to 700 ml H_2O, adjust the pH with formic acid, filter the solution using a 0.45 μm filter, then add 300 ml of acetonitrile. The solution needs to be degassed for 20 min.
12. Vivapure C18 microspin columns (Vivascience).
13. NanoLC capillary column (75 μm ID × 150 mm long) packed with Polaris C18-A resin (Varian Inc.).
14. Mass spectrometer: QSTAR XL MS (AB Sciex) (*see* **Note 4**).
15. Nano LC solvent A: 2 % acetonitrile, 0.1 % formic acid in H_2O.
16. Nano LC solvent B: 98 % acetonitrile, 0.1 % formic acid in H_2O.

2.4 Database Searching for Protein Identification and Quantification

1. LC-MS/MS data extraction: instrument specific scripts from the manufacturer.
2. Protein identification and quantitation software: Protein Prospector (University of California, San Francisco).

2.5 Validation of Dynamic Interactions Using Quantitative Western Blotting

1. Horseradish peroxidase (HRP)-conjugated secondary antibody.
2. Super Signal West Pico chemiluminescent substrate (Pierce/Thermo Fisher Scientific).
3. Anti-Rpt6 antibody (BioMol).
4. Anti-FLAG antibody (Sigma-Aldrich).
5. Odyssey infrared scanning system (LI-COR Biosciences).
6. pcDNA/FRT-ADRM1-FLAG.
7. TuboFect transfection reagent (Thermo scientific).
8. Protein assay kit (Bio-Rad).
9. 10 % SDS-PAGE gel.
10. PVDF membrane (Bio-Rad).
11. Wet/tank blotting system (Bio-Rad).
12. Stripping buffer: 5 mM KH_2PO_4, 0.1 % formic acid. For 1 l of the solution, add 0.68 g KH_2PO_4 and 1 ml of formic acid to 999 ml water.
13. Cy5-conjugated anti-mouse IgG (Invitrogen).

3 Methods

3.1 Cell Culture and Metabolic Labeling

1. Culture $293^{Rpn11\text{-}HTBH}$ cells in light SILAC medium. When cell culture changes from regular medium to SILAC medium, cells need to be grown for more than seven cell doublings to ensure complete labeling. Then grow cells to about 90 % confluence prior to cell lysis.
2. Culture 293^{HTBH} cells (control cell line) in heavy SILAC medium. Cells need to be grown for more than seven cell doublings in heavy SILAC medium to ensure complete labeling. Then grow cells to about 90 % confluence prior to cell lysis.
3. For label-switch experiments, culture $293^{Rpn11\text{-}HTBH}$ cells in heavy SILAC medium and 293^{HTBH} cells in light SILAC medium.

3.2 HB-tag Based Affinity Purification

1. Trypsinize cells and wash them three times with 1× PBS buffer.
2. Collect cell pellets and lyse cells using lysis buffer A by pushing the lysate ten times through a 20 gauge needle.
3. Centrifuge the lysates at maximum speed of a microcentrifuge for 15 min to remove cell debris, and incubate the supernatant with 25 μl of Streptavidin resin per plate for the desired amount of time at 4 °C (*see* **Note 5**).
4. Wash the Streptavidin beads with 20 bed volumes of lysis buffer A without protease and phosphatase inhibitors (*see* **Note 6**).
5. Wash the beads with 10 bed volumes of TEB buffer.
6. Incubate the beads in 2 bed volumes of TEB buffer with 1 % TEV at 30 °C for 1 h with rotation (*see* **Note 7**).
7. Elute the human 26S proteasome complex from the beads by passing the mixture through a Bio-Rad microspin column (*see* **Note 8**).

3.3 Protein Digestion and LC-MS/MS Analysis

To avoid keratin contamination in your samples, you need to wear a hair net, sleeves, and clean gloves for the following procedure.

1. Precipitate purified complexes by adding TCA to a final concentration of 25 %, place the mixture on ice for 1 h. Spin at maximum speed for 15 min. Remove the supernatant. Wash the pellet in 1 ml of ice-cold acetone and centrifuge for 15 min, repeat the washing step two more times (*see* **Note 9**).
2. Redissolve the pellet with a minimal volume of 50 mM NH_4HCO_3 in 8 M urea (*see* **Note 10**).
3. Add 1 μl of endopeptidase LysC stock solution to the protein complex and incubate for 4 h at 37 °C (*see* **Note 11**).

4. Decrease urea concentration to <1.5 M by adding an adequate volume of 50 mM NH_4HCO_3. Add trypsin to a final concentration of 5–10 ng/μl and incubate overnight at 37 °C (*see* **Note 12**).

   Recovery of digested peptides:

5. Add 10 % formic acid to a final concentration of 1 % to stop the digestion.
6. Dry the resulting digest in a SpeedVac. Add 100 μl of water and dry again. Repeat this step one more time. Dissolve the peptide mixture in AKTA buffer A for SCX chromatography (*see* **Note 13**).
7. Separate peptides by SCX chromatography using a PolySULFOETHYL A column at a flow rate of 200 μl/min using an AKTA Basic 10.
8. Elute peptides applying a salt gradient of buffer B: 0–5 % in 2 min, 5–25 % in 20 min, 25–100 % in 10 min.
9. Collect 10–15 fractions manually based on UV absorbance at 215 nm.
10. Desalt collected SCX fractions using Vivapure C18 microspin columns following the manufacturer's instruction.
11. Analyze peptide mixtures by LC-MS/MS using nanoflow reverse phase liquid chromatography (NanoLC) coupled online to a QSTAR XL MS instrument. Elute peptides with a linear gradient of 0–35 % nano LC solvent B in 80 min at a flow of 250 nl/min. LC-MS/MS is operated in an information-dependent mode in which each full MS analysis is followed by three MS/MS acquisitions where the three most abundant peptide molecular ions are dynamically selected for collision induced dissociation (CID) to generate tandem mass spectra (*see* **Note 14**).

3.4 Protein Identification and Quantification Using Protein Prospector

1. Obtain monoisotopic masses of both parent ions and corresponding fragment ions, parent ion charge states, and ion intensities from the MS/MS by using an automated version of the Mascot script from Analyst QS within Protein Prospector.
2. Use the Batch-tag program within Protein Prospector for database searching. Select trypsin as the enzyme and set the maximum number of missed tryptic cleavage sites as 2. Chemical modifications such as protein amino-terminal acetylation, methionine oxidation, amino-terminal pyroglutamine, and deamidation of asparagine residues are selected as variable modifications. These modifications, except for protein amino-terminal acetylation, need to be chosen because of their frequent occurrence during sample preparation. For SILAC experiments, $^{13}C_6{}^{15}N_4$-arginine and $^{13}C_6{}^{15}N_2$-lysine need to be

chosen as variable modifications as well. Set the mass accuracy for parent ions and fragment ions as ±200 ppm and 300 ppm, respectively. Any annotated protein databases such as SwissProt and UniProt can be used for database searching. A concatenated database composed of a normal and its reverse database can be generated in Protein Prospector for database searching. Because we purify the samples from human cell lines, *Homo sapiens* is selected as the restricted species.

3. General protein identification is based on at least two peptides with an expectation value cutoff of 0.01.
4. The SILAC ratios are calculated using the Search Compare program by calculating the relative abundance ratios of arginine/lysine-containing peptides based on ion intensities of monoisotopic peaks observed in the MS spectra at the time when the peptides are sequenced and subsequently identified during database searching. Signal to noise ratio >2 is required for peaks to be considered for quantitation. The SILAC ratios can be further validated by checking all of the raw spectra within the Protein Prospector Search Compare program. The ratio outliers are easily visualized on the ratio plots in Protein Prospector. If the peptide peaks are mixed with other peptide peaks or buried in the noise peaks, they cannot be used for quantification. The SILAC ratios are often reported as average values plus standard deviations. Only reproducible data should be reported as final results.

3.5 Identification of Dynamic and Stable PIPs Using PAM-SILAC and MAP-SILAC

The general workflow for PAM-SILAC and MAP-SILAC experiments is outlined in Fig. 1. For each experiment, use ten 150 mm plates of each type of cells. Perform each experiment at least twice to make sure the results are reproducible.

3.5.1 PAM-SILAC Experiment

1. Lyse $293^{Rpn11\text{-}HTBH}$ cells (grown in light SILAC medium) and 293^{HTBH} (grown in heavy SILAC medium) using lysis buffer A.
2. Mix equal amounts of the two differentially labeled cell lysates.
3. Carry out affinity purification using mixed lysates as described in Subheading 3.2. Use the optimal incubation time, i.e., 2 h (*see* **Note 5**).
4. Perform protein digestion, SCX separation, desalting, and LC-MS/MS analysis as described in Subheading 3.3.
5. Protein identification and quantitation as described in Subheading 3.4.

3.5.2 Time-controlled (Tc)-PAM-SILAC Experiment

Three separate PAM-SILAC experiments are performed by selecting three incubation times. Since the optimal incubation time is 2 h, two shorter incubation times, 20 min and 1 h, are selected. This allows the identification of dynamic proteins based on changes in their relative abundance ratios with incubation times (Fig. 2).

3.5.3 MAP-SILAC Experiment

1. Lyse $293^{Rpn11\text{-}HTBH}$ cells (grown in light SILAC medium) and 293^{HTBH} cells (grown in heavy SILAC medium) using lysis buffer A.
2. Carry out affinity purification as described in Subheading 3.2 from equal amounts of two differentially labeled cell lysates separately. Use the optimal incubation time, i.e., 2 h (*see* **Note 5**).
3. Mix the two purified samples (*see* **Note 15**).
4. Perform protein digestion, SCX separation, desalting, and LC-MS/MS analysis as described in Subheading 3.3.
5. Carry out protein identification and quantitation as described in Subheading 3.4.

3.5.4 Identifying Dynamic and Stable Interactors of Proteasome Complexes Based on MAP-SILAC and PAM-SILAC Ratios

All putative proteasome-specific interacting proteins should have MAP-SILAC ratios >1.5, but not all of them have PAM-SILAC ratios >1.5. The characteristic PAM-SILAC and MAP-SILAC ratio profiles for dynamic and stable interacting proteins are illustrated in Fig. 2.

Dynamic proteasome-interacting proteins are identified when:

1. Their MAP-SILAC ratios are above a selected threshold (>1.5) [14] and are at least twofold higher than their PAM-SILAC ratios (*see* **Note 16**).
2. Their PAM-SILAC ratios increase with decreased incubation time in Tc-PAM-SILAC experiments (*see* **Note 17**).

Stable proteasome interacting proteins are identified when:

1. Their MAP-SILAC and PAM-SILAC ratios are very similar and >1.5.
2. Their PAM-SILAC ratios do not change with incubation time in Tc-PAM-SILAC experiments.

3.6 Validation of Dynamic Interactions Using Quantitative Western Blotting

An alternative strategy to confirm dynamic interactions identified by MAP-SILAC and PAM-SILAC experiments is by protein co-expression, affinity purification, and quantitative immunoblotting. To illustrate the process, we choose to use the validation of a selected proteasome dynamic interactor, ADRM1, as an example. As shown in Fig. 3a, ADRM1 has the characteristic PAM-SILAC and MAP-SILAC ratio profiles for dynamic interactors. To confirm the dynamic interaction between ADRM1 and the proteasome, we examine the interaction exchange of ADRM1 during purification by expressing FLAG-tagged ADRM1 only in control 293^{HTBH} cells and not in $293^{Rpn11\text{-}HTBH}$ cells and by carrying out HB-based affinity purification using the Tc-PAM and MAP methods. Because proteasomes are only purified from $293^{Rpn11\text{-}HTBH}$ cells that express no FLAG-tagged proteins, any co-purification of ADRM1-FLAG using the PAM method should be the result of interactions formed in the mixed lysates during the incubation.

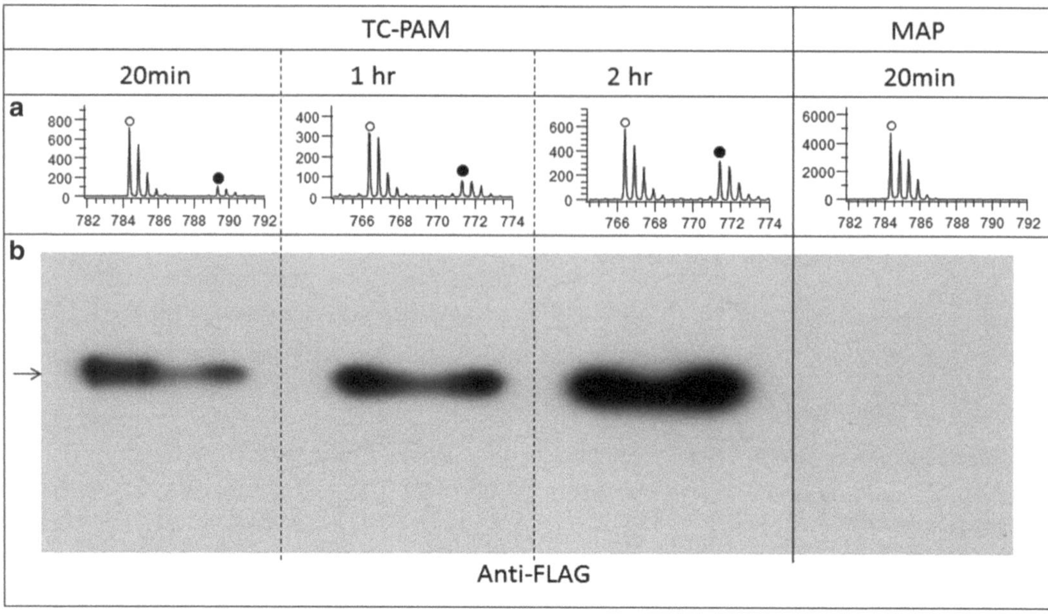

Fig. 3 (**a**) TOF MS spectra of a tryptic peptide (m/z 766.39^{2+}, Acetyl-TTSGALFPSLVPGSR) matched to ADRM1/hRpn13 (a dynamic interactor). "*closed circle*" and "*filled circle*" represent the light and heavy forms of the peptide, respectively. The SILAC ratios for the peptide are shown in the corresponding spectra. As shown, its PAM-SILAC ratios increases when incubation time decreases, while its MAP-SILAC ratio is high (no heavy labeled form detected). (**b**) Validation of the dynamic interaction of ADRM1 with the proteasome using transfection, affinity purification, and quantitative western blot analysis. The band represents ADRM1-FLAG. Comparison of incorporation of ADRM1-FLAG expressed in the control cells into the purified Rpn11-HTBH containing proteasome complexes during the purification with the Tc-PAM approach at three different incubation times (20 min, 1 h, and 2 h) and with the MAP approach (2 h)

Therefore, co-purification of ADRM1-FLAG would be expected only in PAM-purified samples but not in MAP-purified samples (Fig. 3b).

In addition, the amount of co-purified ADRM1-FLAG should increase with increased incubation time during Tc-PAM experiments. Together, this would confirm the dynamic nature of ADRM1 interaction determined by PAM-SILAC and MAP-SILAC experiments (Fig. 3a).

3.6.1 Transfection of ADRM1-FLAG into Control Cell Lines

1. Transiently transfect 293^{HTBH} cells with pcDNA/FRT-ADRM1-FLAG [9]. Twenty-four hours after transfection, wash the cells three times in PBS and lyse the cells in lysis buffer A. Centrifuge the lysate at maximum speed of a microcentrifuge for 15 min to obtain a cleared lysate (lysate A).
2. Grow 293$^{Rpn11\text{-}HTBH}$ cells similarly without transfection and lyse the cells the same way as described above to obtain a cleared lysate (lysate B).
3. Measure protein concentrations of lysates A and B, and divide equal amounts of lysates A and B into four aliquots.

3.6.2 HB-tag Based Affinity Purification Using the Tc-PAM Strategy

1. Take three aliquots of lysates A and B.
2. Mix equal amounts of lysates A and B to make three aliquots of mixed lysates for PAM experiments.
3. Follow the general purification protocol described in Subheading 3.2. The incubation times for the three PAM experiments are 20 min, 1 h, and 2 h.

3.6.3 HB-tag Based Affinity Purification Using MAP Strategy

1. Take one aliquot of lysates A and B.
2. Perform affinity purification as described in Subheading 3.2 from lysates A and B separately. Use the optimal incubation time (i.e., 2 h).
3. Mix the two purified samples for subsequent immunoblotting analysis.

3.6.4 Quantitative Western Blotting

1. Load the four purified samples from Tc-PAM (Subheading 3.6.2) and MAP (Subheading 3.6.3) experiments for one-dimensional SDS-PAGE. Transfer proteins to a PVDF membrane and analyze the proteins by immunoblotting.
2. Probe ADRM1-FLAG protein in the four purified samples using a mouse anti-FLAG antibody (1:2,000) followed by HRP-conjugated anti-mouse IgG (1:10,000).
3. Strip the blots by incubating the membrane in striping buffer for 30 min and re-probe with mouse anti-Rpt6 (1:1,000) followed by HRP-conjugated anti-mouse IgG (1:10,000) to detect the presence of the proteasome in the purified samples. The Rpt6 signal is used as the internal standard for normalization of proteasome loading.
4. Perform quantitative immunoblotting analysis using Cy5-conjugated anti-mouse IgG (1:10,000) as the secondary antibody. Quantify fluorescence intensities of the ADRM1-FLAG and Rpt6 bands using an Odyssey infrared scanning system (*see* **Note 18**).
5. Plot the ratios of ADRM1-FLAG to Rpt6 against incubation times to determine whether interaction exchange between endogenous ADRM1 from $293^{\text{Rpn11-HTBH}}$ cells and ADRM1-FLAG from 293^{HBTH} control cells during Tc-PAM experiments occurred. No interaction exchange should be observed in the sample purified from MAP experiment.

4 Notes

1. The HTBH-tag consists of two hexahistidine tags, a TEV cleavage site, and a signal sequence for in vivo biotinylation, which allows efficient purification of proteasome complexes in a single step by binding to streptavidin resins and specific elution by cleavage with TEV protease [13].

2. SILAC media from other brands such as Thermo Scientific should work as well.
3. Make fresh 1 mM TFA each time from a 100 mM TFA stock solution.
4. Any tandem mass spectrometer that can produce MS1 spectra with a resolution high enough to determine SILAC ratios can be used.
5. For the HTBH-tag, use 10 μl of Streptavidin beads per 150 mm plate of 293 cells for maximum specific binding efficiency with minimal background binding. Purification efficiency should be followed by western blot analysis. The optimal binding for proteasome complexes to Streptavidin beads is 2 h. The binding efficiency decreases when the incubation time decreases.
6. Effective washing steps can be achieved in micro-columns from Bio-Rad, for example, to minimize the bead loss.
7. A Rotator in a 30 °C incubator works best for this step. Alternatively, you can perform this step at 4 °C overnight.
8. The elute can be stored at −80 °C at this point if subsequent analysis will not be carried out immediately.
9. For best results, clear siliconized tubes should be used to visualize the pellet and minimize sample loss.
10. Gradually add a small volume (e.g., 25 μl) of the buffer to dissolve the pellet and keep the volume to the minimum. It is the best not to exceed the final volume of 100 μl.
11. LysC digestion can go from 4 h to overnight.
12. Trypsin digestion can go from 8 h to overnight.
13. It is critical to minimize the salt concentration in the sample before SCX separation. It is the best not to exceed 25 mM salt before loading. If needed, desalting with C18 ZipTips or spin columns can be performed.
14. For MS instruments with fast scanning rates such as the LTQ-Orbitrap, top ten peaks can be sequenced in each LC-MS/MS acquisition cycle.
15. The SILAC ratios of background proteins should be about 1. If not, it suggests that the mixing is not equivalent and the final protein SILAC ratios need to be adjusted accordingly.
16. Comparison of MAP-SILAC and PAM-SILAC ratios alone is sufficient to identify dynamic interactors.
17. For interactors with very fast on/off rates, Tc-PAM-SILAC ratios alone cannot determine whether they are dynamic interactors. This requires MAP-SILAC experiments for unambiguous identification.
18. The Fuji imaging system works well.

Acknowledgments

This work was in part supported by National Institutes of Health Grant GM074830-06A1 and R21CA161807-01A1 (to L. H.).

References

1. Alberts B (1998) The Cell as a collection of protein machines: preparing the next generation of molecular biologists. Cell 92:291–294
2. Charbonnier S, Gallego O, Gavin AC (2008) The social network of a cell: recent advances in interactome mapping. Biotechnol Annu Rev 14:1–28
3. Kar G, Gursoy A, Keskin O (2009) Human cancer protein-protein interaction network: a structural perspective. PLoS Comput Biol 5(12):e1000601
4. Kuhner S, van Noort V, Betts MJ et al (2009) Proteome organization in a genome-reduced bacterium. Science 326(5957):1235–1240
5. Gingras AC, Gstaiger M, Raught B et al (2007) Analysis of protein complexes using mass spectrometry. Nat Rev Mol Cell Biol 8(8):645–654
6. Kaake RM, Wang X, Huang L (2010) Profiling of protein interaction networks of protein complexes using affinity purification and quantitative mass spectrometry. Mol Cell Proteomics 9(8):1650–1665
7. Mann M (2006) Functional and quantitative proteomics using SILAC. Nat Rev Mol Cell Biol 7(12):952–958
8. Ong SE, Kratchmarova I, Mann M (2003) Properties of 13C-substituted arginine in stable isotope labeling by amino acids in cell culture (SILAC). J Proteome Res 2(2):173–181
9. Wang X, Huang L (2008) Identifying dynamic interactors of protein complexes by quantitative mass spectrometry. Mol Cell Proteomics 7(1):46–57
10. Fang L, Wang X, Yamoah K et al (2008) Characterization of the human COP9 signalosome complex using affinity purification and mass spectrometry. J Proteome Res 7(11): 4914–4925
11. Pickart CM, Cohen RE (2004) Proteasomes and their kin: proteases in the machine age. Nat Rev Mol Cell Biol 5(3):177–187
12. Tagwerker C, Flick K, Cui M et al (2006) A tandem affinity tag for two-step purification under fully denaturing conditions: application in ubiquitin profiling and protein complex identification combined with in vivo cross-linking. Mol Cell Proteomics 5(4): 737–748
13. Wang X, Chen CF, Baker PR et al (2007) Mass spectrometric characterization of the affinity-purified human 26S proteasome complex. Biochemistry 46(11):3553–3565
14. Guerrero C, Tagwerker C, Kaiser P et al (2006) An integrated mass spectrometry-based proteomic approach: quantitative analysis of tandem affinity-purified in vivo cross-linked protein complexes (QTAX) to decipher the 26 S proteasome-interacting network. Mol Cell Proteomics 5(2):366–378

Chapter 15

Identifying Nuclear Protein–Protein Interactions Using GFP Affinity Purification and SILAC-Based Quantitative Mass Spectrometry

H. Irem Baymaz, Cornelia G. Spruijt, and Michiel Vermeulen

Abstract

Many cellular proteins assemble into macromolecular protein complexes. Therefore, identifying protein–protein interactions (PPIs) is essential to gain insight into the function of proteins. Recently established quantitative mass spectrometry-based techniques have significantly improved the unbiased search for PPIs. In this chapter, we describe a single-step GFP affinity purification method combined with SILAC-based quantitative mass spectrometry that can be used to identify nuclear PPIs in mammalian cells.

Key words Protein–protein interactions, GFP affinity purification, SILAC, Quantitative mass spectrometry

1 Introduction

Proteins drive all processes in a cell through a complex and dynamic network of protein–protein interactions (PPIs). Identifying these interactions is therefore essential to gain insight into the function of proteins. There are several approaches available to identify cellular protein–protein interactions. When putative interactors are known, co-immunoprecipitation experiments followed by Western blot analysis can be used to validate true interactors among the list of candidates. However, it is quite common that potential interactors of a protein are not known. In this case, unbiased interaction screening approaches are needed. One such approach is the yeast-2-hybrid (Y2H) system which identifies interactors of a protein of interest using a library of "prey" proteins [1]. The major disadvantage of this method, however, is that Y2H tends to result in many false positive interactions. Another limitation is that mammalian proteins expressed in yeast may lack some posttranslational modifications that mediate PPIs. Furthermore, Y2H only detects direct protein–protein interactions and thus cannot be used to determine

Bettina Warscheid (ed.), *Stable Isotope Labeling by Amino Acids in Cell Culture (SILAC): Methods and Protocols*, Methods in Molecular Biology, vol. 1188, DOI 10.1007/978-1-4939-1142-4_15, © Springer Science+Business Media New York 2014

all components of larger protein complexes. An alternative to Y2H is the (tandem) affinity purification of the protein of interest followed by mass spectrometry (AP-MS) [2]. This method requires extensive purification of the protein of interest (the bait) under high stringency conditions to minimize nonspecifically interacting proteins. However, even when using very stringent conditions, high-abundant background binders are not completely removed. This problem is particularly relevant when making use of modern mass spectrometers, which are very sensitive and which can sequence proteins when present in a sample in femtomole amounts [3]. As a consequence, scientists still need to resort to other methods such as those mentioned above to distinguish true interactors from background binders. Therefore, although AP-MS is robust in identifying the proteins that are co-purified with the bait, the fact that interactions are not quantified compromises the ability to discriminate true interactors from background binders. In addition, the high stringency conditions can result in the loss of relatively weak but biologically relevant interactions. To overcome these problems, quantitative mass spectrometry methods have been established in recent years [4]. In most of these methods, differential stable isotopic labeling, either on protein or peptide level, is used in the specific and the control affinity purification. Prior to mass spectrometry analysis, the specific and control pull-downs are combined. Each peptide then has a "light" and a "heavy" intensity, and the ratio between these two states indicates the relative abundance of a peptide and the corresponding protein in the specific and control affinity purification. As a result, the bait and its interactors have a high ratio, whereas background proteins have a ratio close to one. Here we describe an application of this principle, a SILAC-based GFP affinity purification method from mammalian nuclear extracts. The method starts with SILAC labeling of cells that (either stably or transiently) express the GFP-tagged protein of interest [5]. As a control, wild-type cells lacking the GFP-tagged bait are labeled in parallel. Nuclear extracts generated from these cells are then used for GFP affinity purifications followed by quantitative mass spectrometry to identify and quantify protein–protein interactions.

2 Materials

For preparing buffers, ultrapure water (18.5 MΩ cm resistance, total organic carbon <12 parts per billion) which will be referred to as Milli-Q is used. In addition, to avoid polymer accumulation in samples, do not use autoclaved pipette tips and keep all buffers in high-quality glass bottles. Tabletop centrifuges with cooling capacity for 50-ml tubes and microcentrifuge tubes are required throughout the protocol.

2.1 SILAC Labeling

1. SILAC Dulbecco's Modified Eagle Medium (DMEM) lacking arginine and lysine.
2. L-arginine monohydrochloride light and heavy ($^{13}C_6{}^{15}N_4$) each dissolved to a final concentration of 84 mg/ml in Milli-Q.
3. L-lysine monohydrochloride light and heavy ($^{13}C_6{}^{15}N_2$) each dissolved to a final concentration of 146 mg/ml in Milli-Q.
4. Dialyzed fetal bovine serum (D-FBS).
5. 200 mM L-glutamine.
6. 100 U/ml Penicillin–Streptomycin.
7. 50 ml syringes and 0.22 μm filters.
8. Dulbecco's Phosphate Buffered Saline (PBS): 0.2 g/L KCl, 0.2 g/L KH_2PO, 8 g/L NaCl, 2.16 g/L $HNa_2PO_4{\cdot}7H_2O$.
9. Trypsin–EDTA: 200 mg/L EDTA, 170.000 U/L Trypsin.
10. 94 × 16 mm and 145 × 20 mm cell culture dishes (referred to as 10-cm and 15-cm dishes respectively).

    Optional (*see* Subheading 3.1):
11. SILAC RPMI medium lacking arginine and lysine.
12. 100× nonessential amino acids.
13. 2i inhibitors (Axon Medchem, CHIR99021 and PD0325901).
14. Leukemia inhibitory factor (LIF, 1,000,000 U/ml).
15. β-Mercaptoethanol.
16. 100 mM Sodium pyruvate.
17. Accutase.

2.2 Transient Transfection of SILAC-Labeled Cells

Polyethylenimine, linear (PEI, Polysciences), dissolved to a final concentration of 1 mg/ml in Milli-Q and neutralized with HCl.

2.3 Nuclear Extraction

1. PBS.
2. SILAC DMEM.
3. Trypsin–EDTA.
4. Glass dounce homogenizer with type B pestle (tight); depending on the amount of cells, different sizes can be used: 500 μl, 2 ml, 7 ml, or 40 ml.
5. Buffer A: 10 mM KCl, 1.5 mM $MgCl_2$, 10 mM Hepes–KOH pH 7.9.
6. Buffer C: 420 mM NaCl, 20 mM Hepes–KOH pH 7.9, 20 % (v/v) glycerol, 2 mM $MgCl_2$, 0.2 mM EDTA.
7. Nonidet P40 (NP-40), 10 % stock solution.
8. Dithiotreitol (DTT), 500 mM stock solution.

9. Complete EDTA-free protease inhibitor cocktail tablets (CPI; Roche, 1 tablet/ml = 50× stock solution).
   Optional (*see* Subheading 3.3):
10. 5 M NaCl.
11. 100 % Glycerol.

2.4 Bradford Assay for Protein Quantification

1. Albumin from bovine serum (BSA) dissolved in Milli-Q to a concentation of 1 mg/ml.
2. BioRad Protein Assay Dye Reagent Concentrate (BioRad, 5×).

2.5 GFP Affinity Purification

1. GFP-binder beads (e.g., GFP-Trap_A from Chromotek).
2. Buffer C*: 300 mM NaCl, 20 mM Hepes–KOH pH 7.9, 20 % (v/v) glycerol, 2 mM $MgCl_2$, 0.2 mM EDTA.
3. PBS.
4. NP-40, 10 % stock solution.
5. DTT, 500 mM stock solution.
6. Complete EDTA-free protease inhibitor cocktail tablets.
7. Ethidium bromide, 10 mg/ml stock solution.
8. Gel-loader tips.
   Optional (*see* Subheading 3.5):
   Blocked agarose beads (e.g., from Chromotek).

2.6 On-Bead Digestion of Proteins

1. Digestion buffer: 2 M urea dissolved in 100 mM Tris–HCl pH 7.5.
2. DTT, 500 mM stock solution.
3. Iodoacetamide (IAA) or chloroacetamide (CAA) dissolved to a final concentration of 550 mM in 50 mM ammonium bicarbonate.
4. Trypsin dissolved to a final concentration of 0.1 mg/ml in 50 mM acetic acid.
5. Thermoshaker.

2.7 Desalting and Purification of Peptides for Mass Spectrometry

1. Blunt-ended syringe needle (1.2 mm diameter) with a nano-tubing end inserted as a plunger.
2. C18 material (Empore).
3. Methanol (ultrapure).
4. Buffer A: 0.5 % (v/v) acetic acid in Milli-Q.
5. Buffer B: 0.5 % (v/v) acetic acid, 80 % (v/v) acetonitrile (HPLC grade) in Milli-Q.
6. 10 % (v/v) trifluoroacetic acid (TFA) in Milli-Q.

2.8 Elution of Peptides

1. Buffer A: 0.5 % (v/v) acetic acid in Milli-Q.
2. Buffer B: 0.5 % (v/v) acetic acid, 80 % (v/v) acetonitrile in Milli-Q.
3. Eppendorf Combitip plus 2.5 ml.
4. Speed vacuum concentrator (with 96-well plate compatible rotor).
5. HPLC autosampler 96-well plate.

2.9 Mass Spectrometry

1. EASY nLC (Thermo Scientific).
2. High performance mass spectrometer (e.g., LTQ-Orbitrap Velos; Thermo Scientific).

2.10 Data Analysis

1. Windows-operated PC (at least 8 GB of RAM and multiple cores are recommended).
2. MaxQuant software package.

3 Methods

3.1 SILAC Labeling

The most commonly used metabolic labeling strategy is SILAC (Stable Isotope Labeling by Amino acids in Cell culture) [5]. In this method, cells are grown in the presence of normal (light) or heavy stable isotopic versions of certain amino acids (usually arginine and lysine). During cell culture, these light and heavy amino acids are incorporated into the proteins, enabling relative quantification of proteins between two functional states. The workflow described in this chapter consists of a so-called "forward" and "reverse" pull-down. In the forward pull-down, the cell line stably expressing a GFP-tagged transgene (referred to as GFP cells) is labeled "heavy" and the corresponding wild-type (WT) cell line is labeled "light." In the reverse pull-down, the light and heavy labels are swapped. Therefore, four different cultures have to be labeled and expanded (Fig. 1a, *see* **Notes 1** and **2**).

1. Prepare a bottle of "heavy" and a bottle of "light" medium. Once made, SILAC media can be kept for up to 6 weeks at 4 °C. For each condition (light and heavy):
   - (a) Take a bottle of DMEM lacking arginine, lysine, and glutamine (*see* **Note 3**).
   - (b) Transfer 20–30 ml of this medium into a 50-ml tube and add 15 mg of light or heavy arginine and 36.5 mg of light or heavy lysine (*see* **Note 4**).
   - (c) Filter-sterilize these aliquots using a syringe and a 0.22 μm filter back into the DMEM bottle.
   - (d) Add 50 ml of D-FBS, 2 mM of L-glutamine, and 550 U of penicillin–streptomycin.

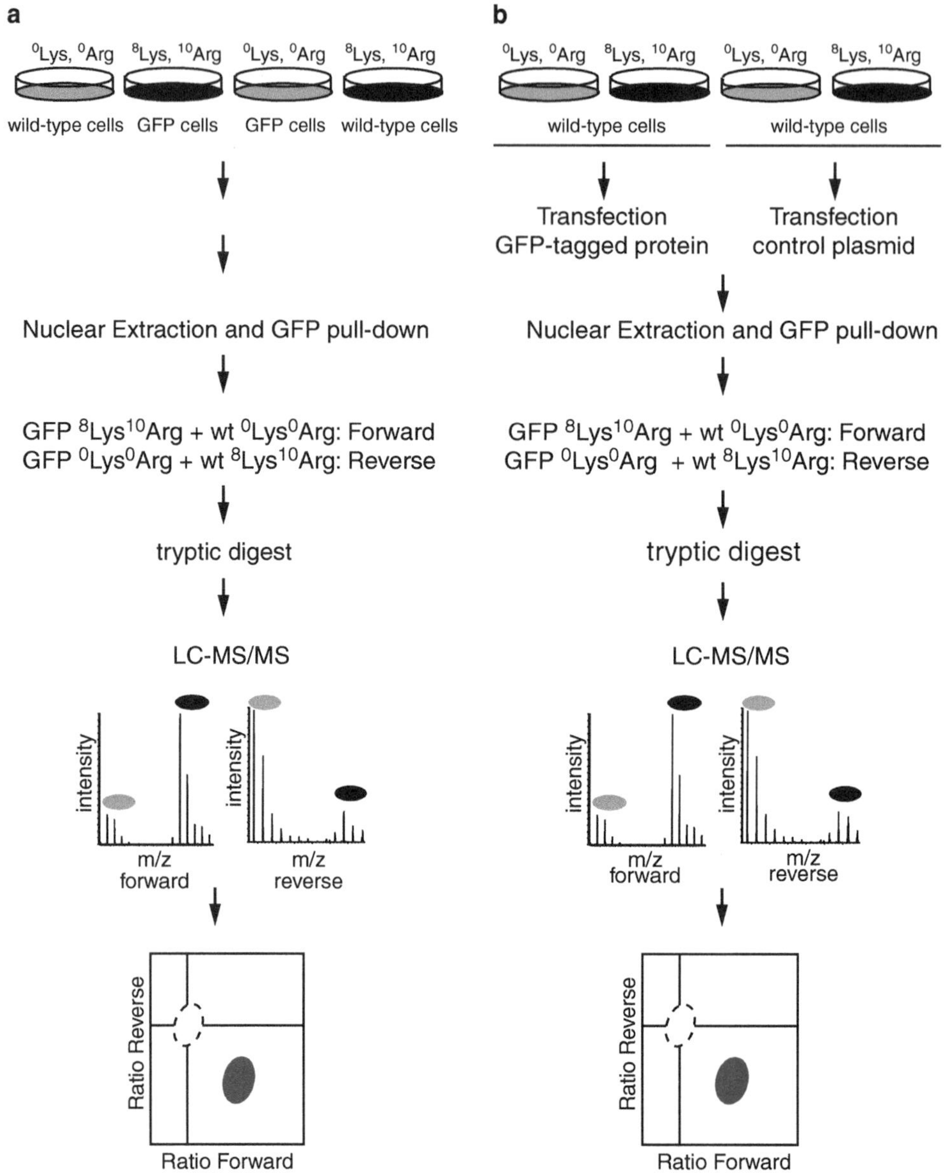

Fig. 1 Schematic representation of the method including the SILAC labeling, nuclear extraction, and GFP-affinity purification followed by quantitative mass spectrometry. (**a**) The workflow for cells stably expressing the GFP-tagged protein of interest and the corresponding wild-type cells. (**b**) The workflow for transient expression of a GFP-tagged bait in SILAC-labeled cells

2. Trypsinize the WT and GFP cells, neutralize trypsin by adding medium and divide each cell suspension into two 15-ml tubes.
3. Spin down the cells at $400 \times g$ for 5 min. This step is necessary to remove trypsin from the cells. Trypsin provides a source of non-labeled amino acids and residual trypsin can therefore compromise SILAC labeling.

4. Aspirate the supernatant and resuspend the cells in light or heavy medium (WT cells and GFP cells are both labeled heavy and light). Transfer 25 % of the cells of each tube into a 10-cm culture dish and add the appropriate amount of medium (heavy or light) to each dish.
5. Culture cells for at least eight doublings in SILAC medium. Keep in mind that some cell lines tend to grow more slowly in SILAC medium compared to normal medium due to the use of D-FBS. While labeling, cells can be expanded in order to end up with the appropriate amount of labeled cells after eight cell doublings. Although it depends on the cell line, a confluent 15-cm dish usually yields around 400 μg of nuclear extract.

3.2 Transient Transfection of SILAC-Labeled Cells

When a stable cell line expressing a GFP-tagged protein of interest is not available, transient transfections can be considered (*see* **Note 5**). In this case, WT cells are light and heavy labeled and expanded. Half of the light and half of the heavy cells are then transfected with the plasmid expressing the GFP-tagged bait, while the other two halves are transfected with a control plasmid (empty GFP plasmid). In the end, there are four cell populations (light control, light GFP, heavy control, heavy GFP; *see* Fig. 1b). The method described below requires a total of 20 × 15-cm dishes, but this scale may be adjusted according to the amount of nuclear extract that is needed.

1. Expand light and heavy labeled cells to the required amount, in this case 10 × 15-cm dishes of light labeled cells and 10 × 15-cm dishes of heavy labeled cells. When cells reach about 60 % confluency (*see* **Note 6**), half of the heavy and half of the light labeled cells (5 × 15-cm dishes of light and 5 × 15-cm dishes of heavy cells) are transfected with the "plasmid expressing the" GFP-tagged bait while the other half is transfected with the control GFP plasmid as follows:
2. Transfer 15 ml of DMEM without lysine, arginine, and glutamine (i.e., SILAC DMEM without anything added to it) into a 50-ml tube for each plasmid (bait and control).
3. Add 150 μg of the plasmid (bait or control) and 450 μl of PEI to each tube (*see* **Note 7**). Vortex for 10 s and incubate at room temperature (RT) for 30 min.
4. Pipette 1.5 ml of the transfection mix dropwise into each of five 15-cm dishes containing light cells and five 15-cm dishes containing heavy cells. Make sure to label the dishes.
5. Culture the cells for an additional 24–48 h at 37 °C. In the end, there are four batches of cells to harvest: control light and heavy, GFP light and heavy. Prior to harvesting, expression of the GFP transgene can be checked by fluorescence microscopy.

3.3 Nuclear Extraction

The next step in the workflow is the generation of nuclear extracts (*see* **Note 8**). The following method is based on Dignam et al. and is suitable for cells harvested from multiple 15-cm dishes [6]. However, the method can be adjusted to smaller or larger scale cultures. The generation of nuclear extracts is a critical step in the procedure. Differential nuclear extraction of different batches of cells results in a variation of individual protein abundance and this introduces more noise into the experiment (i.e., background protein ratios strongly deviating from 1). Therefore, the different batches of cells should be extracted equally.

1. Wash the cells once with 15 ml of PBS.
2. Add 2 ml of trypsin–EDTA to each dish and incubate at 37 °C for about 5 min. Long incubation with trypsin results in cell lysis. Do not trypsinize more than 10 dishes simultaneously.
3. Neutralize trypsin with 10 ml of medium and collect the cells of each batch in a 50-ml tube. Keep the cells on ice from now on.
4. Rinse the dishes with PBS to collect the remaining cells.
5. Centrifuge the cells at 400 × *g* for 5 min at 4 °C and wash them with 20 ml of PBS. Centrifuge at 400 × *g* for 5 min at 4 °C again.
6. Resuspend the pellet with 15 ml of ice-cold PBS and transfer the cell suspension to a 15-ml tube. It is possible to leave the cells in this state for about 30 min.
7. Centrifuge at 400 × *g* for 5 min at 4 °C.
8. Aspirate the supernatant and estimate the volume of the pellet. Resuspend the pellet in 5 volumes of ice-cold buffer A.
9. Incubate the cell suspension on ice for 10 min. Cells swell during this incubation due to the osmotic uptake of water. The extent of swelling, however, varies between different cell lines. HeLa cells, for example, almost double their volume whereas HEK293T cells hardly swell.
10. Centrifuge for 5 min at 400 × *g* at 4 °C.
11. Aspirate the supernatant, resuspend cells in 2 volumes of buffer A containing 1× CPI and 0.15 % NP-40 and transfer the suspension to a dounce homogenizer. Keep the dounce homogenizer on ice at all times.
12. Dounce for 30–40 times. Wait for 30–60 s after every ten strokes of douncing to minimize the temperature increase due to friction.
13. Transfer the suspension into a 15-ml tube and centrifuge for 15 min at 3,200 × *g* at 4 °C.
14. The supernatant is the cytoplasmic extract, which can be aliquoted (*see* **Note 9**), snap-frozen and kept for other purposes. The pellet consists of crude nuclei.

15. Gently add 5 volumes of ice-cold PBS to the pellet and detach the pellet from the tube by flicking it a few times. Do not resuspend the pellet since this will result in partial lysis of the nuclei.
16. Centrifuge for 5 min at 3,200 × *g* at 4 °C. Discard the supernatant and remove any residual liquid by placing the tubes upside down on a tissue paper for 1 min.
17. Estimate the volume of the nuclei and add 2 volumes of buffer C containing 1× CPI, 0.1 % NP-40, and 0.5 mM DTT. Resuspend the nuclei and transfer the suspension to a microcentrifuge tube.
18. Homogenize the nuclei by pipetting up and down 10–15 times.
19. Incubate the suspension at 4 °C on a rotating wheel for 60 min.
20. Centrifuge the microcentrifuge tubes for 40 min in a pre-cooled tabletop centrifuge at maximum speed (about 17,900 × *g*) at 4 °C.
21. Aliquot the supernatant (100–200 μl fractions), which is the nuclear extract, snap-freeze in liquid nitrogen and store at −80 °C. Multiple freeze–thaw cycles should be avoided to preserve the quality of the nuclear extract. The pellet represents the insoluble chromatin fraction which could also be snap-frozen and kept for other purposes.

3.4 Bradford Assay for Protein Quantification

1. Dilute the extracts 1:10 in Milli-Q to a final volume of 20 μl.
2. Take 4 and 10 μl aliquots from the 1:10 dilution and add 1 ml of 1× Bradford protein assay solution to each sample.
3. Take 0, 1, 2, 5, 7, and 10 μl aliquots of 1 mg/ml BSA and add 1 ml of 1× Bradford protein assay solution to each sample.
4. Measure the samples at 595 nm using a spectrophotometer.
5. Generate a standard curve using the BSA samples and calculate the protein concentration of the extracts using linear regression. Note that the extracts were diluted 1:10 in **step 1**.

3.5 GFP Affinity Purification

The next step in the workflow is the GFP affinity purification (*see* **Notes 10** and **11**). Four extracts have been generated in Subheading 3.3: light and heavy WT extracts and light and heavy GFP extracts (*see* **Note 12**). These four extracts are incubated with GFP-binder beads and then combined after the incubation and wash steps to generate the forward (WT light + GFP heavy) and reverse (WT heavy + GFP light) experiment.

1. Add 15 μl of GFP-binder beads (from a 50 % slurry) into 4 microcentrifuge tubes. Cut the tips of 200 μl pipette tips when handling the agarose beads. Centrifuge the beads at 1,500 × *g* for 2 min after each wash and never vortex the beads.

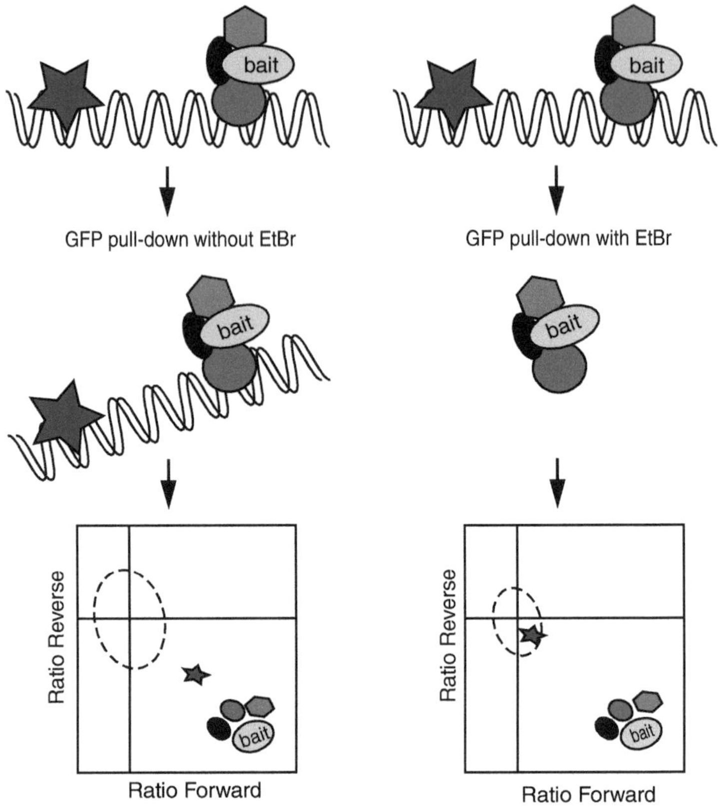

Fig. 2 Preventing DNA-mediated protein interactions. The presence of ethidium bromide during the GFP affinity purification eliminates co-purification of proteins bound to DNA in close proximity to the bait and its interactors. In addition, the cloud of background proteins is less tight in the absence of ethidium bromide

2. Wash the beads three times with 1 ml of buffer C* containing 1× CPI, 0.1 % NP-40, and 0.5 mM DTT (*see* **Note 13**). Beads are washed by adding buffer, inverting the tubes five times and centrifugation.
3. Aspirate the supernatant. For each of the four extracts, calculate the volume for 1 mg of protein. The affinity purification is performed in a final volume of 400 μl that contains the nuclear extract, buffer C* and ethidium bromide at a final concentration of 50 μg/ml (*see* **Notes 14** and **15,** Figs. 2 and 3). First add the necessary amount of buffer C* to the beads, followed by the ethidium bromide and finally the nuclear extract. If the amount of nuclear extract to be added exceeds 400 μl, the final volume can be increased accordingly but it should be kept constant for all the samples.
4. Incubate the samples on a rotating wheel for 90 min at 4 °C.
5. Centrifuge. Wash the beads twice with 1 ml of buffer C* containing 1× CPI, 0.5 % NP-40, and 0.5 mM DTT (*see* **Note 16**).

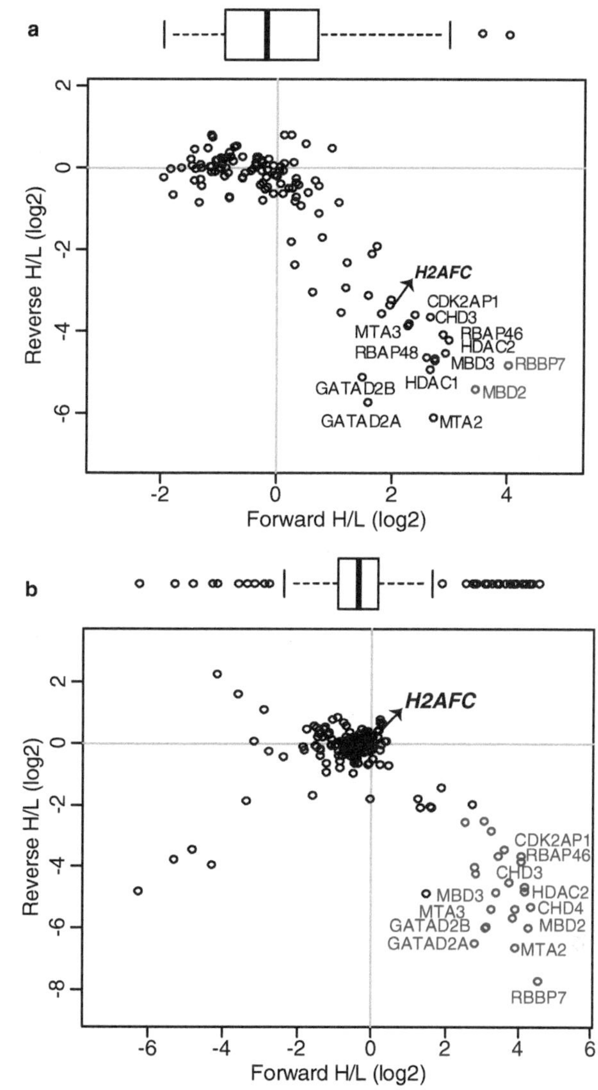

Fig. 3 GFP-CDK2AP1 affinity purifications in the absence or presence of ethidium bromide using wild-type and GFP-tagged CDK2AP1-expressing HeLa cells. In both panels, known interactors of CDK2AP1 (NuRD complex subunits) are indicated and statistically significant interactors are shown in *blue*. (**a**) The affinity purification was performed without ethidium bromide. A wide spreading of the background cloud and interactors is observed in the scatterplot. This spreading affects the boxplot statistics used for the significance calculation of interactors. In this pull-down, H2AFC, a histone variant, shows a high forward and low reverse ratio clustering with known CDK2AP1 interactors. (**b**) The affinity purification was performed in the presence of ethidium bromide. In this case, the background cloud is more compact and tightly clustered around the origin of the scatterplot. As a consequence, NuRD complex subunits, which are known interactors of CDK2AP1, are significant outliers. Note that in this pull-down, H2AFC is identified as a background protein

6. Wash the beads twice with 1 ml of PBS containing 0.5 % NP-40.
7. Wash the beads once with 1 ml of PBS.
8. Add 1 ml of PBS to the light WT beads and transfer the beads to the microcentrifuge tube containing the GFP heavy beads. This is the forward experiment. Combine the heavy WT and light GFP beads in a similar way. This is the reverse experiment.
9. Centrifuge the samples and aspirate the supernatant completely using a gel-loader tip.

3.6 On-Bead Digestion of Proteins

At the end of Subheading 3.5, two samples remain (forward pull-down and reverse pull-down). The next step is digesting the proteins off the beads with trypsin. This method is adapted from Hubner et al. [7, 8].

Always prepare urea solutions fresh before use and never cool the solutions to avoid urea precipitation. Furthermore, do not heat urea-containing samples to avoid adduct formation. Thaw and keep trypsin on ice at all times to minimize self-digestion.

1. Add 50 μl of digestion buffer to the samples.
2. Add DTT to a final concentration of 10 mM and shake the samples at 1,400 rpm on a thermoshaker for 20 min at RT.
3. Add IAA or CAA to a final concentration of 50 mM and incubate the samples for 20 min shaking at 1,400 rpm on a thermoshaker. Note that the stock solutions and IAA-containing samples need to be kept in the dark. CAA should be used if a posttranslational modification analysis (of ubiquitin in particular) will be performed afterwards. IAA can generate adducts on lysine residues that mimic ubiquitination [9].
4. Add 2.5 μl of trypsin to the beads. Incubate for 2 h on the shaker at 1,400 rpm at RT.
5. Centrifuge the samples at 2,000 × *g* for 2 min and transfer the supernatants into new microcentrifuge tubes (avoid taking up any beads). To the remaining beads, add 50 μl of digestion buffer and incubate at RT for 5 min while shaking.
6. Centrifuge the beads, collect the supernatant and add it to the one collected in the previous step for each sample.
7. Add 1 μl of trypsin to the combined supernatant and incubate overnight at RT.

3.7 Desalting and Purification of Peptides for Mass Spectrometry

Following digestion of the proteins, the tryptic peptides have to be desalted prior to mass spectrometry analysis. This method is adapted from Rappsilber et al. [10, 11]. In brief, C18 material inserted into a p200 pipette tip (referred to as StageTips) is used to capture and purify peptides. After each centrifugation step, check

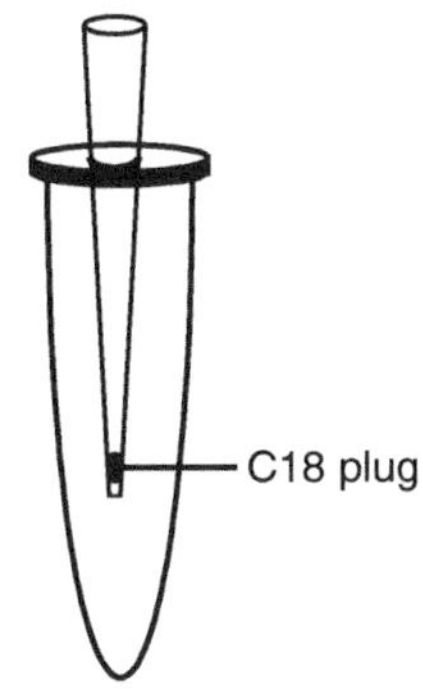

Fig. 4 Schematic representation of a p200 pipette tip with a C18 plug inserted (StageTip) mounted on a 2-ml microcentrifuge tube

the StageTips to ensure that all the liquid has flowed through before proceeding with the next step.

1. For each sample, puncture out a small disk of double-layered C18 material using a blunt-ended syringe needle and transfer the C18 material into a p200 pipette tip. Push the disk into the tip and fix it at the end but do not apply too much pressure.
2. Insert the tip into a microcentrifuge tube with a hole in its cap (*see* Fig. 4). The pipette tip on the tube should be stable enough but it should not touch the bottom of the tube where the flow-through solvents will be collected.
3. Activate the C18 material by adding 50 μl of methanol on top of it and centrifuge at 1,500 × *g* for 3–5 min in a tabletop centrifuge. Check the tip to make sure that all the methanol has flowed through.
4. Add 50 μl of buffer B to the tip and centrifuge at 1,500 × *g* for 3–5 min. Discard the flow-through.
5. Load 50 μl of buffer A to the tip and centrifuge at 1,500 × *g* for 3–5 min.
6. Repeat **step 5**.
7. To each of the samples that were digested overnight with trypsin (Subheading 3.6), add 10 μl of 10 % TFA and resuspend. Centrifuge the samples at 2,000 × *g* for 2 min. Make sure not to take up any residual beads that may have been carried over when supernatants were collected in the previous section.
8. Load the samples onto the (labeled) StageTips and centrifuge at 400 × *g* for 10–15 min. It is important to centrifuge more slowly during this step to ensure efficient binding of peptides to the C18 plug.
9. When all the sample has flowed through the StageTip, load 50 μl of buffer A and centrifuge at 1,500 × *g* for 3–5 min. The StageTips can be stored at 4 °C for months at this point.

3.8 Elution of Peptides

1. When proceeding directly after desalting and concentration of peptides using StageTips, start from **step 2** of this section. If the samples have been stored at 4 °C on StageTips, rehydrate StageTips by loading 30 μl of buffer A and centrifuge at 1,500 × *g* for 3–5 min.
2. Elute the peptides into a 96-well autosampler plate that is compatible with the nanoHPLC connected to the mass spectrometer. To elute, add 30 μl of buffer B and fit an air-filled Eppendorf combitip to the back of the StageTip. Apply pressure on the combitip to slowly force the solvent through the StageTip directly into the autosampler plate.
3. Concentrate the eluted samples until the volume is about 5 μl using the speed vacuum concentrator. Add 7 μl of buffer A to the samples and transfer them to the autosampler plate of the nanoHPLC.

3.9 Mass Spectrometry

The following section provides guidelines for measuring GFP affinity purification samples by nanoHPLC-MS/MS, e.g., an EASY nLC coupled to an LTQ-Orbitrap Velos mass spectrometer. Expertise is required to operate these machines, therefore the following steps should be performed under the supervision of an experienced mass spectrometrist.

1. Program for an injection volume of 5 μl of the sample into the nanoHPLC.
2. Peptides are eluted from the nano-column packed with C18 using a 5–30 % acetonitrile (v/v) gradient followed by a sharp increase to 60 % acetonitrile in 10 min with a flow rate of 250 nl/min. Total elution time is around 120 min, but this can be increased if the sample is very complex.
3. The recommended settings for data acquisition are: for precursor MS spectra, *m/z* range 300–1,750 with a resolution of 60,000 and a target value of 1 million ions per full scan. For MS/MS spectra, CID is selected as the fragmentation method and the 15 most intense precursor ions are selected for fragmentation from each full MS scan with a minimal ion count target value of 500. Dynamic exclusion is set to 30 s (both repeat duration and exclusion duration), list size 500, early expiration enabled (count 2, S/N threshold 2). MS/MS scans are acquired in the centroid mode in the dual pressure linear ion trap with a normalized collision energy of 35 %.

3.10 Data Analysis

After data acquisition is complete, raw data are transferred to a Windows-operated PC. We use the MaxQuant software package to analyze the raw data which can be freely downloaded at www.maxquant.org [11]. The software also contains a module for downstream data analysis (statistical tests, filtering, clustering, GO term enrichments, etc.) called Perseus. New versions of the software are continously being generated and made available at the Web site. There is also an active Google group (MaxQuant) for posting questions.

1. Open the MaxQuant.exe program.
2. Load the raw data files (using load files option) that were copied to a local disk in a separate data folder containing only the files that are to be analyzed together (in this case, only the raw data files of the forward and the reverse experiment). Do not use spaces when creating data folders and raw files since this generates an error while running MaxQuant.
3. Click on "Exp. Design". Then use Excel to open the "experimentalDesignTemplate.txt" that is written into the "combined" folder within the data folder. Specify the forward and reverse experiments in the "experiment" column and save the changes.
4. Go to the "Group-specific parameters" tab and check the protease that is used (trypsin(P)), the multiplicity (2) and the labeled amino acids (Arg10 and Lys8). If necessary, adjust these options according to the experiment.
5. Go to the "MS/MS & Sequences" tab and upload the FASTA file database of the appropriate organism. These FASTA files can be downloaded at the MaxQuant Web site.
6. Go to the "Identification and quantification" tab and upload the "experimentalDesignTemplate.txt".
7. Go to the "Misc." tab and click the "Match between runs" and "Re-quantify" options.
8. Start the analysis by clicking on "Start".

At the end of the analysis, the forward and reverse ratios of proteins are reported in "ProteinGroups.txt" (within the txt folder that is written into the data folder). Using Perseus (or other programs such as R, MatLab etc.), the contaminants, reverse hits and proteins with less than 3 reported peptide ratios (Ratio H/L count) can be filtered out. After this step, log-transform (base 2) the protein ratios and calculate significance B (in Perseus). Significance B indicates the probability of a protein being a significant outlier from the background cloud based on intensity and ratio. After these steps, plot the log2-transformed normalized forward and reverse ratios against each other. The nonspecific background binders (with 1:1 ratios in both experiments) cluster around the origin of the graph while the specific interactors have high forward and low reverse ratios (*see* Fig. 1). Contaminant proteins (such as keratins and serum proteins) are easily distinguished since they have a low forward and a low reverse ratio (as they are not SILAC-labeled).

4 Notes

1. When a cell line is being labeled for the first time, it is recommended to perform a SILAC label incorporation check before proceeding with large scale experiments. This can be done by digesting a small amount of heavy-labeled nuclear or whole cell extract (10 μg) with trypsin using standard in solution digestion protocols or the filter-aided sample preparation (FASP) method followed by LC-MS/MS [12]. If incorporation efficiency is lower than 95 %, the maximum observable ratios decrease significantly (i.e., 90 % incorporation results in a maximum ratio of 9, at 80 % the maximum ratio is 4, etc.). The minimum recommended incorporation efficiency is 95 %. Note that the labeling efficiency will never reach 100 % due to trace amounts of non-labeled amino acids in the culture medium. Another potential problem is arginine-to-proline conversion, which can be investigated during the incorporation check. When a heavy arginine (Arg10) is converted to proline, this proline is 6 Da heavier compared to normal proline. For heavy SILAC-labeled peptides containing one or more internal prolines, a third isotope cluster appears in the mass spectrum. The intensity of this third isotope cluster is not taken into account during quantification and this results in an underestimation of the peptide ratio. A small percentage of arginine-to-proline conversion can easily be normalized for, but as a rule of thumb, heavy proline peaks should not have an intensity of more than 5 % of the normal heavy peak. Some cell lines are more prone to this problem than others and in some cases, titrating the amount of arginine and proline in the SILAC medium can reduce the amount of conversion. However, proline to arginine conversion also occurs and this potentially compromises arginine labeling efficency. Therefore, the titration should be performed carefully. As a last resort, lysine-only labeling can be used to circumvent this problem. In this case, LysC instead of trypsin is used to digest the proteins.
2. When working with a stable cell line expressing a GFP-tagged protein of interest, it is recommended to SILAC-label this cell line and the WT control cells simultaneously. Both lines will be labeled light and heavy. In the end, four batches of cells are harvested: light and heavy WT cells and light and heavy GFP cells. While culturing and harvesting, it is important to handle all these batches of cells as reproducibly as possible in order to minimize variations between the extracts.
3. For suspension cells, SILAC RPMI medium lacking lysine, arginine and glutamine is available. For labeling of mouse embryonic stem cells (mESCs), SILAC DMEM (with 15 %

D-FBS, L-glutamine, and penicillin–streptomycin) can be used with the necessary additions: 1× nonessential amino acids, 1 mM sodium pyruvate, lysine and arginine (light or heavy in amounts described), 1,000 U/ml LIF, 4.2 μl of 99 % (v/v) β-mercaptoethanol, and 2i inhibitors (3 μM and 1 μM of CHIR99021 and PD0325901 respectively). 2i inhibitors consist of small molecule inhibitors of GSK-3 and ERK1/2 signaling and thus prevent differentiation [13]. Note that the mESCs should not be grown on feeder cells since these are not SILAC-labeled. Several mESC lines such as IB10 and R1 can grow in the absence of feeder cells when culture dishes are coated with 0.15 % gelatin. Instead of trypsin–EDTA, accutase is used for detaching the cells which is a mixture of proteolytic and collagenolytic enzymes effective in detaching primary fibroblasts, neurons, endothetial cells, and ESCs [14].

4. The amount of labeled amino acids to be added to the medium is optimized for SILAC labeling of commonly used cell lines such as HeLa, HEK293T, and MCF7. These amounts may need to be altered for labeling other cell lines.

5. We prefer to use cell lines stably transfected with a bacterial artificial chromosome (BAC) construct containing the gene of interest [15]. These BACs have been recombined to express the protein of interest with a GFP tag. Since the BAC is expected to contain the proximal regulatory regions for expression, the expression level of the GFP-tagged bait is at near endogenous levels. In this workflow, the parental WT cells are used as control cell line. An alternative approach is to generate a stable cell line in which the expression of the GFP-tagged bait is inducible. In this case, uninduced cells serve as the control. If no stable cell line is available, the GFP-tagged bait can be transiently over-expressed. However, the disadvantage of transient transfection is that the expression level of the GFP-tagged bait cannot be controlled. Although the expression level of the bait depends on the strength of the promoter in the plasmid, it usually exceeds endogenous levels. While this may not necessarily be problematic, a gross over-expression may induce false positive interactions. In addition, when the majority of the bait is not associated with interactors due to over-expression, this may compromise the depth of sample sequencing, thereby reducing the identification of substoichiometric interactors.

6. The confluency required at the time of transient transfection depends on the proliferation rate of the cells. It is recommended to keep the cells in culture after transfection for at least 24 h and at most 48 h. The confluency at the time of transfection should be adjusted accordingly to prevent overgrowth of the cells.

7. Usually 10–20 μg of DNA per 15-cm dish is used for transfection. The amount may be reduced in order to lower the amount of GFP-tagged bait in the nuclear extract in case of very strong over-expression.
8. When working with nuclear proteins, making nuclear extracts is highly recommended since this is an efficient way to get rid of a lot of highly abundant cytoplasmic proteins. This step decreases the background in the pull-down and in the mass spectrometer, facilitating the sequencing and identification of the GFP-tagged bait and interactors. The affinity purification method described here, however, is not restricted to nuclear extracts and can also be applied in combination with cytoplasmic or whole cell extracts. In these cases, the stringency of the buffer used during extraction and of Buffer C* used during the affinity purification should be increased and more input material should be used.
9. The cytoplasmic extract should be snap-frozen after adding glycerol to a final volume of 10 % and NaCl to a final concentration of 150 mM.
10. In this chapter, the GFP tag is used for affinity purification. Although GFP is a large protein tag compared to other commonly used tags such as Flag, HA, etc., tagging proteins with GFP rarely results in a non-functional fusion protein [15]. In our hands, the GFP tag works well and does not often interfere with PPIs. This is true even when tagging very small proteins [16]. Depending on the domain structure of the bait, N- or C-terminal tagging may be preferred.
11. The GFP-binder beads used for affinity purification only bear the epitope recognition domain of a high-affinity monoclonal GFP antibody. This small, 13 kDa fragment is expressed in *E. coli* and then covalently cross-linked to agarose beads [17]. The result is a high-affinity enrichment resin lacking the heavy and light immunoglobin chains present in conventional antibodies. When using conventional antibodies for affinity purification, the heavy and light chains are digested together with the affinity-purified proteins and dominate the MS spectra due to their abundance. In that case, a specific elution protocol (peptide elution or acidic elution) or in-gel digestion is recommended instead of on-bead digestion [18].
12. When a "wild-type" cell line (the parental cell line that was used to generate the transgenic cell line) is not available, it is possible to use the nuclear extracts from cells expressing the GFP-tagged transgene in combination with blocked agarose beads (BAB) as a negative control. These BAB are commercially available (*see* Subheading 2.5). This may also be necessary when expression of the GFP-tagged transgene significantly changes the growth rate and/or morphology of the parental cell line. In this case, the experimental setup changes. Instead of four different cultures, only the GFP-tagged cell line is

labeled light and heavy but expanded to twice the amount. Four affinity purifications are then performed as follows: light and heavy nuclear extracts incubated with GFP-binder beads and light and heavy nuclear extracts incubated with BAB. The forward experiment becomes heavy GFP-binder pull-down + light BAB pull-down and the reverse experiment becomes light GFP-binder pull-down + heavy BAB pull-down.

13. Note that the buffer C* used for affinity purification contains 300 mM NaCl as opposed to the buffer C (420 mM NaCl) used for nuclear extraction. At the last step of nuclear extraction, addition of two volumes of buffer C with 420 mM NaCl to the crude nuclei results in a final salt concentration of approximately 300 mM. Since the protein concentrations of different batches of nuclear extracts may vary, the volume of nuclear extract to be used for affinity purification may also vary (in order to use the same amount of protein). By using a buffer with 300 mM salt for the GFP pull-down, the final salt concentration is kept equal in all affinity purifications.
14. Using 1 mg of nuclear extract is usually sufficient to identify the bait and interactors. For low-expressed bait proteins or when using whole cell lysates, this amount can be increased to a maximum of 5 mg. The amount of beads used during the pull-down need not be changed.
15. Using ethidium bromide is very critical to eliminate DNA-mediated indirect interactions. Although the bulk of the DNA ends up in the insoluble chromatin fraction during nuclear extraction, the nuclear extract still contains a fair amount of double-stranded DNA. In the absence of ethidium bromide, the pull-downs are more "noisy" and more proteins are identified as interactors of the bait, many of which are false positive, DNA-mediated secondary interactions (*see* Figs. 2 and 3). As an alternative, DNA and RNA in the extract can be digested with an endonuclease such as MNase prior to the pull-down.
16. The amount of detergent and salt in the wash buffer may be altered. To identify low affinity or sub-stoichiometric interactors, the final NP-40 concentration can be decreased to 0.2 %; to identify very strong interactors, it can be raised to 1 %. The key point is to find a balance between specificity and sensitivity, and this may vary depending on the bait.

Acknowledgments

We thank members of the Vermeulen group and Nina Hubner for critical reading of the manuscript and fruitful discussions. Work in the Vermeulen group is supported by grants from the Netherlands Organisation for Scientific Research (VIDI and Cancer Genomics Centre (CGC)) and the Dutch Cancer Society (KWF).

References

1. Fields S, Song O (1989) A novel genetic system to detect protein-protein interactions. Nature 340:245–246
2. Gingras AC, Gstaiger M, Raught B et al (2007) Analysis of protein complexes using mass spectrometry. Nat Rev Mol Cell Biol 8: 645–654
3. Altelaar AF, Heck AJ (2012) Trends in ultrasensitive proteomics. Curr Opin Chem Biol 16:206–213
4. Vermeulen M, Hubner NC, Mann M (2008) High confidence determination of specific protein-protein interactions using quantitative mass spectrometry. Curr Opin Biotechnol 19:331–337
5. Ong SE, Blagoev B, Kratchmarova I et al (2002) Stable isotope labeling by amino acids in cell culture, SILAC, as a simple and accurate approach to expression proteomics. Mol Cell Proteomics 1:376–386
6. Dignam JD, Lebovitz RM, Roeder RG (1983) Accurate transcription initiation by RNA polymerase II in a soluble extract from isolated mammalian nuclei. Nucleic Acids Res 11: 1475–1489
7. Hubner NC, Bird AW, Cox J et al (2010) Quantitative proteomics combined with BAC TransgeneOmics reveals in vivo protein interactions. J Cell Biol 189:739–754
8. Hubner NC, Mann M (2011) Extracting gene function from protein-protein interactions using Quantitative BAC InteraCtomics (QUBIC). Methods 53:453–459
9. Nielsen ML, Vermeulen M, Bonaldi T et al (2008) Iodoacetamide-induced artifact mimics ubiquitination in mass spectrometry. Nat Methods 5:459–460
10. Rappsilber J, Mann M, Ishihama Y (2007) Protocol for micro-purification, enrichment, pre-fractionation and storage of peptides for proteomics using StageTips. Nat Protoc 2: 1896–1906
11. Cox J, Matic I, Hilger M et al (2009) A practical guide to the MaxQuant computational platform for SILAC-based quantitative proteomics. Nat Protoc 4:698–705
12. Wisniewski JR, Zougman A, Nagaraj N et al (2009) Universal sample preparation method for proteome analysis. Nat Methods 6: 359–362
13. Lee KH, Chuang CK, Guo SF et al (2012) Simple and efficient derivation of mouse embryonic stem cell lines using differentiation inhibitors or proliferation stimulators. Stem Cells Dev 21:373–383
14. Zhang P, Wu X, Hu C et al (2012) Rho kinase inhibitor Y-27632 and Accutase dramatically increase mouse embryonic stem cell derivation. In Vitro Cell Dev Biol Anim 48:30–36
15. Poser I, Sarov M, Hutchins JR et al (2008) BAC TransgeneOmics: a high-throughput method for exploration of protein function in mammals. Nat Methods 5:409–415
16. Spruijt CG, Bartels SJ, Brinkman AB et al (2010) CDK2AP1/DOC-1 is a bona fide subunit of the Mi-2/NuRD complex. Mol Biosyst 6:1700–1706
17. Rothbauer U, Zolghadr K, Muyldermans S et al (2008) A versatile nanotrap for biochemical and functional studies with fluorescent fusion proteins. Mol Cell Proteomics 7:282
18. Paul FE, Hosp F, Selbach M (2011) Analyzing protein-protein interactions by quantitative mass spectrometry. Methods 54:387

Chapter 16

Analyzing the Protein Assembly and Dynamics of the Human Spliceosome with SILAC

Carla Schmidt, Monika Raabe, Reinhard Lührmann, and Henning Urlaub

Abstract

Quantitative mass spectrometry has become an indispensable tool in proteomic studies. Numerous methods are available and can be applied to approach different issues. In most studies these issues include the quantitative comparison of different cell states, the identification of specific interaction partners or determining degrees of posttranslational modification. In this chapter we describe a SILAC-based quantification in order to analyze dynamic protein changes during the assembly of the human spliceosome on a pre-mRNA in vitro. We provide protocols for assembly of spliceosomes on pre-mRNA (including generation of pre-mRNAs and preparation of nuclear extracts), quantitative mass spectrometry (SILAC labeling, sample preparation), and data analysis to generate timelines for the dynamic protein assembly.

Key words Spliceosome, Metabolic labeling, Assembly timelines, Protein dynamics

1 Introduction

1.1 The Spliceosome

1.1.1 Eukaryotic Pre-mRNAs

Eukaryotic pre-mRNAs consist of protein-coding sequences (exons) and noncoding sequences (introns). The introns are defined by very short, conserved sequences at the 5′ and 3′ splice sites (i.e., the exon/intron and intron/exon junctions) as well as the branch-point site, which contains a conserved adenosine (branch-point adenosine) and in most cases a polypyrimidine tract (Y_n; Fig. 1a).

During pre-mRNA splicing, the introns are excised and the exons are ligated to yield mature mRNA. This proceeds by two consecutive transesterification reactions (Fig. 1b): First, the 2′ hydroxyl group of the branch-point adenosine attacks the phosphodiester bond at the 5′ splice site (5′ss) resulting in a phosphodiester bond between the branch-point adenosine and the first nucleotide of the intron. In the second step of splicing, the free 3′ hydroxyl group of exon1 attacks the phosphodiester bond of the 3′ splice site (3′ss); in this way, exon 1 and exon 2 become ligated and the intron is released in the form of a lariat (Fig. 1b) [1–5].

Bettina Warscheid (ed.), *Stable Isotope Labeling by Amino Acids in Cell Culture (SILAC): Methods and Protocols*, Methods in Molecular Biology, vol. 1188, DOI 10.1007/978-1-4939-1142-4_16, © Springer Science+Business Media New York 2014

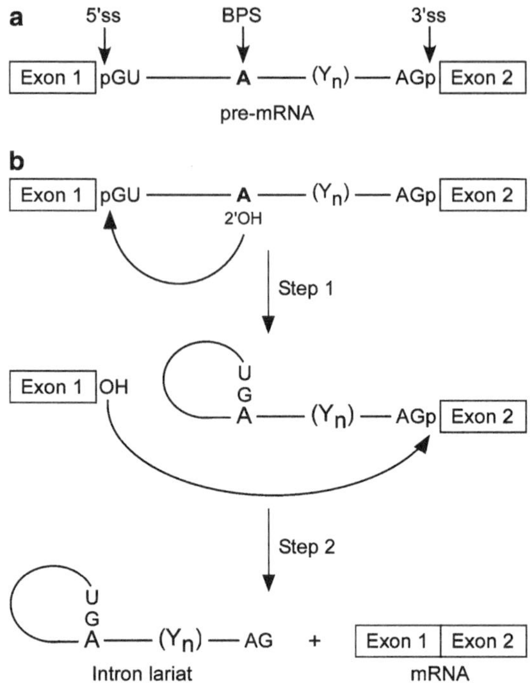

Fig. 1 The two steps of eukaryotic pre-mRNA splicing. (**a**) The arrangement of exons and introns in eukaryotic pre-mRNA (for details see text). (**b**) The splicing reaction follows two transesterification steps, after which the spliced mRNA (Exon1–Exon2) and the intron-lariat are released

1.1.2 U snRNPs and Non-snRNP Splicing Factors

The process of eukaryotic pre-mRNA splicing is highly dynamic and is catalyzed by the spliceosome—a multi-megadalton machine that assembles from the so-called U snRNPs (uridine-rich small nuclear ribonucleoprotein particles) as well as non-snRNP proteins. There are five spliceosomal U snRNPs: the U1, U2, U4, U5, and U6 snRNPs and also the U4/U6 di-snRNP and the U4/U6.U5 tri-snRNP. All U snRNPs consist of a specific uridine-rich RNA (U snRNA) and a particle-specific set of proteins (for review *see* [6]).

Common to all U snRNPs (except U6) are seven Sm proteins (E, F, G, D1, D2, D3, and B/B′), which form a ring-shaped heptamer and bind the U snRNA at the Sm site via a Sm motif [7–9]. The U6 snRNA instead associates with a group of related proteins, called Sm-like proteins (LSm2–LSm8). They also form a heptameric ring and bind the U6 snRNA at the 3′ end [10, 11].

In addition, every U snRNP contains a set of particle specific proteins: The U1 snRNP contains the U1-A, U1-70K, and U-1C proteins. Of these, U1-C is important for splicing activity, as it directly contacts the pre-mRNA near the 5′ss (*see* Fig. 1a) stabilizing snRNA–pre-mRNA interactions [12, 13]. The 12S U2 snRNP contains in addition to the Sm proteins two additional proteins U2-A′ and U2-B″, and the splicing active U2 snRNP, was found two contain two further heteromeric splicing factors—SF3a and

SF3b, composed of three and five proteins, respectively [14–16]. SF3a and SF3b proteins contact the pre-mRNA near the branch point site (*see* Fig. 1a) and are essential for the spliceosomal assembly [17, 18]. The U5 snRNP contains eight U5 specific proteins with an apparent molecular weight of 15, 40, 52, 100, 102, 116, 200, and 220 kDa [19]; most of these are involved in structural rearrangements in the first step of splicing (reviewed by [20]. The U5 snRNP is recruited to the spliceosome after tri-snRNP formation with U4/U6 di-snRNP comprising U4 and U6 snRNAs, Sm and LSm proteins, and five U4/U6 specific proteins (Fig. 2). The thus formed U4/U6.U5 tri-snRNP contains all U4/U6 and U5 proteins (except U5-52K) and three additional proteins (110K, 65K, and 27K proteins), which are required for integration into the spliceosome [21]. In addition to the snRNP-specific proteins, additional non-snRNP protein components play important roles in pre-mRNA splicing. The hPrp19/CDC5L complex is one of them. It consists of seven proteins (CDC5L, Hsp70, CTNNBL1, PRL1, hPrp19, AD-002, and SPF27; [22, 23]) and associates with additional related proteins with the U5 snRNP to form the remodeled 35S U5 [23] during activation of the spliceosome (Fig. 2). It is suggested to play a crucial role in the assembly of a catalytically active spliceosome, presumably by stabilizing the RNA interaction network in the catalytic core [22].

Several other splicing factors belong to the DExD/H-box protein family: These proteins are able to rearrange RNP and RNA–RNA interactions and are therefore required for structural rearrangements. Some belong to U snRNP specific proteins (e.g., U5-220K, U5-100K, see above), whereas others are non-snRNP specific (e.g., hPrp5, UAP56, hPrp2). Furthermore, other proteins that have been found to be specific for the different spliceosomal transition states. For instance, Prp16, Prp17, Prp18, Prp22, and Slu7 (reviewed by [24]) have been reported to bind the spliceosome after the first step of splicing to function at the second step of splicing—these are so-called second step splicing factors.

Complete pre-mRNA splicing in higher eukaryotes hence requires several processes: (1) the ordered assembly of the U snRNPs on the pre-mRNA in concert with the recruitment of additional non-snRNP splicing factors, (2) the partial dissociation of these factors upon rearrangement of the gross structure of the spliceosome and the accompanying disruption and formation of protein–protein, RNA–RNA, and protein–RNA binding, (3) the completion of the two transesterification steps, and (4) the release of the mature mRNA generated (for detailed review *see* [25]).

1.1.3 Assembly of the Spliceosome

The spliceosome assembles on the pre-mRNA in a stepwise manner, passing through a series of functional intermediates. The various states are outlined for the human spliceosome in Fig. 2. In the first assembly step, the U1 snRNP binds to the 5′ss of the

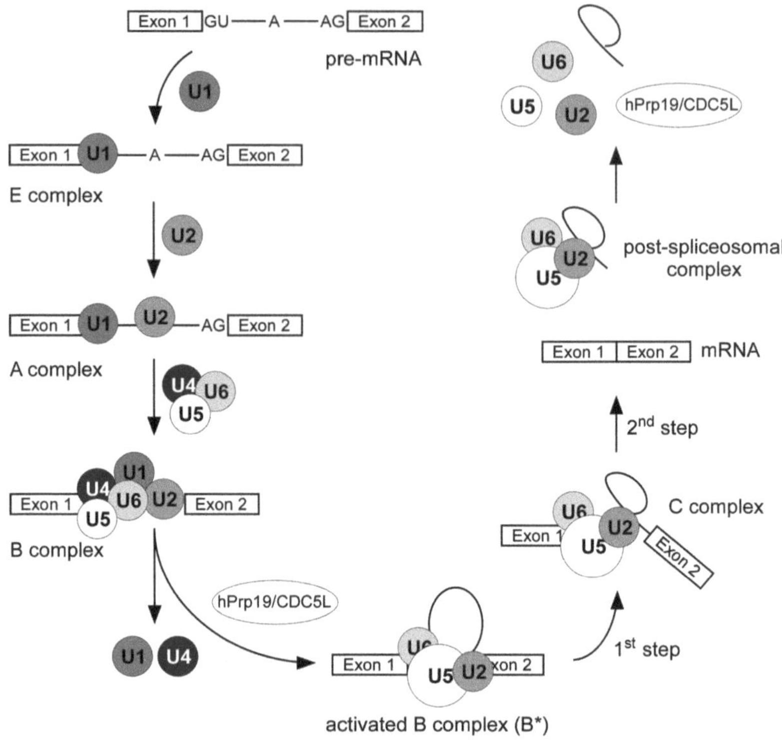

Fig. 2 The stepwise assembly of the spliceosome during pre-mRNA splicing. First, U1 snRNP binds to the 5′ss forming the E complex followed by binding of U2 snRNP (A complex formation). Recruitment of the pre-assembled tri-snRNP (U4/U6.U5) leads to formation of the pre-catalytic B complex. Upon structural rearrangements U1 and U4 snRNPs dissociate and incorporation of the hPrp19/CDC5L complex leads to remodeling of U5 generating the activated B complex. The first step of splicing occurs in this intermediate assembly yielding the C complex, in which the second step of splicing is carried out. The generated mRNA and the post-spliceosomal complex are released and the splicing factors are reconstituted. In addition to the protein complex hPrp19/CDC5L, numerous non-snRNP specific proteins, not shown here, join and leave the spliceosome at various points during the cycle

pre-mRNA, forming the E ("early") complex [12, 26]. The recruitment of U2 snRNP leads then to formation of the A complex, which is also called the pre-spliceosome [27, 28]. Upon integration of the U4/U6.U5 tri-snRNP and additional splicing factors, called B specific proteins, the pre-catalytic spliceosome (B complex) is developed [29]. Structural RNA and protein rearrangements within the B complex induced by RNA helicases Brr2 (U5-200K) and Snu114 (U5-116K) cause the dissociation of U1 and U4 snRNPs. Dissociation of U1 and U4 together with U4/U6 specific proteins and remodeling of U5 initiated by the binding of the hPrp19/CDC5L complex generate the activated spliceosome (B*), in which the first catalytic step of splicing occurs [30, 31]. The complex that forms during this process is the catalytically active C complex, which goes on to perform the second step of splicing for which the second step splicing factors (see above) are required [32]. The final steps are the release of the mature mRNA product, dissociation of the post-spliceosomal intron complex, and recycling of the splicing factors (Fig. 2).

1.2 Analyzing Dynamic Protein Changes by Quantitative Mass Spectrometry

So far, only few studies have applied quantitative mass spectrometry combined with metabolic labeling using stable isotopes to describe dynamic protein changes in ribonucleoprotein complexes. In a first study, the SILAC strategy was applied to analyze the proteome of the nucleolus from differentially labeled cells after different durations of treatment [33]. The time-dependent composition profiles of protein subunits from RNA polymerase I, snRNPs and ribosomes were recorded [33]. In a similar manner, the assembly kinetics of the 30S ribosomal subunit of *Escherichia coli* have been studied by quantitative pulse-chase MS (PC/QMS). ^{15}N-labeled proteins were incubated with 16S rRNA and, after assembly had taken place for various times, chased with an excess of ^{14}N-labeled proteins. 30S subunits were then completely assembled and purified, and the ^{15}N/^{14}N ratio in their proteins was used to reveal the binding kinetics [34]. In this chapter we describe the quantitative analysis of the dynamic protein changes that occur during pre-mRNA splicing by using stable isotope labeling and subsequent mass spectrometry.

1.2.1 Protein Assembly and Dynamics of the Human Spliceosome

The distinct assembly states of the human spliceosome (i.e., A, B, B*, and C complexes, see above) in vitro have been analyzed in previous studies and compared in a semi-quantitative manner to determine differences in their protein compositions [27, 29, 30, 32, 35]. However, this approach only monitors the quantitative changes of the protein composition during the transition of one purified state of the spliceosomes to another. Yet no description of the dynamic protein changes that occur during assembly of proteins pre-mRNA splicing in a time dependent manner has been applied. We therefore used SILAC quantification to monitor the protein assembly on a pre-mRNA in a time-dependent manner.

We used MS2-tagged PM5 pre-mRNA, which in previous studies had been successfully applied to purify catalytically active spliceosomes [32], and a splicing-inactive variant of this pre-mRNA, which was generated by deletion of the 5′ss. A direct comparison was made between the assembled proteins on the splicing-active and on the splicing-inactive pre-mRNA at different time points during pre-mRNA splicing (Fig. 3).

For this purpose, we prepared HeLa nuclear extracts from differentially labeled HeLa cells (light and heavy SILAC cells; Table 1).

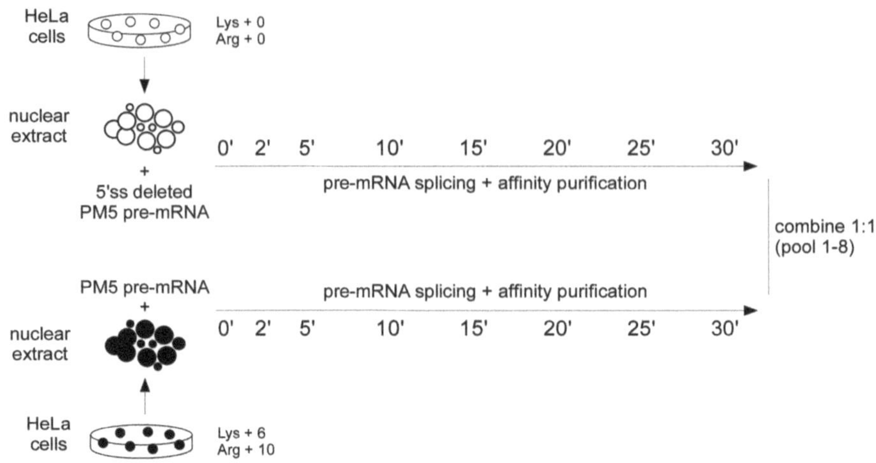

Fig. 3 Experimental setup to monitor protein dynamics during pre-mRNA splicing. SILAC nuclear extracts were prepared from differentially labeled HeLa cells (light and heavy; the additional masses arising from the isotopically labeled lysine and arginine are indicated on the *left*). Splicing reactions were assembled on PM5 and 5′ss deleted PM5 pre-mRNA using light and heavy nuclear extracts, respectively. Assembled complexes from the same time point but assembled on the different pre-mRNAs were pooled in equal amounts

Table 1
Amino acids required in order to prepare duplex SILAC nuclear extracts

		L-Arginine	Δm	L-Lysine	Δm
duplex SILAC	Light	–	0	–	0
	Heavy	$^{13}C_6{}^{15}N_4$	+10 Da	$^{13}C_6$	+6 Da

The light cells (and corresponding nuclear extracts) are prepared by using "normal" (non-labeled, i.e., ^{12}C-, ^{14}N-, ^{1}H-containing) amino acids

Pre-mRNA splicing was then performed using the two pre-mRNAs (splicing-active and splicing-inactive) and the two SILAC nuclear extracts (Fig. 3). The splicing reaction was stopped at different time points and the assembled complexes were affinity-purified by using the MS2-tag that was present on both pre-mRNAs. The samples to be compared were pooled in the same amounts and the proteins were separated by gel electrophoresis. The proteins were digested in-gel and the peptides generated were analyzed by LC-online MS/MS. Peptide and protein ratios were obtained by using the MaxQuant software [36]. Protein ratios were normalized by calculating a normalization factor for each time point from background protein ratios, as these is supposed to be present in equimolar amounts on both pre-mRNAs. The protein assembly of distinct spliceosomal protein groups was displayed by plotting their average normalized protein ratios against time.

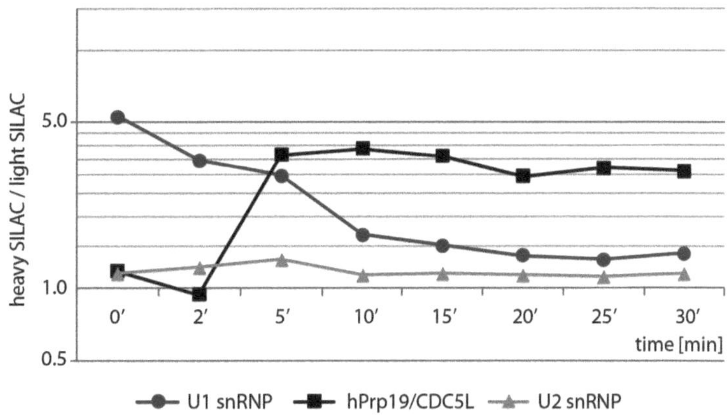

Fig. 4 Protein assembly timelines for spliceosomal protein groups. Assembly on the splicing-active (heavy SILAC nuclear extract) and splicing-inactive (light SILAC nuclear extract) PM5 pre-mRNA was directly compared. Assembly timelines for U1 snRNP, U2 snRNP, and hPrp19/CDC5L specific proteins are shown. Differences in the assembly on the two pre-mRNAs are observed. The U1 snRNP proteins are more abundant on the splicing-active PM5 pre-mRNA than on the splicing-inactive pre-mRNA during the first time points of pre-mRNA splicing. Their SILAC protein ratios decrease after approximately 5 min clearly demonstrating that they dissociate from the pre-mRNA. The U2 snRNP proteins instead do not show any differences in their assembly on the two pre-mRNAs. They bind to the branch-point site of the pre-mRNA and are thus not affected by the 5′ss deletion. They show constant protein ratios of approximately 1:1 over the whole time frame, meaning that they are present to the same extent on both pre-mRNAs. The hPrp19/CDC5L complex proteins do not show differences in their assembly on the two pre-mRNAs for the first time points represented by their protein ratios of 1:1. After 5 min of splicing their protein ratios increase significantly showing that they associate with the spliceosome after 5 min. This is in agreement with their role in the assembly of the activated spliceosome. Increased protein ratios of members of the hPrp19/CDC5L also reveal that they are more abundant on the splicing-active pre-mRNA indicating that catalytically active spliceosomes are not formed on the splicing-inactive pre-mRNAs

In this chapter, we show some examples of the kinetics of the protein assembly on the two pre-mRNAs analyzed for spliceosomal protein groups that are affected by deletion of the 5′ss. We compared the assembly dynamics for the U1, U2 snRNPs and the hPrp19/CDC5L complex proteins, all of which are components of different spliceosomal intermediate states (see above). Our results demonstrate that, differences in the protein assembly on the two pre-mRNAs were identified by direct comparison (Fig. 4). The U1 snRNP is specific for the pre-catalytic spliceosomal complexes (Fig. 2) and, at early time points, its proteins are more abundant on the splicing-active pre-mRNA than on the splicing-inactive pre-mRNA as obtained from their high SILAC protein ratios between zero and 5 min (Fig. 4). For the U2 snRNP proteins, which are present in all intermediate states (Fig. 2), no differences were observed between the two pre-mRNAs (Fig. 4).

Interestingly, the hPrp19/CDC5L complex, which the spliceosome incorporates during its activation (see above and Fig. 2), assembles at later time points only on the splicing-active pre-mRNA; the SILAC protein ratios increased significantly after 2 min of incubation for this group of proteins. Thus, our quantitative MS results demonstrate that the different protein groups indeed show different assembly kinetics, and they also show which proteins are affected by the deletion on the 5′ss. The timelines generated for the assembly of whole spliceosomal protein groups during pre-mRNA splicing thus contribute substantially toward gaining an understanding of this dynamic process.

2 Materials

2.1 SILAC Labeling of HeLa Cells

1. HeLa S3 cells (wt).
2. DMEM, high glucose, w/o arginine, w/o lysine.
3. Dialyzed fetal bovine serum (FBS).
4. 100× penicillin–streptomycin.
5. 50 mg/l L-arginine, 50 mg/l $^{13}C_6{}^{15}N_4$-L-arginine.
6. 50 mg/l L-lysine, 50 mg/l $^{13}C_6$-L-lysine.
7. 2.0 L spinner flasks.
8. 2.5 L fermenter (bioreactor).
9. *See* also **Notes 1–3** and Table 1.

2.2 Preparation of SILAC HeLa Nuclear Extracts

1. Phosphate-buffered saline (PBS): 130 mM NaCl, 0.2 mM K-PO_4, ice-cold.
2. MC buffer: 10 mM HEPES–KOH pH 7.6, 10 mM KOAc, 0.5 mM $Mg(OAc)_2$, ice-cold.
3. 0.25 M dithioerythritol (DTE).
4. EDTA-free protease inhibitor cocktail (Roche).
5. Roeder C buffer: 25 % (v/v) glycerol, 20 mM HEPES–KOH pH 7.6, 420 mM NaCl, 1.5 mM $MgCl_2$, 0.2 mM EDTA, ice-cold.
6. 0.1 M phenylmethylsulfonyl fluoride (PMSF), dissolved in isopropanol (*see* **Note 4**).
7. Roeder D buffer: 10 % (v/v) glycerol, 20 mM HEPES–KOH pH 7.6, 100 mM KCl, 1.5 mM $MgCl_2$, 0.2 mM EDTA, 0.5 mM DTT (dithiothreitol), 0.5 mM PMSF, ice-cold.
8. Dounce homogenizer.
9. Dialysis tubing (MWCO 6,000–8,000 Da).
10. *See* also **Note 5**.

2.3 Transcription of pre-mRNA

1. Transcription-optimized 5× buffer.
2. 0.1 M ATP, 0.1 M UTP, 0.1 M CTP, 0.01 M GTP.
3. ^{32}P-αUTP.
4. m^{7}GpppG cap (Kedar, Poland).
5. Stock solutions of $MgCl_2$ (1 M) and DTT (1 M).
6. 10 mg/ml BSA.
7. RNAsin (40 U/μl).
8. SP6 RNA polymerase (2 U/μl).
9. DNA template.
10. RNA extraction buffer: 20 mM Tris–HCl pH 7.5, 150 mM NaCl, 0.5 % (w/v) SDS, 0.2 mM EDTA pH 8.0.
11. RQ1 DNase.
12. 5 % polyacrylamide gels containing 8 M urea.
13. X-ray film.
14. Phenol–chloroform–isoamyl alcohol (25:24:1).
15. Chloroform.
16. 10 μg/μl glycogen.
17. 100 % (v/v) ethanol, ice-cold.
18. 80 % (v/v) ethanol, ice-cold.
19. 3 M sodium acetate (NaOAc) pH 5.3.
20. RNAse-free water.
21. Vortex mixer.
22. Vacuum centrifuge.
23. *See* also **Notes 6** and 7.

2.4 Splice Reaction

1. m^{7}G(5′)ppp(5′)G-capped and MS2-tagged pre-mRNA (^{32}P-labeled and unlabeled).
2. MS2-MBP fusion protein [37].
3. SILAC-labeled HeLa nuclear extracts.
4. Stock solutions of KCl (2 M), $MgCl_2$ (1 M), ATP (0.1 M), creatine phosphate (0.5 M).
5. Scintillation counter.
6. *See* also **Notes 8–10**.

2.5 Affinity Purification of Assembled Spliceosomal Complexes

1. Disposable chromatography columns.
2. Amylose resin.
3. 20 mM HEPES–KOH pH 7.6, 1.5 mM $MgCl_2$, 150 mM NaCl.
4. 50 mM maltose dissolved in 20 mM HEPES–KOH pH 7.6, 1.5 mM $MgCl_2$, 150 mM NaCl.

2.6 Quantification and LC-MS Analysis

2.6.1 Mixing "Light" and "Heavy" Assembled Complexes and Gel Electrophoresis

1. 100 % ethanol, ice-cold.
2. 80 % (v/v) ethanol, ice-cold.
3. 0.3 M NaOAc, pH 5.3.
4. NuPAGE 4–12 % Bis-Tris pre-cast gels, 4× sample buffer, 10× reducing agent, 20× MOPS SDS running buffer, antioxidant (Life Technologies).
5. *See* also **Note 11**.

2.6.2 In-Gel Hydrolysis of Proteins and Extraction of Peptides

1. Ultrapure water.
2. Acetonitrile (ACN).
3. 100 mM ammonium bicarbonate (NH_4CO_3), pH 8.0.
4. 10 mM DTT in 100 mM NH_4CO_3.
5. 55 mM iodoacetamide (IAA) in 100 mM NH_4CO_3.
6. 5 % (v/v) formic acid (FA).
7. Trypsin (sequencing grade, 0.1 μg/μl).
8. Buffer 1: 50 μl H_2O, 50 μl of 100 mM NH_4CO_3, 5 μl of 100 mM $CaCl_2$, 15 μl trypsin.
9. Buffer 2: 50 μl H_2O, 50 μl of 100 mM NH_4CO_3, 5 μl of 100 mM $CaCl_2$.
10. Gel loader pipette tips.
11. Thermomixer (Eppendorf).
12. Vacuum centrifuge.
13. *See* also **Notes 12** and **13**.

2.6.3 LC-MS Analysis

1. 1 % (v/v) FA (loading buffer).
2. *See* **Note 14**.

2.7 Data Analysis

1. Computer system (Intel Pentium III/800 MHz or higher, 2 GB RAM minimum).
2. MaxQuant software package.
3. GProx software platform.
4. *See* also **Notes 15** and **16**.

3 Methods

3.1 SILAC Labeling of HeLa Cells

1. Prepare custom-made DMEM containing the following ingredients (*see* Table 1 for combinations of light and heavy L-arginine and L-lysine to obtain duplex SILAC medium):
   (a) 500 ml DMEM w/o arginine, w/o lysine.
   (b) 50 ml dialyzed FBS.

(c) 5 ml of 100× penicillin–streptomycin.
(d) 5.55 ml of 50 mg/l l-arginine.
(e) 5.55 ml of 50 mg/l l-lysine.

2. Grow HeLa S3 cells in small volumes for at least six passages and then expand to 2.0 L in spinner flasks (0.5–1.0 × 10^6 cells/ml).
3. Transfer the cells to a 2.5 L fermenter and grow under standard conditions (2.5–5.0 × 10^6 cells/ml).

3.2 Preparation of SILAC HeLa Nuclear Extracts

1. Harvest cells from the fermenter by centrifugation for 5 min at 1,200–1,600 × *g* and wash cells with ice-cold PBS.
2. Resuspend the cells in 1.25 volumes of MC buffer supplemented with 1/500 volumes of 0.25 M DTE and 1/100 volumes of EDTA-free protease-inhibitor cocktail.
3. Incubate on ice for 5 min.
4. Lyse in a Dounce homogenizer (18 strokes) at 4 °C.
5. Pellet the nuclei by centrifugation for 5 min at 18,000 × *g*.
6. Dounce (20 strokes) at 4 °C in 1.3 volumes of Roeder C buffer supplemented with 1/500 volumes of 0.25 M DTE and 1/200 volumes of 0.1 M PMSF.
7. Stir for 40 min at 4 °C.
8. Centrifuge for 30 min at approx. 30,000 × *g*.
9. Dialyze the supernatant three times for 2 h against 50 volumes of Roeder D buffer.
10. Centrifuge the dialysate for 2 min at 9,000 × *g*.
11. Prepare aliquots of the supernatant and freeze in liquid nitrogen. Store nuclear extracts at –80 °C.

3.3 In Vitro Transcription of Pre-mRNA

Synthesize pre-mRNA by in vitro transcription using RNA polymerase and linearized DNA template. To synthesize ^{32}P-labeled pre-mRNA, add a certain amount of ^{32}P-αUTP.

1. For in vitro transcription, use 1× transcription buffer, 7.5 mM ATP, 7.5 mM CTP, 7.5 mM UTP, 1.3 mM GTP, 5 mM m^{7}GpppG cap, 20 mM $MgCl_2$, 10 mM DTT, 0.1 μg/ml BSA, 1 U/μl RNasin, 0.1 μg/μl DNA template, and 2 U/μl SP6 RNA polymerase. Adjust the volume to 50 (^{32}P-labeled pre-mRNA) or 150 μl (non-labeled pre-mRNA) with RNAse-free water.
2. Incubate for approx. 4 h at 40 °C.
3. Digest the DNA template using 1 U of RQ1 DNase/μg template and incubate for 20 min at 37 °C.
4. Purify RNA transcripts by gel purification using 5 % polyacrylamide gels containing 8 M urea.

5. Visualize unlabeled RNA by UV-shadowing (254 nm) and ^{32}P-αUTP-labeled RNA by exposure of an X-ray film.
6. Excise bands from the gel.
7. Extract RNA by incubation with RNA extraction buffer overnight.
8. Purify extracted RNA further by Phenol–Chloroform–Isoamyl alcohol PCI extraction and ethanol precipitation (see below).
9. Resuspend the purified RNA in RNase-free water.
10. *See* also **Notes 6**, **7**, **17**, and **18**.

PCI extraction:

1. Mix the sample with 1 volume of PCI and 1 μl of 10 μg/μl glycogen.
2. Vigorously agitate on a vortex (15 min).
3. Separate aqueous and organic phases by centrifugation for 5 min at 13,000 rpm at room temperature.
4. Transfer the aqueous RNA containing phase (upper phase) to a new tube.
5. Add 1 volume of chloroform.
6. Vigorously agitate on a vortex (15 min).
7. Separate aqueous and organic phases by centrifugation for 5 min at 13,000 rpm at room temperature.
8. Transfer the aqueous phase to a new tube and precipitate RNA with ethanol.

Ethanol precipitation:

1. Add 3 volumes of ice-cold 100 % ethanol and 1/10 volumes of 3 M NaOAc, pH 5.3.
2. Incubate at −20 °C for at least 2 h.
3. Centrifuge for 30 min at 16,200 × *g* at 4 °C.
4. Remove the supernatant and wash the pellet with 1 ml ice-cold 80 % (v/v) ethanol.
5. Spin down for 30 min at 16,200 × *g* at 4 °C.
6. Remove the supernatant and dry the protein pellet in a vacuum centrifuge.

3.4 Spliceosome Assembly

To perform in vitro splicing and subsequently purify assembled protein–RNA complexes, use m^{7}G(5′)ppp(5′)G-capped and MS2-tagged pre-mRNA. In our laboratory, we use a mixture of ^{32}P-labeled (radioactive) and non-labeled pre-mRNA, i.e., the non-labeled pre-mRNA is spiked with a small amount of radioactive ^{32}P-labeled pre-mRNA to allow for determination of the concentration. The amount of pre-mRNA and, thus the molar amounts of assembled protein complexes can then be determined by using a scintillation counter. Use duplex SILAC nuclear extracts to compare directly the protein assembly on splicing-active pre-mRNAs with the assembly on splicing-inactive pre-mRNAs (*see* Fig. 3).

1. Pre-incubate the pre-mRNA with a 20-fold molar excess of MS2-MBP fusion protein for approx. 30 min on ice.
2. Prepare several splicing reactions, each containing 20 pmol of pre-mRNA and 50 % (v/v) HeLa nuclear extract, 65 mM KCl, 3 mM $MgCl_2$, 2 mM ATP, and 20 mM creatine phosphate.
3. Incubate for different time intervals at 30 °C.
4. Stop the assembly by placing the reaction vessel on ice.

3.5 Affinity Purification of Assembled Spliceosomal Complexes

Affinity-purify assembled complexes on amylose beads:

1. Use disposable chromatography columns and add the amylose beads.
2. Wash the beads three times with 20 mM HEPES–KOH pH 7.6, 1.5 mM $MgCl_2$, 150 mM NaCl.
3. Add the assembled complexes to the beads.
4. Wash again three times.
5. Elute complexes with 50 mM maltose (dissolved in 20 mM HEPES–KOH pH 7.6, 1.5 mM $MgCl_2$, 150 mM NaCl).
6. Perform all steps at 4 °C.

3.6 Quantification and LC-MS Analysis

3.6.1 Mixing "Light" and "Heavy" Assembled Complexes

1. Determine the molar amounts of assembled complexes within the samples from different time points by measuring the radioactivity of the pre-mRNA.
2. Pool samples from different time points to be compared in equal molar amounts.
3. Precipitate proteins with ethanol (see above).
4. Redissolve the proteins in SDS-PAGE sample buffer and perform gel electrophoresis.

3.6.2 In-Gel Hydrolysis of Proteins and Extraction of Peptides

In-gel hydrolysis: Carry out all incubation steps at 26 °C in a thermomixer at 1,050 rpm for 15 min unless otherwise stated. Remove the solutions after incubation steps using gel loader pipette tips.

1. Cut gel slices from entire gel lanes and cut the slices into small pieces.
2. Wash the gel pieces with 150 μl of water.
3. Dehydrate with 150 μl of ACN.
4. Dry the gel pieces in a vacuum centrifuge.
5. Reduce disulfide bonds of proteins by addition of 100 μl of 10 mM DTT and incubation at 56 °C for 50 min.
6. Dehydrate with 150 μl of ACN.

7. Alkylate reduced cysteine residues by addition of 100 µl of 55 mM IAA and incubation at 26 °C for 20 min.
8. Incubate the gel pieces with 150 µl of 100 mM NH_4CO_3 for 15 min.
9. Add of 150 µl ACN and incubate for 15 min.
10. Add of 150 µl ACN and incubate for 15 min.
11. Dry the gel pieces in a vacuum centrifuge.
12. Rehydrate gel pieces on ice with buffer 1.
13. Cover the gel pieces with buffer 2 and carry out the tryptic digestion overnight at 37 °C.

Extraction of peptides: Carry out all incubation steps at 37 °C in a thermomixer at 1,050 rpm for 15 min.

1. Incubate gel pieces with 50 µl of water.
2. Add 50 µl of ACN.
3. Remove the supernatant containing tryptic peptides and collect it in a new microcentrifuge tube.
4. Add 50 µl of 5 % (v/v) FA to the gel pieces.
5. Add 50 µl of ACN.
6. Remove the supernatant and pool it with the first supernatant.
7. Evaporate supernatants to dryness in a vacuum centrifuge and store the peptide pellets at −20 °C.

3.6.3 LC-MS Analysis

1. Dissolve the samples in loading buffer and analyze them by LC-MS/MS.
2. Analyze samples in technical replicates.
3. *See* also **Note 14**.

3.7 Data Analysis

3.7.1 Max Quant Data Analysis

1. Analyze the raw data using the MaxQuant software package.
2. Define labeled amino acids and appropriate settings for database search.
3. Use protocols provided [38, 39].
4. *See* also **Notes 15** and **19**.

3.7.2 Normalization of Protein Ratios

Normalize the obtained protein ratios by using the ratios of background proteins.

1. Choose a multitude of background (approx. 10–20 proteins).
2. Calculate a normalization factor for every time point in the assembly.
3. Apply the normalization factor to the protein ratios obtained.
4. *See* also **Note 20**.

Here, we used the protein ratios of ribosomal proteins which have been found to be present in equal amounts within the different SILAC nuclear extracts.

3.7.3 Clustering of Protein Groups

Clustering of proteins into protein groups can help with the interpretation of the results and is thus an important step during data analysis. In this study, we used GProx for clustering of protein groups. Please refer to protocols provided [39]. *See* also **Note 16**.

4 Notes

1. DMEM, FBS, penicillin–streptomycin, and heavy-labeled amino acids are available from different commercial sources in different purity grades.
2. Dissolve amino acids in DMEM and if necessary adjust the pH of the solution by using filtered sodium hydroxide solution.
3. As heavy-labeled amino acids are in most cases high-priced, the concentration of lysine and arginine can be reduced compared with normal DMEM. However, the concentration of both amino acids should be adjusted to ensure normal cell growth. The differently labeled cells should in all three cases—i.e., light, medium, and heavy cells—be the same to ensure comparability of the cells (nuclear extracts).
4. PMSF is not very soluble in water and is usually dissolved in isopropanol or ethanol. PMSF should be added freshly, below the liquid surface, to avoid precipitation.
5. DTT, DTE and PMSF solutions should be prepared freshly and added before use.
6. Depending on the promoter, each DNA template requires the use of the appropriate RNA polymerase (e.g., SP6 or T7). This protocol describes the use of SP6 polymerase. When a different polymerase is used, the protocol may need to be adjusted.
7. When working with RNA, the use of RNase-free water is highly recommended to avoid RNA hydrolysis.
8. For preparation of ^{32}P-labeled and unlabeled m^{7}G(5′)ppp(5′) G-capped and MS2-tagged pre-mRNA, *see* Subheading 3.3.
9. For preparation of SILAC-labeled HeLa nuclear extracts, *see* Subheading 3.2 and Table 1.
10. For preparation of MS2-MBP fusion protein, *see* ref. 37.
11. In our laboratory, the NuPAGE gel system has been found to be well-suited for subsequent MS analysis. In principle, any other gel system can be applied to separate the purified protein complexes.

12. For all buffers and solutions, p.a. grade water and solvents should be used.
13. All buffers for in-gel digestion of proteins and extraction of peptides should be prepared freshly before use.
14. Since every laboratory has its own individual setup for LC-MS/MS, we do not provide a specific protocol for this. However, it is worth mentioning that the MaxQuant software [36], which has been proven to be well suited for the analysis of SILAC experiments, is only compatible with data acquired on high-resolution mass spectrometers (i.e., LTQ-Orbitraps, Exactive and Q-Exactive, and FT-ICR; Thermo Fisher Scientific).
15. The MaxQuant software package is freely available (www.maxquant.org). Visit the Web site for additional information and support.
16. The software GProx is freely available (http://gprox.sourceforge.net/). Visit the Web site for additional information and support.
17. Addition of ^{32}P-αUTP will generate radioactively labeled pre-mRNA. The incorporation of ^{32}P-αUTP will be random. The specific activity of the labeled pre-mRNA can be calculated from the mixing ratio of UTP to ^{32}P-αUTP, the number of uridines within the pre-mRNA and the radioactivity of the ^{32}P-αUTP.
18. For further information, please see standard molecular biology protocols.
19. In theory, any other software can be used to analyze the raw data on the assembly kinetics of protein(-RNA) complexes. However, the MaxQuant software package is well-suited to the analysis of the large SILAC datasets that are generated when protein dynamics are analyzed. In addition, it can be applied fully automated and provides additional tools, e.g., for statistical analysis of the data obtained.
20. It is highly recommended that one performs an initial experiment pooling the differentially labeled nuclear extracts in a 1:1 ratio. All proteins should be present in equal amounts and should not be upregulated or downregulated in the different extracts. Background proteins that deviate from the overall 1:1 protein ratios should not be selected for normalization of the data.

Acknowledgements

We thank Klaus Hartmuth for fruitful discussions and Uwe Plessmann and Johanna Lehne for excellent technical assistance.

References

1. Burge CB, Tuschl T, Sharp PA (1999) Splicing of precursors to mRNA by the Spliceosome. In: Gesteland RF, Cech TR, Atkins TR (eds) The RNA world, 2nd edn. Cold Spring Harbor Laboratory Press, Cold Spring Harbour, pp 525–560
2. Green MR (1991) Biochemical mechanisms of constitutive and regulated premRNA splicing. Annu Rev Cell Biol 7:559–599
3. Moore MJ, Query CC, Sharp PA (1993) Splicing of precursors to mRNA by the spliceosome. In: Gesteland RF, Atkins TR (eds) The RNA world, 1st edn. Cold Spring Harbor Laboratory Press, Cold Spring Harbour, pp 303–357
4. Moore MJ, Sharp PA (1993) Evidence for two active sites in the spliceosome provided by stereochemistry of pre-mRNA splicing. Nature 365:364–368
5. Nilsen TW (1998) RNA-RNA interactions in nuclear pre-mRNA splicing. In: Simmons RW, Grunerg-Manaro M (eds) RNA structure and function, 1st edn. Cold Spring Harbor Laboratory Press, Cold Spring Harbour, pp 1793–1809
6. Wahl MC, Will CL, Luhrmann R (2009) The spliceosome: design principles of a dynamic RNP machine. Cell 136:701–718
7. Raker VA, Hartmuth K, Kastner B et al (1999) Spliceosomal U snRNP core assembly: Sm proteins assemble onto an Sm site RNA nonanucleotide in a specific and thermodynamically stable manner. Mol Cell Biol 19:6554–6565
8. Raker VA, Plessel G, Luhrmann R (1996) The snRNP core assembly pathway: identification of stable core protein heteromeric complexes and an snRNP subcore particle in vitro. EMBO J 15:2256–2269
9. Urlaub H, Raker VA, Kostka S et al (2001) Sm protein-Sm site RNA interactions within the inner ring of the spliceosomal snRNP core structure. EMBO J 20:187–196
10. Achsel T, Brahms H, Kastner B et al (1999) A doughnut-shaped heteromer of human Sm-like proteins binds to the 3′-end of U6 snRNA, thereby facilitating U4/U6 duplex formation in vitro. EMBO J 18:5789–5802
11. Vidal VP, Verdone L, Mayes AE et al (1999) Characterization of U6 snRNA-protein interactions. RNA 5:1470–1481
12. Heinrichs V, Bach M, Winkelmann G et al (1990) U1-specific protein C needed for efficient complex formation of U1 snRNP with a 5′ splice site. Science 247:69–72
13. Pomeranz Krummel DA, Oubridge C, Leung AK et al (2009) Crystal structure of human spliceosomal U1 snRNP at 5.5 A resolution. Nature 458:475–480
14. Behrens SE, Tyc K, Kastner B et al (1993) Small nuclear ribonucleoprotein (RNP) U2 contains numerous additional proteins and has a bipartite RNP structure under splicing conditions. Mol Cell Biol 13:307–319
15. Brosi R, Groning K, Behrens SE et al (1993) Interaction of mammalian splicing factor SF3a with U2 snRNP and relation of its 60-kD subunit to yeast PRP9. Science 262:102–105
16. Will CL, Urlaub H, Achsel T et al (2002) Characterization of novel SF3b and 17S U2 snRNP proteins, including a human Prp5p homologue and an SF3b DEAD-box protein. EMBO J 21:4978–4988
17. Gozani O, Feld R, Reed R (1996) Evidence that sequence-independent binding of highly conserved U2 snRNP proteins upstream of the branch site is required for assembly of spliceosomal complex A. Genes Dev 10:233–243
18. Kramer A, Gruter P, Groning K et al (1999) Combined biochemical and electron microscopic analyses reveal the architecture of the mammalian U2 snRNP. J Cell Biol 145: 1355–1368
19. Bach M, Winkelmann G, Luhrmann R (1989) 20S small nuclear ribonucleoprotein U5 shows a surprisingly complex protein composition. Proc Natl Acad Sci U S A 86:6038–6042
20. Staley JP, Guthrie C (1998) Mechanical devices of the spliceosome: motors, clocks, springs, and things. Cell 92:315–326
21. Makarova OV, Makarov EM, Luhrmann R (2001) The 65 and 110 kDa SR-related proteins of the U4/U6.U5 tri-snRNP are essential for the assembly of mature spliceosomes. EMBO J 20:2553–2563
22. Ajuh P, Kuster B, Panov K et al (2000) Functional analysis of the human CDC5L complex and identification of its components by mass spectrometry. EMBO J 19:6569–6581
23. Makarova OV, Makarov EM, Urlaub H et al (2004) A subset of human 35S U5 proteins, including Prp19, function prior to catalytic step 1 of splicing. EMBO J 23:2381–2391
24. Umen JG, Guthrie C (1995) The second catalytic step of pre-mRNA splicing. RNA 1: 869–885
25. Will CL, Luhrmann R (2011) Spliceosome structure and function. Cold Spring Harb Perspect Biol 3: pii: a003707
26. Will CL, Rumpler S, Klein GJ et al (1996) In vitro reconstitution of mammalian U1 snRNPs active in splicing: the U1-C protein enhances the formation of early (E) spliceosomal complexes. Nucleic Acids Res 24:4614–4623

27. Behzadnia N, Golas MM, Hartmuth K et al (2007) Composition and three-dimensional EM structure of double affinity-purified, human prespliceosomal A complexes. EMBO J 26:1737–1748
28. Hartmuth K, Urlaub H, Vornlocher HP et al (2002) Protein composition of human prespliceosomes isolated by a tobramycin affinity-selection method. Proc Natl Acad Sci U S A 99:16719–16724
29. Deckert J, Hartmuth K, Boehringer D et al (2006) Protein composition and electron microscopy structure of affinity-purified human spliceosomal B complexes isolated under physiological conditions. Mol Cell Biol 26:5528–5543
30. Bessonov S, Anokhina M, Krasauskas A et al (2010) Characterization of purified human Bact spliceosomal complexes reveals compositional and morphological changes during spliceosome activation and first step catalysis. RNA 16:2384–2403
31. Makarov EM, Makarova OV, Urlaub H et al (2002) Small nuclear ribonucleoprotein remodeling during catalytic activation of the spliceosome. Science 298:2205–2208
32. Bessonov S, Anokhina M, Will CL et al (2008) Isolation of an active step I spliceosome and composition of its RNP core. Nature 452:846–850
33. Andersen JS, Lam YW, Leung AKL et al (2005) Nucleolar proteome dynamics. Nature 433:77–83
34. Talkington MW, Siuzdak G, Williamson JR (2005) An assembly landscape for the 30S ribosomal subunit. Nature 438:628–632
35. Agafonov DE, Deckert J, Wolf E et al (2011) Semiquantitative proteomic analysis of the human spliceosome via a novel two-dimensional gel electrophoresis method. Mol Cell Biol 31:2667–2682
36. Cox J, Mann M (2008) MaxQuant enables high peptide identification rates, individualized p.p.b.-range mass accuracies and proteome-wide protein quantification. Nat Biotechnol 26:1367–1372
37. Jurica MS, Licklider LJ, Gygi SR et al (2002) Purification and characterization of native spliceosomes suitable for three-dimensional structural analysis. RNA 8:426–439
38. Cox J, Matic I, Hilger M et al (2009) A practical guide to the MaxQuant computational platform for SILAC-based quantitative proteomics. Nat Protoc 4:698–705
39. Rigbolt KT, Vanselow JT, Blagoev B (2011) GProX, a user-friendly platform for bioinformatics analysis and visualization of quantitative proteomics data. Mol Cell Proteomics 10:701–718

Chapter 17

Identification and Validation of Protein-Protein Interactions by Combining Co-immunoprecipitation, Antigen Competition, and Stable Isotope Labeling

Frederik Sommer, Timo Mühlhaus, Dorothea Hemme, Daniel Veyel, and Michael Schroda

Abstract

Co-immunoprecipitation (coIP) in combination with mass spectrometry (MS) is a powerful tool to identify potential protein-protein interactions. However, unspecifically precipitated proteins usually result in large numbers of false-positive identifications. Here we describe a detailed protocol particularly useful in plant sciences that is based on ^{15}N stable isotope labeling of cells, ^{14}N antigen titration, and coIP/MS to distinguish true from false protein-protein interactions.

Key words Co-immunoprecipitation (coIP), ^{15}N stable isotope metabolic labeling, Mass spectrometry (MS), Protein-protein interaction

1 Introduction

Most proteins in the cell do not function alone but within complexes with other proteins whose composition often is dynamic and depending, e.g., on environmental conditions, tissue type, or the presence of specific effector molecules. Accordingly, knowledge of a protein's interaction partners provides important insights into its function. A frequently used method for the identification of protein-protein interactions is co-immunoprecipitation (coIP) [1]. Here, a bait protein is precipitated together with its interaction partners from cell lysates by the use of specific antibodies coupled to beads. The precipitates are then analyzed, ideally by mass spectrometry (MS), to identify coprecipitated proteins which potentially represent the interaction partners of the bait protein.

The high sensitivity of today's mass spectrometers allows for a thorough analysis of the precipitated proteins and usually results in lists of several hundreds of proteins. A fraction of them are true interaction partners while the remaining majority represent

Bettina Warscheid (ed.), *Stable Isotope Labeling by Amino Acids in Cell Culture (SILAC): Methods and Protocols*, Methods in Molecular Biology, vol. 1188, DOI 10.1007/978-1-4939-1142-4_17,

"contaminants" which were coprecipitated because they unspecifically bind to the beads or cross-react with the antibodies used [1, 2]. To discriminate between true interaction partners and contaminants, several experimental designs and methods have been described [2–8] using appropriate controls, reciprocal approaches, tandem affinity tags, epitope competition, stable isotope labeling by amino acids in cell culture (SILAC), strains differentially expressing the bait proteins (QUICK and QUICK-X), affinity modulation, or combinations thereof. The by far most commonly used approach is a combination of reciprocal coIP and SILAC. Unfortunately, SILAC approaches require the target organism to be auxotrophic, ideally for lysine and/or arginine, which is not the case for most organisms including autotrophic ones like plants and algae, unless auxotrophic mutants exist. Moreover, when heavy arginine is fed, arginine-to-proline interconversion results in additional mass shifts for each proline in a peptide and slightly dilutes heavy with light arginine, thereby rendering quantification more tedious and less accurate [3, 9]. Finally, labeled amino acids are costly. For autotrophic organisms these problems are overcome by growing them on comparably cheap $^{15}NO_3^-$ or $^{15}NH_4^+$ as nitrogen source, leading to full ^{15}N labeling of the proteome [10].

Another problem inherent to the commonly used strategies for the identification of protein-protein interactions employing stable isotope labeling is that they require the bait to be present in one cell type and absent, or at least diminished, in another. This may be achieved by the overexpression of a tagged version of the bait [8] or by knocking down bait expression using, e.g., RNA interference [3, 5]. The latter may require the often tedious construction of appropriate transgenic organisms. Moreover, changing the expression levels of the bait protein may lead to alterations in the composition of the cell's proteome or of the bait's interaction partners, thus potentially rendering the interpretation of quantitative proteomics results difficult [3].

We aimed at developing a coIP-MS approach for the identification of protein-protein interactions that is suitable for autotrophic organisms and does not require an alteration of the expression levels of the bait protein. Moreover, we strove for an approach that allows for distinguishing true from unspecific interaction partners of a bait protein by assigning probability values. The method described here takes advantage of the fact that the epitope that was used for antibody production is usually available as purified protein, e.g., from overexpressing *Escherichia coli* strains, or is easily commercially available (e.g., as HA, FLAG, or GFP epitope) in case tagged proteins and antibodies specific for the respective tag are used. Prior to the coIP, an inactive form of the antigen that is not able to interact with its interaction partners is spiked as non-labeled antigen (i.e., as ^{14}N-labeled bait) at increasing concentrations into lysates of metabolically ^{15}N-labeled cells (for an overview,

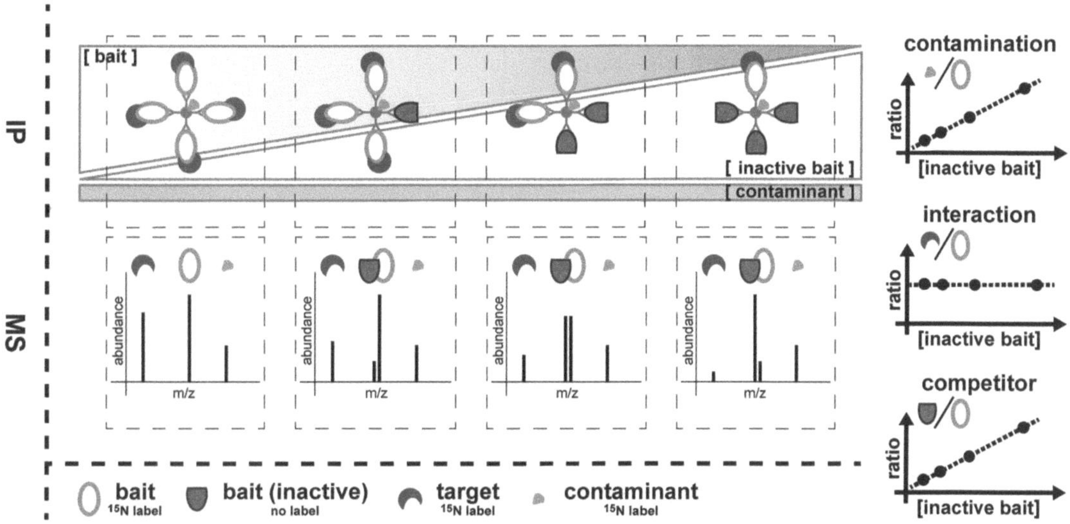

Fig. 1 Workflow for competition coIP. Different amounts of inactivated ^{14}N bait protein are added to lysates of ^{15}N-labeled cells. CoIP is performed using sub-stoichiometric amounts of affinity-purified antibodies coupled to beads and precipitates are analyzed by nanoLC-MS/MS. After peptide identification, peak volumes of the eluting peptides are calculated and assembled to a protein abundance value. Data can be inspected manually by plotting the abundance ratios of ^{15}N-labeled co-purified proteins and ^{15}N-labeled bait against added ^{14}N bait. The resulting slopes should have high positive values for contaminations and ^{14}N bait, while they should have values around zero for specific interaction partners

see Fig. 1). Thus, the inactive, ^{14}N-labeled bait competes with the ^{15}N-labeled bait (and its interaction partners) for binding sites on the antibodies in a dose-dependent manner. As a result, with increasing concentrations of the inactive ^{14}N-labeled bait, less of the ^{15}N-labeled bait and its interaction partners are present in the immunoprecipitates, while the amount of precipitated contaminants is unaffected. By plotting the intensity ratios of precipitated proteins against the amount of added ^{14}N bait, unspecific interaction partners can be distinguished from specific ones by a simple manual inspection of the data (Fig. 1).

This way of data presentation also eliminates variations in coIP efficiency. Moreover, statistical analyses like Pearson correlation can be employed to estimate the significance with which a precipitated protein can be assigned as a true interaction partner of the respective bait protein.

2 Materials

2.1 Buffers and Chemicals

1. 1 M Tris–HCl (pH 7.5).
2. 0.1 M sodium phosphate buffer (pH 7.4): Mix 0.1 M NaH_2PO_4 and 0.1 M Na_2HPO_4 at a 19:81 (v/v) ratio.

3. 0.1 M boric acid adjusted to pH 9.0 using 5 M NaOH.
4. Dimethyl pimelimidate (DMP).
5. H_2O bidest, used for all buffers.
6. PBS pH 7.4: 137 mM NaCl, 2.7 mM KCl, 4.3 mM Na_2HPO_4, 1.4 mM KH_2PO_4. Add 0.1 % (v/v) Tween-20 to obtain PBS-T.
7. Bovine serum albumin (BSA).
8. 6 M urea in PBS.
9. pH-shock buffer: 137 mM NaCl, 0.1 M glycine, adjust to pH 2.5 with HCl, add 0.1 % (v/v) Tween-20.
10. 1 M Na_2CO_3 buffer.
11. 2× SDS-sample buffer: 4 % SDS, 125 mM Tris–HCl (pH 6.8), 20 % glycerol, 10 % 2-mercaptoethanol.
12. pH-indicator paper.

2.2 Chlamydomonas Cell Culture

1. *Chlamydomonas reinhardtii* strain CF185 (cw_d, mt^+, *arg7*$^-$), complemented with plasmid-borne *ARG7* wild-type gene [11].
2. TAP minus N, prepared as described in [12] without NH_4Cl.
3. TAP-^{15}N: Dilute 3.7 M $^{15}NH_4Cl$ (Cambridge Isotopes, purity >98 %) 1:500 (v/v) in TAP minus N (*see* **Note 1**).

2.3 CoIP

1. Sonicator Sonoplus HD2070 with a KE76 tip.
2. Tabletop centrifuge for 15 ml tubes and 1.5 ml microcentrifuge tubes.
3. Lysis buffer: 20 mM HEPES, 1 mM $MgCl_2{\cdot}6H_2O$, 10 mM KCl, 154 mM NaCl; adjust to pH 7.2 with KOH.
4. 10 % (v/v) Triton X-100.
5. Ice-cold acetone.
6. ABC buffer: 50 mM NH_4HCO_3.
7. ABC buffer containing 2 % SDS.
8. Protein A-Sepharose beads.
9. Purified antigen: The protein isolated, e.g., from overexpressing *E. coli* strains that was used to generate the antiserum (*see* **Note 2**).
10. Inactivation buffer: 20 mM sodium phosphate buffer pH 7.4, 6 M urea, 10 mM GSSG.
11. Hybond-ECL nitrocellulose membrane.
12. Millipore Amicon Ultra 0.5 ml centrifugal ultrafiltration device, molecular weight cutoff (MWCO) of 10,000 Da.
13. Protease inhibitors.
14. Gel electrophoresis and protein transfer system for SDS-PAGE and immunoblotting.

2.4 Mass Spectrometry and Data Analysis

1. Reduction buffer: 6.5 mM dithiothreitol (DTT).
2. Alkylation buffer: 27 mM iodoacetamide.
3. LysC in ABC buffer at 0.1 μg/μl.
4. Trypsin sequencing grade at 0.1 μg/μl in ABC buffer.
5. ACN: Acetonitrile, HPLC/MS grade.
6. Glacial acetic acid, HPLC/MS grade.
7. HPLC buffers A and B: A, 2 % ACN, 0.4 % acetic acid; B, 90 % ACN, 0.4 % acetic acid.
8. SpeedVac.
9. High-mass-accuracy LC-MS system, e.g., nanoAQUITY UPLC system (Waters) coupled to an LTQ-Orbitrap XL (Thermo Scientific) with nanoACQUITY UPLC Symmetry C18 Trapping Column, 5 μm, 180 μm × 20 mm, and nanoACQUITY UPLC™ BEH C18 75 μm × 150 mm 1.7 μm for separation.
10. 2 M thiourea, 6 M urea in ABC buffer.
11. MaxQuant version 1.2.2.5.
12. MSQuant version 1.5 with helper application including DTAsupercharge version 1.37.
13. Mascot version 2.3.01.

3 Methods

3.1 Affinity Purification of Antibodies

Unless purified antibodies were purchased, the immune serum should be affinity-purified to reduce cross-reactivity and, thus, contaminants in the coIPs. Here, HSP70B antibodies are purified from immune serum as follows [13]:

1. Incubate 500 μg of purified HSP70B antigen in 1 ml of PBS containing 6 M urea for 1 h at 25 °C. Cut a 10 cm × 6 cm nitrocellulose membrane and incubate it with the HSP70B protein, diluted with 10 ml of PBS, on a rocking shaker for 1 h at 25 °C.
2. Discard the supernatant and wash the membrane three times for 10 min with 10 ml of PBS-T.
3. Block the membrane for 1 h with 10 ml of PBS-T containing 1 % (w/v) BSA.
4. Discard the supernatant and wash the membrane three times for 10 min with 10 ml of PBS-T.
5. Incubate the membrane with 1 ml of antiserum in 10 ml of PBS-T containing 1 % BSA for 1 h.
6. Discard the supernatant and wash the membrane three times for 10 min with 10 ml of PBS-T.

7. Elute the antibodies from the membrane twice by applying 3 ml of pH-shock buffer for 30 s.
8. Immediately adjust the pooled eluate to pH 8 by adding some μl of 1 M Na_2CO_3 and monitor the pH change with pH-indicator paper.

3.2 Cell Culture and Cell Lysis

Chlamydomonas reinhardtii cells are grown in TAP medium [12] in 1 l Erlenmeyer flasks on an orbital shaker (120 rpm) in constant light (40 μE/m^2/s) to logarithmic phase (5×10^6 cells/ml) using $^{15}NH_4Cl$ as nitrogen source for at least ten generations to ensure complete (>97 %) metabolic labeling [14]. Cell lysis is performed as follows (described here for four IPs):

1. Harvest 400 ml of *Chlamydomonas* cells by centrifugation for 5 min at 4,000 × *g*.
2. Resuspend the cells in 6 ml of ice-cold lysis buffer by pipetting up and down and add protease inhibitors.
3. Break cells by sonifying them four times for 20 s on ice with output control of 75 % and duty cycle of 60 %. Include breaks of 20 s between the sonication cycles for cooling.
4. Split the broken cells between several 1.5 ml microcentrifuge tubes and centrifuge in a tabletop centrifuge at 4 °C for 15 min at highest speed to pellet insoluble material (*see* **Note 3**).
5. Carefully transfer the supernatants into a 15 ml tube.
6. Determine the protein concentration of the sample, e.g., by Lowry's assay [15] (we usually obtain approximately 10 mg/ml).
7. For each IP, transfer 1 ml of soluble protein into a 15 ml tube and add lysis buffer containing 0.1 % Triton X-100 to a final volume of 6 ml.

3.3 Preparation of Affinity Beads

To prepare affinity beads, IgGs are adsorbed to Protein A-Sepharose and cross-linked to reduce contamination of the MS samples with IgGs. The protocol described here is for four IPs [4]:

1. Rehydrate 60 mg of Protein A-Sepharose in a 15 ml tube in 4 ml of 0.1 M phosphate buffer (pH 7.4) for 30 min at 25 °C.
2. Centrifuge for 60 s at 1,000 × *g* and 25 °C to pellet the swollen Protein A-Sepharose. Carefully remove the supernatant and wash the beads in the same buffer at least three times.
3. Fill up to 4 ml with phosphate buffer and transfer equal volumes of bead suspension to four 1.5 ml microcentrifuge tubes. Fill up each to 1.3 ml with phosphate buffer and the required amounts of purified antibodies as described and determined in Subheading 3.4.

4. Allow Protein A-Sepharose beads to adsorb IgGs for 1 h at 25 °C on an end-over-end mixer.
5. Centrifuge for 60 s at 5,000 × *g* and 25 °C to pellet the Protein A-Sepharose beads. Carefully remove the supernatant and resuspend the beads in 1.3 ml of 0.1 M sodium borate buffer (pH 9.0). Repeat this step three times and resuspend the affinity beads in 1.3 ml of borate buffer.
6. Dissolve 28 mg of DMP in 1 ml of borate buffer and add 200 μl to each tube of affinity beads to obtain a final concentration of 20 mM DMP.
7. Allow IgGs to cross-link to Protein A for 30 min at 25 °C on an end-over-end mixer.
8. Centrifuge for 60 s at 5,000 × *g* and 25 °C to pellet the beads. Carefully remove the supernatant and resuspend the beads in 1.3 ml of 1 M Tris–HCl (pH 7.5) each to quench free cross-linker. Repeat this step once and incubate for 2 h at 25 °C on an end-over-end mixer.
9. Wash the affinity beads with lysis buffer as described in **step 8** (*see* **Note 4**).

3.4 Determination of the Amounts of Antibodies Needed for CoIP

Nonnative ^{14}N bait added to the ^{15}N cell lysate prior to the IP will compete with the ^{15}N bait for the antibody-binding sites only when the latter are limiting. If antibody-binding sites are in excess, the additional ^{14}N bait will simply be precipitated together with the ^{15}N bait from the cell lysate and the method will not work. To test whether indeed the antibodies are limiting, different amounts of antibodies are coupled to the beads, the IP is performed, and the amount of bait protein pulled down should increase proportionally with increased amounts of antibodies added (the protocol given here is for four IPs):

1. Prepare affinity beads as mentioned in Subheading 3.3 using different amounts of affinity-purified antibodies (here 5, 10, 20, and 50 μl of affinity-purified anti-HSP70B).
2. Add 1 ml affinity beads to the 6 ml cell lysate in 15 ml tubes obtained in **step** 7 of Subheading 3.2 and incubate for 1 h at 25 °C on a tube roller.
3. Pellet the beads by centrifugation for 1 min at 25 °C at 2,000 × *g* and take off the supernatant leaving approximately 0.4 ml on top of the beads. Keep an aliquot of the supernatant for immunoblot analysis (*see* Fig. 2).
4. Resuspend and transfer the beads to 1.5 ml microcentrifuge tubes, rinse the 15 ml tubes with 0.8 ml of lysis buffer containing 0.1 % Triton X-100, and add it to the beads in the microcentrifuge tubes.

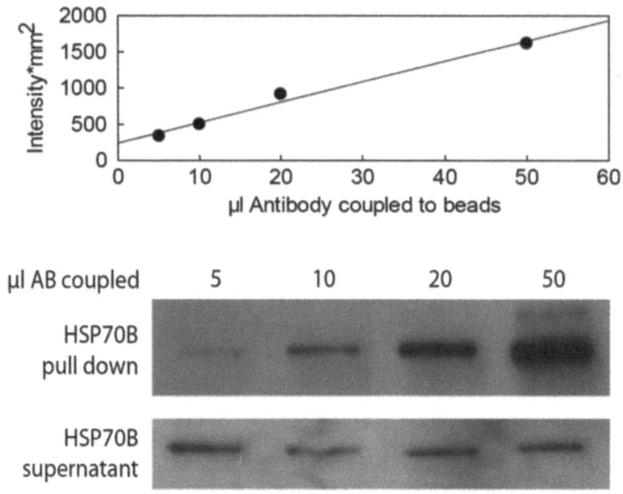

Fig. 2 Antibody titration. For the competition coIP experiment, sub-stoichiometric amounts of antibodies must be used. To determine the amount of antibodies needed, different amounts of them are covalently coupled to Protein A-Sepharose beads and IP is performed. 10 % of the precipitated proteins and 0.025 % of the remaining supernatants were analyzed by immunoblotting. The supernatant is not depleted of bait and the amount of precipitated bait (here HSP70B) increases proportionally with increasing amounts of antibodies added, as seen here by the immunoblot and subsequent densitometry analysis

5. Pellet the beads at 25 °C by centrifugation for 30 s at 16,000 × *g*, discard the supernatant, and wash the beads with 1.3 ml of lysis buffer containing 0.1 % Triton X-100. Repeat this step three times.
6. Transfer the beads to fresh 1.5 ml tubes and perform three washing steps as described in **step 5** using lysis buffer without Triton X-100.
7. Perform two washing steps as described in **step 5** using 20 mM phosphate buffer.
8. Add 100 µl of 2× SDS-sample buffer, mix, incubate at 95 °C for 5 min, pellet the beads, and analyze approx. 10 µl of the eluate by SDS-PAGE and immunoblotting for the bait protein (Fig. 2, here shown for HSP70B).

3.5 Determination of the Amount of Antigen Needed

The amount of ^{14}N-labeled bait added to the coIP to compete for the affinity pull down of ^{15}N bait and interacting proteins should be approximately in a stoichiometric range. This ensures a balance between sensitivity and selectivity of the method. If too much of the ^{14}N bait is added, all ^{15}N-labeled bait and associated interaction partners are replaced from the beads and the subsequent MS analysis will not detect them. If too little of the ^{14}N bait is added, the effect on bound ^{15}N bait and associated ^{15}N contaminants is too small for a robust data analysis. To be able to add ^{14}N bait

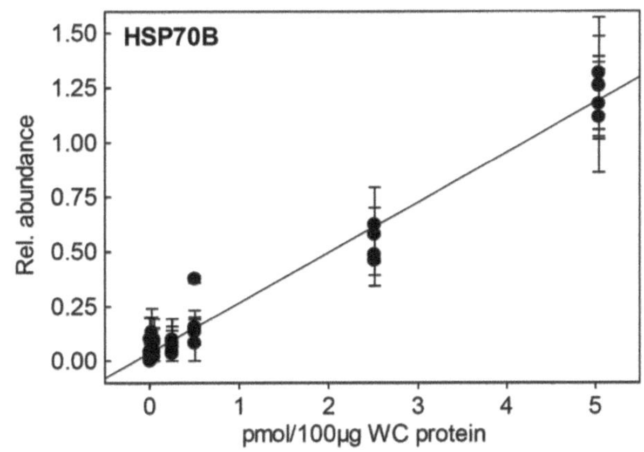

Fig. 3 Estimation of the amount of competitor needed. Different amounts of ^{14}N bait protein (here HSP70B) were spiked into whole-cell (WC) lysates of ^{15}N-labeled cells and analyzed by shotgun MS/MS. $^{14}N/^{15}N$ ratios of bait protein abundances were plotted against the amount of spiked protein. From the linear part of the titration results, the absolute amount of the bait protein in the cell extract can be calculated and the amounts of ^{14}N bait required for coIP competition estimated

protein at about stoichiometric concentrations to ^{15}N-labeled cell lysates, we need to know the absolute amounts of ^{15}N bait protein in the lysate. One approach is to spike different amounts of the ^{14}N antigen into ^{15}N-labeled cell lysates and to determine the absolute amount of the bait protein in the lysate by MS analysis (Fig. 3) (*see* **Note 5**):

1. Different amounts of ^{14}N bait protein (here 33, 16.5, 3.3, 1.65, 0.33, and 0.165 pmol of HSP70B) are spiked into 630 µg of ^{15}N-labeled *Chlamydomonas* whole-cell proteins at a final protein concentration of 2 µg/µl in ABC buffer.
2. 200 µl of the soluble fraction of the protein mix is precipitated by adding 800 µl of ice-cold acetone at −20 °C for >4 h.
3. Samples are digested, desalted, and analyzed by MS as described previously [14].
4. From the linear range of the dilution series (Fig. 3), the absolute amount of bait protein (HSP70B) can be determined. We determined here that HSP70B represents ~0.28 % of whole-cell protein. Cells used here were subjected to heat stress, which is known to lead to an ~2.3-fold increase in HSP70B protein levels [14]. Hence, HSP70B should constitute ~0.12 % of whole-cell proteins in non-stressed cells, which is in good accordance with an average value of ~0.131 % determined previously by immunoblotting [13].

3.6 Inactivation of ^{14}N-Labeled Bait Protein

The added ^{14}N bait protein must be inactivated to ensure that it is not binding to ^{15}N interaction partners in the cell lysate, thereby obscuring the results. Inactivation can be done by denaturation and/or modification. Here, ^{14}N HSP70B is inactivated by denaturation in urea and modification of cysteine residues with GSSG (*see* **Note 6**):

1. Incubate 125 μg of recombinant, purified HSP70B protein in inactivation buffer at a final volume of 400 μl for 30 min at 25 °C.
2. To remove GSSG, exchange the buffer by centrifugation for 5 min at 16,000 × *g* at 25 °C in a 0.5 ml centrifugal ultrafiltration device (MWCO 10,000 Da). Fill up the ultrafiltration device to 0.5 ml with PBS. Repeat this step five times.
3. Resuspend the inactivated HSP70B in 125 μl of lysis buffer containing 0.1 % Triton X-100.

3.7 CoIP Competition

Perform the coIP competition experiment by adding different amounts of inactivated ^{14}N bait (here HSP70B) to the ^{15}N-labeled cell lysates (*see* **Note 7**):

1. Take cell lysates generated in **step 7** of Subheading 3.2 and add different amounts of ^{14}N-labeled bait protein (here 0, 10, 30, and 60 μl corresponding to approximately 0, 1:2, 1.5:1, and 3:1 molar ratios of ^{14}N:^{15}N bait in the sample, as determined in **step 4** of Subheading 3.5 and **step 6** of Subheading 3.2) (*see* **Note 8**).
2. Mix, add equal amounts of affinity beads obtained from **step 9** in Subheading 3.3 (all with the same amount of antibodies coupled), and perform the IP as described in **steps 2–7** of Subheading 3.4.
3. Elute the proteins from the beads by adding 100 μl of ABC buffer containing 2 % SDS, mixing, and incubating for 10 min at 65 °C.
4. Centrifuge at 25 °C for 1 min at 16,000 × *g* to pellet the beads and collect the supernatant.
5. Repeat **step 3** at 95 °C and pool the supernatants.
6. Optional: To control for a successful IP, immunoblot analysis can be performed using 10 μl of the eluates (Fig. 4).

3.8 MS Sample Preparation and Analysis

1. Precipitate the eluted proteins (200 μl) and remove SDS by adding 800 μl of acetone; place the tubes at −20 °C for >4 h.
2. To pellet precipitated proteins, centrifuge at 4 °C for 15 min at 16,000 × *g*. Wash the precipitate two times with ice-cold acetone followed by centrifugation as above.
3. Resuspend precipitated proteins in 20 μl of 6 M urea and 2 M thiourea in ABC buffer and incubate in a sonication bath two times for 10 min.

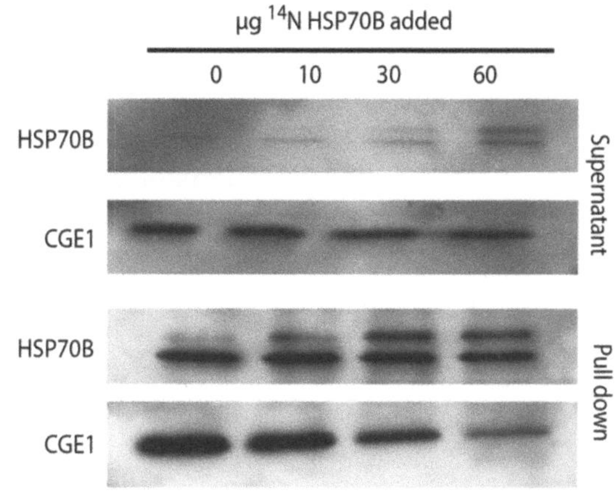

Fig. 4 Test of competition coIP with a known interaction partner. Precipitates and remaining supernatants of four coIPs on lysates from ^{15}N-labeled cells containing different amounts of inactivated ^{14}N HSP70B bait protein were analyzed by immunoblotting. HSP70B and its known interaction partner CGE1 [31] are not depleted from the supernatant fraction. While constant amounts of $^{14/15}$N HSP70B are precipitated, amounts of precipitated ^{15}N CGE1 decline with increasing amounts of inactive ^{14}N HSP70B added. Note that ^{14}N HSP70B contains a hexa-histidine tag and therefore migrates with a larger molecular weight than the native form

4. Add 1 µl of reduction buffer, mix, and incubate at 25 °C for 30 min.
5. Add 1 µl of alkylation buffer, mix, and incubate in the dark for 20 min.
6. Add 16 µl of ABC buffer, 2 µl of ACN, and 1 µl of LysC solution and incubate at 37 °C for >3 h.
7. Add 120 µl of ABC buffer and 2.5 µl of trypsin solution, mix, and incubate for >6 h at 37 °C.
8. Quench the reaction by adding 2 µl of glacial acetic acid and mixing.
9. Desalt the resulting peptides on C18 material with StageTips [4, 16].
10. Evaporate the eluate in a SpeedVac and resuspend the peptides in 10 µl of HPLC buffer A.
11. Analyze the samples (3 µl each) on a high-mass-accuracy LC-MS system, e.g., nanoAQUITY UPLC coupled to an LTQ-Orbitrap XL, running in data-dependent mode as described in [14] and employing a constant flow of 300 nl/min and a linear gradient from 0 to 60 min, ramping from 2 to 40 % HPLC buffer B, and then to 90 % buffer B within 10 min. Perform a final wash for 15 min with 90 % buffer B.

3.9 Data Analysis

Raw files only containing information on the obtained spectra are generated by the instrumentation software. By the help of mass spectrometry software (1) "feature" lists containing masses and intensities of the peptide peaks are extracted from the MS spectra. (2) MS/MS fragmentation spectra are used to identify peptides by comparing the spectra against a reference sequence database. (3) Identified peptides and features are assembled into proteins and their corresponding relative abundances. Probabilities for true-positive interaction partners can be estimated by correlating the relative abundance of the different identified proteins with that of the bait protein (here HSP70B) over the different competition coIPs.

1. Build or download a search reference database including all (predicted) protein sequences of the used organism in FASTA format (here, a combined protein sequence database including all translated sequences of Augustus gene models version 10.2 of the *Chlamydomonas* genome sequence as well as all sequences from *Chlamydomonas* mitochondrial and chloroplast proteins; downloaded from http://chlamycyc.mpimp-golm.mpg.de/files/sequences/protein/).
2. Analyze Raw files in a combined search using MaxQuant for peptide identification, protein assembly, and protein quantification [17, 18]. The ^{15}N mass differences per amino acid have to be included as fixed modifications for all amino acids. Using the label-free quantitation and "match between runs" option for mass-retention time correlation and an FDR of <1 % for peptides and proteins, 238 proteins were identified (*see* **Note 9**).
3. (Optional) Since MaxQuant can handle either ^{14}N or ^{15}N identification and quantitation, MSQuant version 1.5 [19] can be used for quantifying $^{14}N/^{15}N$ ratios of added versus cellular HSP70B. Extract Raw files to mgf files using the helper utility DTAsupercharge version 1.37. Use Mascot to search these against the protein FASTA database for peptide identification using the option ^{15}N metabolic labeling. Use the result files from Mascot as input for MSQuant for peptide and protein quantification to obtain $^{14}N/^{15}N$ ratios (*see* Fig. 5a) (*see* **Note 10**).
4. For statistical analysis, calculate the Pearson's correlation coefficient between each identified and quantified protein in the precipitate and the HSP70B bait protein (this is done only for the ^{15}N-labeled proteins).
5. Calculate pairwise *p*-values between the proteins and the bait using Tukey's jackknife method [20]. After filtering according to a significance threshold ($p \leq 0.05$), the list of potential interaction partners can be ranked according to their Pearson's correlation coefficient (*see* **Note 11**).

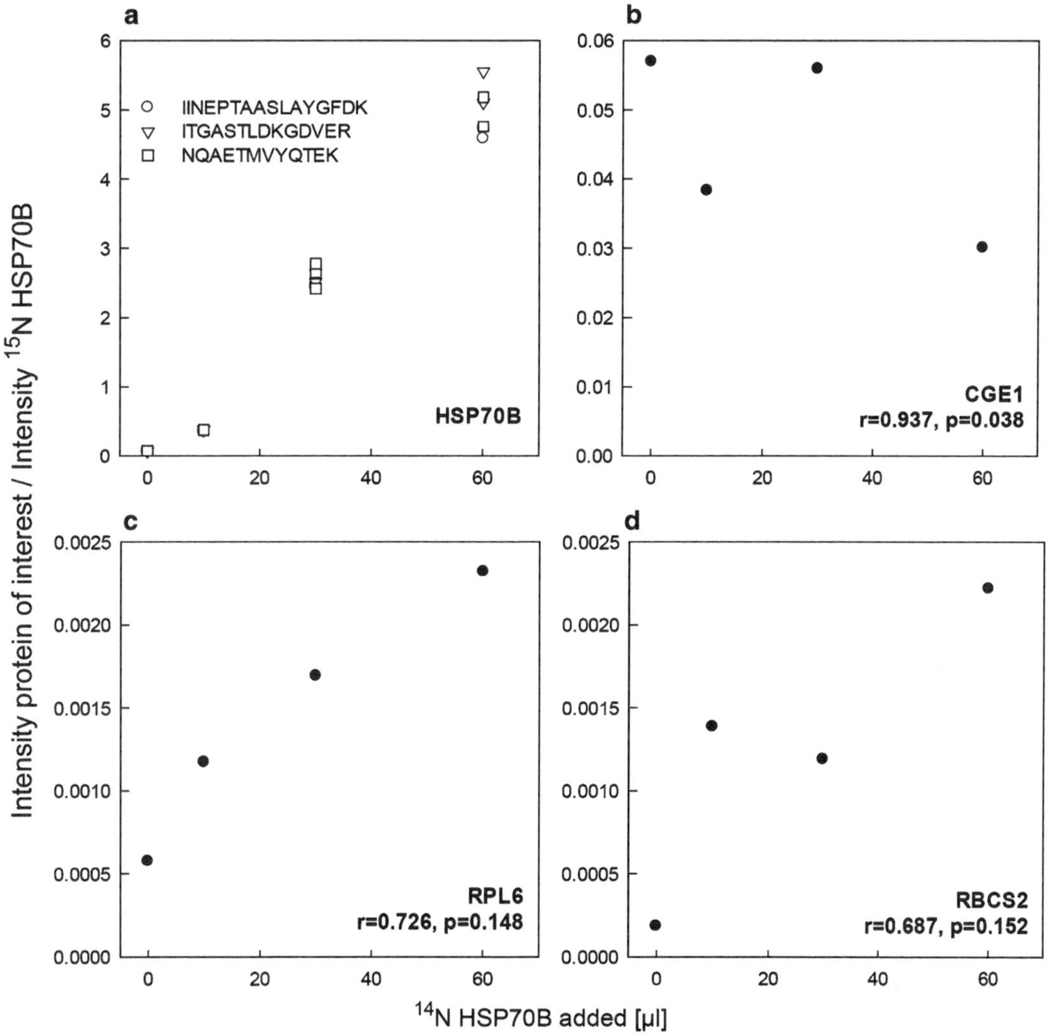

Fig. 5 Graphs for manual inspection and results from statistical analysis for selected coprecipitated proteins. (**a**) Ratios of ^{14}N/^{15}N-labeled HSP70B peptides as determined with MSQuant. (**b**–**d**) Intensity ratios of coprecipitated proteins to HSP70B (both ^{15}N labeled) were plotted against the amount of added inactive ^{14}N HSP70B bait. Correlation coefficients (r) and p-values indicating how likely the respective protein is an interaction partner of HSP70B are given in the diagrams. At a significance cutoff of 5 %, CGE1 is confirmed as an interaction partner of HSP70B [31], while RBSC2 and RPL6 are not. This is also reflected by the zero slope value for CGE1 and the high, positive slope values for RPL6 and RBCS2

6. For visual inspection of the results, the ratio between the abundance of a potential interaction partner and the ^{15}N-labeled bait can be plotted against the amount of added ^{14}N-labeled bait. While high positive slope values are expected for contaminants (Fig. 5c, d), the slope values for true interaction partners should be around zero (Fig. 5b).

4 Notes

1. The method described here is also applicable for SILAC. On the one hand, SILAC might be beneficial if tools for the identification and quantitation of ^{15}N-labeled peptides are not readily available. On the other hand, SILAC (1) requires strains that are auxotrophic for the labeled amino acids added to the medium, (2) is much more cost intensive than ^{15}N stable isotope labeling, and (3) using arginine as labeled amino acid may lead to arginine-to-proline interconversion which renders quantification much more tedious [9, 21]. As Chlamydomonas strains auxotrophic for arginine exist, SILAC using $^{13}C_6$-L-arginine has been applied to Chlamydomonas previously [22–24]. However, as Chlamydomonas strains auxotrophic for lysine do not exist, only tryptic arginine peptides can be quantified. Moreover, significant arginine-to-proline interconversion was observed in Chlamydomonas [3]. Therefore, full metabolic labeling with ^{15}N is considered to represent the preferred labeling method for this organism [14].
2. Depending on the experimental design, this may also be an epitope tag like the GFP moiety, HA, or FLAG peptides if the bait protein to be investigated is a fusion protein with the respective tag. Neither stable isotope labeling nor protein inactivation would be required if the experiment was done with affinity tags.
3. For a more rigorous separation between soluble and membrane proteins, ultracentrifugation may be employed.
4. Protein A-Sepharose beads coupled to IgGs may be stored until further use at 4 °C after adding sodium azide to a final concentration of 0.02 % (v/w).
5. Several other possibilities exist to determine the absolute amount of the bait in cell lysates: (1) absolute amounts might be reported in the literature, especially for well-studied species like *Saccharomyces cerevisiae* [25]. (2) Absolute protein amounts can be estimated by immunoblotting using dilution series of whole-cell extracts and the purified antigen [13]. (3) Synthetic ^{14}N proteotypic peptides may be spiked into tryptic digests of ^{15}N-labeled whole-cell extracts to estimate the absolute amount by MS.
6. Reactivation of denatured unlabeled bait by cellular chaperones is unlikely as cell lysates, at least in *Chlamydomonas*, are rapidly depleted from ATP [26].
7. Remember that in this step the same amount of antibodies is coupled to the beads for all IPs. The amount of antibodies used should be sub-stoichiometric with respect to the ^{15}N-labeled bait in the sample. It is important that the affinity

beads are prepared in one batch to ensure their homogeneity. For this, the preparation of the affinity beads is performed in a single 15 ml tube.

8. For the competition coIP, inactivated ^{14}N bait should be used at stoichiometries of approximately 0.1–10 relative to the ^{15}N-labeled bait to provide a balance between sensitivity and selectivity.
9. Different search engines, commercial and free, for the identification and quantitation of peptides/proteins are available and they differ in their capability to identify or quantify ^{14}N and ^{15}N peptides in one run [MaxQuant [17], MSQuant [19], Sequest [27], Mascot [28], X!Tandem [29], ASAPRatio [30], just to name a few]. Most programs can be easily adjusted to do at least ^{15}N identification/quantification for the data analysis needed here. It is advisable to use an identification/quantitation program like MaxQuant or MSQuant that uses mass-retention time correlation to minimize missing identification/quantification values in the titration series.
10. This step is only necessary if quantitation of the ratio of ^{14}N to ^{15}N baits is required for estimating the efficiency with which the native ^{15}N bait is competed from antibody-binding sites by the added ^{14}N bait. In case MaxQuant is used for two separate searches on ^{14}N and ^{15}N amino acids using label-free quantitation, obtained results can be used for the analysis as well.
11. More than the four coIPs done here, including additional variations of the amount of ^{14}N bait added and biological replicates, would improve judging the significance of a potential protein-protein interaction and, thus, the sensitivity and specificity of the method.

Acknowledgements

This work was supported by grants from the Deutsche Forschungsgemeinschaft (Schr 617/5-1) and the Bundesministerium für Bildung und Forschung (Systems Biology Initiative FORSYS, project *Go*FORSYS), and by the Max Planck Society.

References

1. Markham K, Bai Y, Schmitt-Ulms G (2007) Co-immunoprecipitations revisited: an update on experimental concepts and their implementation for sensitive interactome investigations of endogenous proteins. Anal Bioanal Chem 389:461–473
2. ten Have S, Boulon S, Ahmad Y et al (2011) Mass spectrometry-based immuno-precipitation proteomics - the user's guide. Proteomics 11:1153–1159
3. Heide H, Nordhues A, Drepper F et al (2009) Application of quantitative immunoprecipitation

combined with knockdown and cross-linking to *Chlamydomonas* reveals the presence of vesicle-inducing protein in plastids 1 in a common complex with chloroplast HSP90C. Proteomics 9:3079–3089

4. Schmollinger S, Strenkert D, Offeddu V et al. (2012) A protocol for the identification of protein-protein interactions based on ^{15}N metabolic labeling, immunoprecipitation, quantitative mass spectrometry and affinity modulation. J Vis Exp 67, pii: 4083
5. Selbach M, Mann M (2006) Protein interaction screening by quantitative immunoprecipitation combined with knockdown (QUICK). Nat Methods 3:981–983
6. Rigaut G, Shevchenko A, Rutz B et al (1999) A generic protein purification method for protein complex characterization and proteome exploration. Nat Biotechnol 17:1030–1032
7. Wan LK, Chen N, Klemmer P et al (2012) Identifying true protein complex constituents in interaction proteomics: the example of the DMXL2 protein complex. Proteomics 12: 2428–2432
8. Paul FE, Hosp F, Selbach M (2011) Analyzing protein-protein interactions by quantitative mass spectrometry. Methods 54:387–395
9. Ong SE, Mann M (2006) A practical recipe for stable isotope labeling by amino acids in cell culture (SILAC). Nat Protoc 1:2650–2660
10. Engelsberger WR, Erban A, Kopka J et al (2006) Metabolic labeling of plant cell cultures with $K(^{15})NO_3$ as a tool for quantitative analysis of proteins and metabolites. Plant Methods 2:14
11. Schroda M, Vallon O, Wollman FA et al (1999) A chloroplast-targeted heat shock protein 70 (HSP70) contributes to the photoprotection and repair of photosystem II during and after photoinhibition. Plant Cell 11:1165–1178
12. Harris EH (2008) The *Chlamydomonas* sourcebook: introduction to *Chlamydomonas* and its laboratory use, vol 1, 2nd edn. Elsevier, San Diego, CA
13. Willmund F, Schroda M (2005) HEAT SHOCK PROTEIN 90C is a bona fide Hsp90 that interacts with plastidic HSP70B in *Chlamydomonas reinhardtii*. Plant Physiol 138: 2310–2322
14. Mühlhaus T, Weiss J, Hemme D et al (2011) Quantitative shotgun proteomics using a uniform 15N-labeled standard to monitor proteome dynamics in time course experiments reveals new insights into the heat stress response of *Chlamydomonas reinhardtii*. Mol Cell Proteomics 10(M110):004739
15. Lowry OH, Rosebrough NJ, Farr AL et al (1951) Protein measurement with the Folin phenol reagent. J Biol Chem 193:265–275
16. Rappsilber J, Ishihama Y, Mann M (2003) Stop and go extraction tips for matrix-assisted laser desorption/ionization, nanoelectrospray, and LC/MS sample pretreatment in proteomics. Anal Chem 75:663–670
17. Cox J, Mann M (2008) MaxQuant enables high peptide identification rates, individualized p.p.b.-range mass accuracies and proteome-wide protein quantification. Nat Biotechnol 26:1367–1372
18. Cox J, Neuhauser N, Michalski A et al (2011) Andromeda: a peptide search engine integrated into the MaxQuant environment. J Proteome Res 10:1794–1805
19. Mortensen P, Gouw JW, Olsen JV et al (2010) MSQuant, an open source platform for mass spectrometry-based quantitative proteomics. J Proteome Res 9:393–403
20. Sokal RR, Rohlf FJ (1995) Biometry: the principles and practice of statistics in biological research, 3rd edn. W.H. Freeman, New York, NY
21. Van Hoof D, Pinkse MW, Oostwaard DW et al (2007) An experimental correction for arginine-to-proline conversion artifacts in SILAC-based quantitative proteomics. Nat Methods 4:677–678
22. Naumann B, Busch A, Allmer J et al (2007) Comparative quantitative proteomics to investigate the remodeling of bioenergetic pathways under iron deficiency in *Chlamydomonas reinhardtii*. Proteomics 7:3964–3979
23. Naumann B, Stauber EJ, Busch A et al (2005) N-terminal processing of Lhca3 Is a key step in remodeling of the photosystem I-light-harvesting complex under iron deficiency in *Chlamydomonas reinhardtii*. J Biol Chem 280:20431–20441
24. Terashima M, Specht M, Naumann B et al (2010) Characterizing the anaerobic response of *Chlamydomonas reinhardtii* by quantitative proteomics. Mol Cell Proteomics 9: 1514–1532
25. Ghaemmaghami S, Huh WK, Bower K et al (2003) Global analysis of protein expression in yeast. Nature 425:737–741
26. Liu C, Willmund F, Golecki JR et al (2007) The chloroplast HSP70B-CDJ2-CGE1 chaperones catalyse assembly and disassembly of VIPP1 oligomers in *Chlamydomonas*. Plant J 50:265–277
27. Eng J, McCormack AL, Yates JR (1994) An approach to correlate tandem mass spectral

data of peptides with amino acid sequences in a protein database. J Am Soc Mass Spectrom 5:976–989

28. Perkins DN, Pappin DJ, Creasy DM et al (1999) Probability-based protein identification by searching sequence databases using mass spectrometry data. Electrophoresis 20:3551–3567

29. Craig R, Beavis RC (2004) TANDEM: matching proteins with tandem mass spectra. Bioinformatics 20:1466–1467

30. Li XJ, Zhang H, Ranish JA et al (2003) Automated statistical analysis of protein abundance ratios from data generated by stable-isotope dilution and tandem mass spectrometry. Anal Chem 75:6648–6657

31. Schroda M, Vallon O, Whitelegge JP et al (2001) The chloroplastic GrpE homolog of *Chlamydomonas*: two isoforms generated by differential splicing. Plant Cell 13: 2823–2839

Chapter 18

Protein Correlation Profiling-SILAC to Study Protein-Protein Interactions

Anders R. Kristensen and Leonard J. Foster

Abstract

An interactome describes the global organization of protein interactions within a cell and is typically generated using affinity purification-mass spectrometry (AP-MS), yeast two-hybrid screening, or protein-fragment complementation assays (Gavin et al. Nature 440: 631–636, 2006; Krogan et al. Nature 440: 637–643, 2006; Uetz et al. Nature 403: 623–627, 2000; Tarassov et al. Science 320: 1465–1470, 2008). These techniques have been widely used to depict the interactome as we know it today but current models of interactomes do not contain stoichiometric or temporal information. In this chapter we describe size-exclusion chromatography (SEC) combined with protein correlation profiling-stable isotope labeling by amino acids in cell culture (PCP-SILAC) to generate dynamic chromatographs for thousands of proteins (Kristensen et al. Nat Methods 9: 907–909, 2012). Using the precise co-elution of two proteins as evidence that they interact, it is possible to identify similar numbers of protein interactions without overexpression or creating fusion proteins as other high-throughput techniques require. In addition, triplex SILAC allows us to quantify protein stoichiometry and temporal changes to the interactome following perturbation. Finally, SEC-PCP-SILAC is very time efficient since it generates two orders of magnitude fewer samples for LC-MS analysis and avoids the tedious tagging and purification steps, making it possible for everyone with a single mass spectrometer to study the interactome.

Key words PCP-SILAC, Interactome, Protein-protein interaction network, Co-elution, Size-exclusion chromatography, Matlab

1 Introduction

The study of the interactome (all protein-protein interactions within a system) holds great promise for answering fundamental questions about biological pathways, not only with respect to their basic organization but also, more importantly, how they are rearranged following perturbation. Existing methods for studying interactomes require tagging or creating fusion proteins of all open reading frames of interest to provide a measurable readout or enable purification of the protein complex and subsequent identification of the interacting proteins by mass spectrometry [1, 2].

Bettina Warscheid (ed.), *Stable Isotope Labeling by Amino Acids in Cell Culture (SILAC): Methods and Protocols*, Methods in Molecular Biology, vol. 1188, DOI 10.1007/978-1-4939-1142-4_18, © Springer Science+Business Media New York 2014

A protein tag is problematic because it can be time consuming to introduce and it can disrupt interactions or alter localization of the protein complex [1, 3]. In addition, existing large-scale methods are not easily amenable to addressing how an interactome responds to stimulation, thereby missing the amazing cellular adaptations we know to exist from a multitude of more focused studies.

Protein correlation profiling (PCP)-SILAC has been used to profile organelle proteins across sucrose gradients and thereby assign localizations to those proteins; it does not rely on an organelle being purified to homogeneity, but instead relies on proteins localized to the same organelle displaying similar profiles across a density gradient [4–8].

Recently we combined size-exclusion chromatography (SEC) and PCP-SILAC to generate dynamic chromatographs for thousands of proteins, leading to the assignment of over 7,000 binary protein interactions and 300 protein complexes [9]. This method detects similar numbers of protein interactions without overexpression or creating fusion proteins as other high-throughput techniques but also quantifies protein stoichiometry and temporal changes to the interactome following perturbation. The basic principle starts with SILAC labeling of three populations of cells, mass-encoding them using arginine and lysine isotopologs. The heavy population can subsequently be perturbed while the medium and light populations are left untreated and then all three populations are mechanically lysed. An enrichment of the cytosolic protein complexes from each lysate is performed, to ensure that irrelevant interactions among proteins in different cellular compartments will be minimized, before each lysate is separated on an SEC column with optimal resolution between 150 kDa and 2 MDa. The fractions from the light sample are then pooled together and aliquots are mixed into each of the medium/heavy fractions prior to tryptic digestion and mass spectrometric analysis (Fig. 1a). In this scheme, the light-labeled proteins act as internal standards and any interactome changes following perturbation of the heavy population are monitored with the medium/heavy ratio (Fig. 1b).

2 Materials

2.1 Cell Culture and Lysis

1. Standard culture medium: Dulbecco's modified Eagle medium (DMEM) for PCP-SILAC; custom-made culture medium formulated identically to the standard medium but lacking arginine and lysine [10, 11].
2. Amino acids for SILAC media: Add $^{12}C_6{}^{14}N_4$-arginine and $^{12}C_6{}^{14}N_2$-lysine to the "light," $^{13}C_6{}^{14}N_4$-arginine and $^{2}H_4$-lysine to the "medium," and $^{13}C_6{}^{15}N_4$-arginine and $^{13}C_6{}^{15}N_2$-lysine to the "heavy" medium.

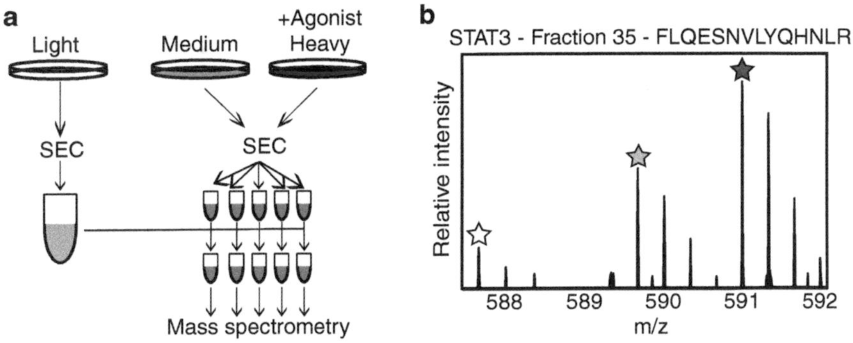

Fig. 1 Principle of the SEC-PCP-SILAC approach. (**a**) Three populations of Hela cells are metabolically labeled with amino acid isotopologs and the heavy population is stimulated with an agonist. The cells are lysed and the high MW fraction enriched by ultrafiltration prior to size-exclusion chromatography (SEC). SEC fractions from the light cells are pooled and subsequently spiked into each fraction from medium/heavy fractions as an internal standard prior to LC-MS/MS. (**b**) Mass spectrum of the triple SILAC-labeled peptide FLQESNVLYQHNLR from STAT3 that displays different spatiotemporal interaction changes following EGF stimulation. The medium:light ratio (M/L) is used to generate chromatograms, whereas the heavy-to-medium (H/M) ratio represents the impact of the agonist on that protein. In this example, STAT3 gets recruited to a complex in fraction 35 following EGF stimulation

3. SILAC media: 10 % dialyzed fetal bovine serum, L-glutamine 2 mM, penicillin/streptomycin 100 µg/ml.
4. Phosphate-buffered saline (PBS): 137 mM NaCl, 2.7 mM KCl, 10 mM Na_2HPO_4, 1.8 mM KH_2PO_4, pH 7.2, adjusted with HCl.
5. Lysis buffer: 20 mM Tris base, 50 mM KCl, 50 mM $NaCH_3CO_2$, including Halt protease and phosphatase inhibitors cocktail (Thermo Scientific) pH 7.2, adjusted with HCl; must be prepared freshly.
6. Dounce tissue grinder with thick pestle.

2.2 Sample Preparation and Protein Separation

1. Molecular weight markers for gel filtration chromatography.
2. SEC running buffer: 20 mM Tris base, 50 mM KCl, 50 mM $NaCH_3CO_2$, pH 7.2, adjusted with HCl.
3. 10 % stock solution of sodium deoxycholate.
4. 100 mM stock solution dithiothreitol (DTT).
5. 550 mM stock solution iodoacetamide (IAA).
6. Sequence grade-modified trypsin.
7. Trifluoroacetic acid (TFA).
8. Molecular weight cutoff filter (100 kDa).
9. HPLC-SEC column (*see* **Note 1**).
10. Preparative HPLC system, with fraction collector and UV detector and degasser.

2.3 LC-MS/MS and Data Handling

1. Hydrophilic buffer: 0.5 % acetic acid in water.
2. Hydrophobic buffer: 0.5 % acetic acid in 80 % acetonitrile.
3. StageTips [12, 13].
4. Benchtop computer with MaxQuant and Matlab installed.

3 Methods

3.1 Sample Preparation

1. Label the cells by letting them grow for at least five generations in SILAC media and then perturb the "heavy" population as desired (*see* **Note 2**).
2. Wash all three cell populations three times with PBS before harvesting them in 1 ml of PBS per 15 cm plate, collect them by centrifugation at 600 × *g* at 4 °C for 5 min, and lyse the cells by douncing in 2 ml of lysis buffer (*see* **Note 3**).
3. Clarify the lysates by centrifugation at 100,000 × *g* at 4 °C for 15 min and reduce the volume of the supernatant by ultrafiltration to 50 μl using spin filters with a molecular weight cutoff of 100 kDa to enrich for macromolecular complexes and concentrate the sample.
4. Combine "medium" and "heavy" samples just prior to loading them onto the SEC column.

3.2 Size-Exclusion Chromatography and Digestion

One of the most important steps in SEC-PCP-SILAC is to separate the protein complexes on SEC with the highest possible resolution. We operate a semipreparative HPLC system using a flow rate of 0.5 ml/min and a column temperature of 12 °C. The optimal number of fractions will come from the resolution of the column, with more fractions needed for higher resolution columns. Typically one should adjust the number of fractions collected such that even the narrowest peaks are covered by multiple fractions.

1. Run an SEC standard and make sure that the peaks are symmetric and have the expected elution volumes.
2. Load the combined "heavy/medium" sample in the smallest possible volume (*see* **Note 4**) and collect the fractions using a fraction collector. Typically 20 to 50 fractions are collected, depending on the column resolution and experimental design.
3. Load the "light" sample and collect the fraction from the same elution window as in **step 2** above using a fraction collector.
4. Pool all the fractions from the "light" sample collected in **step 3** into one tube and mix thoroughly.
5. Spike an aliquot of the recombined "light" fractions into each fraction from the "heavy/medium" fractions at a 1:1 ratio (v/v). In this instance the spiked light sample will be acting as an internal standard for quantitation.

6. Add sodium deoxycholate to a final concentration of 1 % and boil for 5 min. Reduce disulfide bonds with 10 mM DTT for 30 min at 37 °C followed by alkylation with 55 mM IAA for 20 min at 37 °C. Finally, digest the proteins into peptides by adding 1 μg of trypsin to 50 μg of protein overnight at 37 °C.
7. Precipitate the deoxycholic acid by adding TFA to a final concentration of 1 % and pellet the cholic acid at 16,000 × *g* for 10 min before desalting the peptides using StageTips as described [12].
8. Inject peptide mixtures into a nano-HPLC column, run a gradient from 0 to 60 % hydrophobic buffer, and operate the mass spectrometer in a data-dependent mode.

3.3 Data Analysis

The next step is to determine which proteins interact with one another. To this end, we have employed two approaches to systematically address the interactions represented in the chromatograms. First, for every chromatogram we calculated the Euclidian distance to all other chromatograms with the assumption that two proteins that always occur together in the same complex(es) would have similar chromatograms and, thus, small distances. Second, we deconvolved each chromatogram into component Gaussian curves with the assumption that for large complexes, which are made of independent, stable, and observable subcomplexes, the constituent proteins might only show similarities in part of the chromatogram (overlapping Gaussian curves) because they are not all always in a complex together (Fig. 2).

1. Quantify and identify the proteins with a suitable proteomic software package, e.g., MaxQuant [14]. Define each fraction as a separate experiment so that you will generate individual protein ratios for each fraction.
2. Prefilter the chromatograms by (1) adding one missing quantitation value by taking the average of the neighbor fractions, (2) making sure that there are quantitation values for five consecutive fractions or else discard the quantitation values.
3. To determine which proteins are always together in a complex, compare the raw chromatograms to each other, e.g., by clustering or calculating correlation coefficients between the profiles. Those very close in space can be considered to interact but the actual distance cutoff one can use must be determined empirically.
4. To determine if proteins interact but are not always with only the same partners, deconvolve each chromatogram to its component Gaussian curves using, e.g., the curve fitting toolbox in Matlab (www.mathworks.com) (*see* **Note 5**). If two proteins display similar center, height, and width, for a

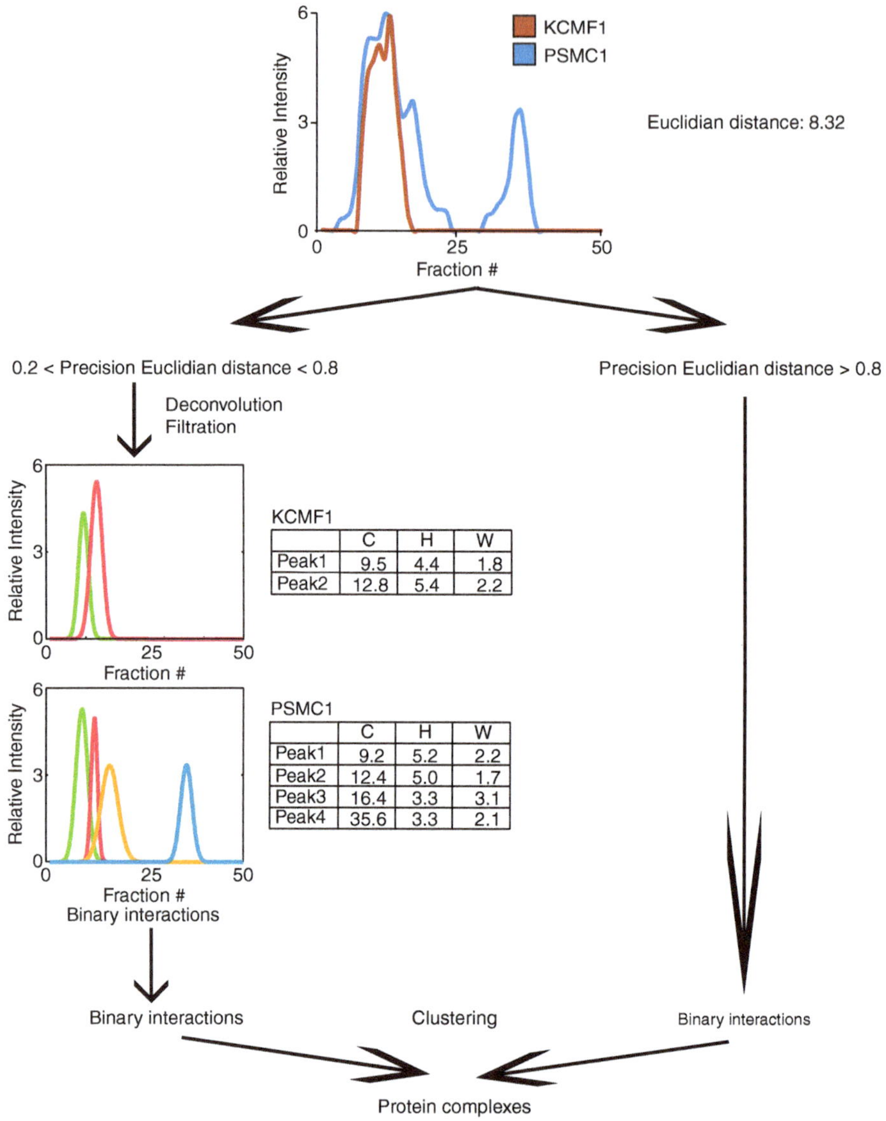

Fig. 2 Data analysis approach. If the precision was larger than 0.8 for a given Euclidian distance between any two chromatograms, the two proteins represented by those chromatograms were considered to be in a binary interaction. For distances below this threshold, the chromatograms were deconvolved into Gaussian components resulting in three coefficients per curve: center (C), height (H), and width (W). Then, if the Gaussian coefficients for any curves of two different proteins fell within a given window of one another, those two proteins were considered to be in a binary interaction. These interactions, represented in a binary matrix, can then be clustered to determine which proteins were in a complex together

deconvoluted Gaussian peak, the proteins are assigned as interacting (*see* **Note 6**).

5. Finally, by clustering a matrix of the binary interactions, protein complexes can be identified (*see* **Note 7**).

4 Notes

1. It is very important that the SEC column has the highest available resolution. We use a 7.8 × 600 mm BioSep4000 column (Phenomenex); however, any column with similar high resolution can be used. The dimensions of the column should be carefully considered: larger pore sizes for larger complexes and smaller column diameter if less protein is applied.
2. The perturbation should be relatively short (<30 min) to avoid protein expression having an effect. If using longer perturbation times, then the interaction change should be normalized to protein expression change, similar to long-time phosphoproteomics studies [15].
3. Check under a microscope if the cells are lysed efficiently. The procedure can also be performed for organelles in the cell. In that case, the organelle of interest should first be isolated before disrupting it to release the protein complexes.
4. It is important to load the protein complexes in the smallest possible volume onto the SEC column since the minimum peak volume is defined by the loading volume where no binding and release from the SEC column are taking place.
5. It is important to perform cross-validation, e.g., the leave-one-out cross-validation approach, to avoid over-fitting the curves to the chromatogram [16].
6. It is important to draw receiver-operator characteristic and precision-recall curve to estimate the performance of the parameters. The parameter performances are a result of the resolution of the SEC column, number of fractions being collected, precision of quantitation, etc.
7. Programs such as Matlab, R, and TM4 are commonly available packages that contain all the clustering tools one would need for this type of data.

References

1. Gavin A-C, Aloy P, Grandi P et al (2006) Proteome survey reveals modularity of the yeast cell machinery. Nature 440:631–636
2. Krogan NJ, Cagney G, Yu H et al (2006) Global landscape of protein complexes in the yeast Saccharomyces cerevisiae. Nature 440:637–643
3. Werner JN, Chen EY, Guberman JM et al (2009) Quantitative genome-scale analysis of protein localization in an asymmetric bacterium. Proc Natl Acad Sci U S A 106: 7858–7863
4. Andersen JS, Wilkinson CJ, Mayor T et al (2003) Proteomic characterization of the human centrosome by protein correlation profiling. Nature 426:570–574
5. Andersen JS, Lam YW, Leung AKL et al (2005) Nucleolar proteome dynamics. Nature 433: 77–83
6. Foster LJ, de Hoog CL, Zhang Y et al (2006) A mammalian organelle map by protein correlation profiling. Cell 125:187–199
7. Jakobsen L, Vanselow K, Skogs M et al (2011) Novel asymmetrically localizing components of human centrosomes identified by complementary proteomics methods. EMBO J 30: 1520–1535

8. Dengjel J, Hoyer-Hansen M, Nielsen MO et al (2012) Identification of autophagosome-associated proteins and regulators by quantitative proteomic analysis and genetic screens. Mol Cell Proteomics 11:M111.014035
9. Kristensen AR, Gsponer J, Foster LJ (2012) A high-throughput approach for measuring temporal changes in the interactome. Nat Methods 9:907–909
10. Ong S-E, Blagoev B, Kratchmarova I et al (2002) Stable isotope labeling by amino acids in cell culture, SILAC, as a simple and accurate approach to expression proteomics. Mol Cell Proteomics 1:376–386
11. Ong S-E, Mann M (2007) A practical recipe for stable isotope labeling by amino acids in cell culture (SILAC). Nat Protoc 1: 2650–2660
12. Rappsilber J, Mann M, Ishihama Y (2007) Protocol for micro-purification, enrichment, pre-fractionation and storage of peptides for proteomics using StageTips. Nat Protoc 2: 1896–1906
13. Rappsilber J, Ishihama Y, Mann M (2003) Stop and go extraction tips for matrix-assisted laser desorption/ionization, nanoelectrospray, and LC/MS sample pretreatment in proteomics. Anal Chem 75:663–670
14. Cox J, Mann M (2008) MaxQuant enables high peptide identification rates, individualized p.p.b.-range mass accuracies and proteome-wide protein quantification. Nat Biotechnol 26:1367–1372
15. Olsen JV, Vermeulen M, Santamaria A et al (2010) Quantitative phosphoproteomics reveals widespread full phosphorylation site occupancy during mitosis. Sci Signal 3:ra3
16. Lachenbruch PA, Mickey MR (1968) Estimation of error rates in discriminant analysis. Technometrics 10:1–11

Chapter 19

Autophagosomal Proteome Analysis by Protein Correlation Profiling-SILAC

Andrea C. Becker and Jörn Dengjel

Abstract

Autophagy is one of the two major degradation pathways within eukaryotic cells. Nevertheless, little is known about the protein composition of autophagosomes, the vesicles shuttling proteins to lysosomes for degradation. Protein correlation profiling in combination with stable isotope labeling by amino acids in cell culture is a stringent method to investigate the dynamics of the autophagosomal proteome. It enables the discrimination between autophagosomal and co-purifying proteins identifying organellar candidate proteins for further investigation.

Key words Autophagy, Organelle, Proteomics, SILAC, Protein correlation profiling, Mass spectrometry

1 Introduction

The autophagosomal-lysosomal system is responsible for degradation of intracellular, cytoplasmic protein complexes and whole organelles [1]. The proteomic analysis of autophagosomes is very challenging as the vesicles can virtually contain all cellular proteins shuttling them to lysosomes for degradation. A purification procedure has to be established that enables the discrimination between organellar proteins and content, and contaminating proteins. Most contaminations in organellar purification methods derive from cellular sub-compartments, which share similar physiochemical properties as the organelle of interest. Organelles may be purified (a) biochemically by cell fractionation or by affinity purification (AP), or (b) by mapping their distribution over a gradient, e.g., by mass spectrometry (MS), also described as protein correlation profiling (PCP) [2]. As modern mass spectrometers are very sensitive even minute amounts of contaminating proteins will be identified [3]. To discriminate contaminations from true hits, quantitative MS-based proteomics experiments have to be performed [4]. In PCP, organelles, in our case vesicles, are separated by density gradient centrifugation followed by quantitative MS-based proteomics

Bettina Warscheid (ed.), *Stable Isotope Labeling by Amino Acids in Cell Culture (SILAC): Methods and Protocols*, Methods in Molecular Biology, vol. 1188, DOI 10.1007/978-1-4939-1142-4_19,

as readout. Distribution profiles of several thousand proteins across a gradient can be determined by using peptide-extracted ion currents (XICs) as a measure of peptide abundance in each gradient fraction [5]. It is assumed that all proteins with profiles that closely follow the profiles of specific organellar marker proteins belong to the respective organelle. Comparing protein profiles allows distinguishing novel organellar candidates from contaminants.

In 2003, PCP was introduced using label-free quantification [5]. The centrosome was purified across a sucrose gradient, resulting in the identification of 41 candidate proteins of which 23 were validated. With the development of relative quantification assays employing stable isotope labeling it became evident that a combination of PCP with, e.g., iTRAQ [6–8] or stable isotope labeling by amino acids in cell culture (SILAC) [9, 10] was preferable as this resulted in more accurate protein profiles. In a recent study we used PCP-SILAC to study autophagosomal protein dynamics during different autophagy-inducing conditions [11]. Amongst others, the proteasome turned out to be one of the "favorite" substrates of autophagosomes. Stimulus-independent, proteasomal proteins were always identified associated to autophagosomes. During active autophagy proteasomal protein abundance as well as proteasome activity decreased in cells, indicating its degradation via autophagy.

In this chapter, we discuss the use of PCP-SILAC to study autophagosome composition and dynamics. SILAC leads to the metabolic incorporation of distinct stable isotope variants of L-arginine and L-lysine allowing relative quantification by MS-based proteomics. In comparison to iTRAQ, the quantitation is performed on the MS level instead of the MS/MS level resulting in higher numbers of data points per peptide and therefore more accurate protein quantification.

2 Materials

2.1 Cell Culture

1. Stable isotope-labeled amino acids: L-arginine-$^{13}C_6$ hydrochloride (Arg6), L-arginine-$^{13}C_6$, $^{15}N_4$ hydrochloride (Arg10), L-lysine-4,4,5,5-D_4 hydrochloride (Lys4) and L-lysine-$^{13}C_6$, $^{15}N_2$ hydrochloride (Lys8).
2. Unlabeled amino acids: L-arginine, L-lysine, and L-proline.
3. SILAC Dulbecco's modified Eagle medium (SILAC-DMEM) lacking arginine and lysine.
4. Dialyzed fetal bovine serum (dFBS).
5. 200 mM L-glutamine (100×).
6. Penicillin/streptomycin (100×; 10,000 U/ml and 10,000 μg/ml, respectively).
7. Trypsin-EDTA solution (200 mg/l trypsin, 500 mg/l EDTA).

8. Cell scraper.
9. Heraeus multifuge 3 SR E (75006445).

2.2 Cell Lysis, Density Gradient Centrifugation, and Protein Fractionation

1. Homogenization medium (HM): 0.25 M Sucrose, 1 mM EDTA, 20 mM Hepes-NaOH pH 7.4, protease inhibitors (Complete™ tablets, Roche Diagnostics).
2. Dounce tissue grinder with "tight" pestle (Wheaton): Depending on the sample amount, 1 and 7 ml grinders are generally used.
3. Iodixanol OptiPrep density gradient medium.
4. Ultracentrifuge with suitable swing-out rotor such as the Sorvall WX Ultra 80 with the rotor TH 641 (Thermo Scientific) and 12 ml tubes (# 06752; Thermo Scientific) for gradient centrifugation.
5. Micro-ultracentrifuge with fixed-angle rotor such as the Sorvall Discovery M150 SE with the rotor S100AT6 (Thermo Scientific) and 4PC tubes (# S404332A; Hitachi Koki) for pelleting of vesicular gradient fractions.
6. Dithiothreitol (DTT) (100 mM stock concentration, stored at −20 °C).
7. Iodoacetamide (555 mM stock concentration, stored at −20 °C).
8. SDS loading buffer [6× loading buffer with 0.35 M TRIS (pH 6.8), 10 % SDS, 30 % glycerol (99.9 %), 0.0005 % bromophenol blue].
9. NuPAGE® Novex 4–12 % Bis-Tris gradient gels (Life Technologies).
10. MOPS running buffer (20× buffer composed of 50 mM MOPS, 50 mM Tris base, 0.1 % SDS, 1 mM EDTA, pH 7.7).
11. Antioxidants.
12. Colloidal Blue Stain.
13. Scalpel.
14. Parafilm.

2.3 In-Gel Digest and LC-MS Analysis

1. ABC buffer: 100 mM ammonium bicarbonate, pH 7.5.
2. Ethanol (HPLC grade).
3. Trifluoroacetic acid (TFA), 2 %.
4. Buffer A: 0.5 % acetic acid in water.
5. Buffer A*: 3 % acetonitrile (ACN) and 0.3 % TFA in water.
6. Buffer B: 0.5 % acetic acid and 80 % ACN in water.
7. StageTip material (Empore Discs, C18 material from 3 M).
8. Modified sequencing-grade trypsin, 12.5 ng/μl in ABC buffer.
9. Reprosil-Pur C18 AQ 3 μm beads (Dr. Maisch GmbH).
10. Silica Tip Emitters (New Objective).

3 Methods

The experimental approach provided below describes the application of PCP-SILAC for the proteomic analysis of autophagosomes. The experimental setup can be used for double- or triple-labeling approaches. The triple labeling has the advantage of producing a biological replicate in a single experiment, which reduces MS measuring time (Fig. 1).

3.1 Stable Isotope Labeling by Amino Acids in Cell Culture

The concentration of SILAC amino acids and proline in the medium is cell line dependent and needs to be adjusted (*see* **Note 1**). This is an exemplified recipe for MCF7 cells.

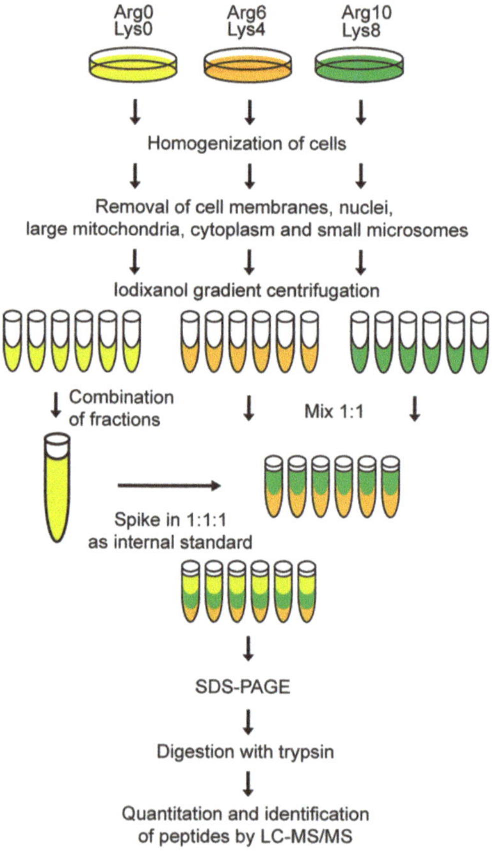

Fig. 1 Experimental setup for a triple-label PCP-SILAC experiment. Autophagosomes are purified by different centrifugation steps, followed by gradient centrifugation. Six fractions of each label are collected and the Lys0/Arg0 fractions are combined yielding an internal standard mixture of proteins over the gradient. The Lys4/Arg6-labeled fractions are mixed with the respective Lys8/Arg10-labeled fractions. The internal standard is distributed in a 1:1:1 ratio to the mixed Lys4/Arg6-Lys8/Arg10-labeled fractions. The combined samples are separated by SDS-PAGE, in-gel digested by trypsin, and analyzed by LC-MS/MS

1. Dissolve amino acids (final concentration of 42 mg/l L-arginine HCl, 73 mg/l L-lysine HCl, and 1.33 mg/l L-proline) in approx. 10 ml of SILAC medium, sterile filter the solution, and add it to the final amount of SILAC medium.
2. Complement SILAC-DMEM with 10 % dFCS, 2 mM L-glutamine, and 1 % Pen/Strep.
3. Cells need to be cultured in SILAC medium containing light (Lys0 and Arg0), medium (Lys4 and Arg6), or heavy (Lys8 and Arg10) labeled amino acids. To gain full incorporation of labeled amino acids, cells should be cultured for at least five cell doublings in SILAC medium. Incomplete incorporation could lead to quantification inaccuracies. The same number of cells should be used for each condition to ensure comparable loading of gradients and avoid inaccurate quantification (*see* **Note 1**).

3.2 Preparation of Density Gradients

1. Prepare gradient steps by mixing HM with iodixanol. Gradient steps consist in total of 1.6 ml. Prepare five solutions of 5 %, 10 %, 16 %, 24 %, and 30 % iodixanol, respectively.
2. Prepare gradient by underlying layers with higher density solutions in 12 ml centrifugation tubes (start with 5 % iodixanol and add the next higher density solution underneath).
3. Seal tubes with parafilm. The stacked gradients need to diffuse for 4–5 h in a horizontal position at RT.

3.3 Organellar Fractionation

1. Harvest at least 3×10^7 cells per condition by scraping. Pellet cells at 4 °C, 1,000 ×*g*, for 5 min (*see* **Note 2**). Resuspend the cell pellet in HM (approx. three times the pellet volume) and dounce 150 times to disrupt the cells mechanically and to keep organelles intact (*see* **Note 3**). The different SILAC-labeled cell lysates need to be prepared in parallel.
2. Use different centrifugation steps to remove cell compartments. Spin for 10 min at 1,000 ×*g* to remove nuclei and plasma membrane. Transfer the supernatant to a new reaction tube and spin for 10 min at 3,000 ×*g* to remove mitochondria. Transfer again the supernatant to a new reaction tube and spin for 15 min at 17,000 ×*g* to pellet the vesicular fraction. Discard the supernatant and resuspend the pellet in 1 ml of HM (*see* **Note 4**; Fig. 1). All centrifugation steps should be performed at 4 °C.
3. Load the pre-purified vesicular fraction, which contains the autophagosomes, on top of a density gradient. The vesicular mixture will get separated according to its density.
4. Centrifuge for 17 h at 100,000 ×*g* and 6 °C in a swing-out rotor.
5. Collect six 1 ml gradient fractions from the top of each gradient using a pipette (*see* **Note 5**). Mix all collected fractions of the light label (*see* **Note 6**; Fig. 1). Add 1 ml of the mixture as an internal standard to each collected fraction of the heavy

SILAC label. In case of triple labeling, mix the respective fractions of medium-heavy and heavy labels and add 1 ml of the internal, light standard.

6. Dilute the sample 1:1 with HM and centrifuge at 40,000 × *g* at 4 °C for 20 min in a fixed-angle rotor to pellet the vesicles. Dissolve the pellets in 25 µl SDS sample buffer.

3.4 MS Sample Preparation

1. Heat the samples to 75 °C for 10 min with 1 mM DTT to reduce the disulfide bonds of cysteine residues. For alkylation, incubate for 30 min with 5.5 mM iodoacetamide at room temperature in the dark.
2. Separate the proteins by SDS-gel electrophoresis on NuPAGE® Novex 4–12 % Bis-Tris gels. Antioxidant should be used to maintain proteins in a reduced state during SDS-gel electrophoresis. Visualize the proteins by staining with Colloidal Blue according to the manufacturer's protocol.
3. Cut the gel lane with a scalpel into ten slices of equal size and cut afterwards each slice into small cubes (approx. 1 mm^3). Transfer cubes into a reaction tube and wash out the remaining Colloidal Blue by incubation for 10 min in ABC buffer followed by incubation for 10 min in ethanol. Repeat washing steps three times.
4. After the last incubation with ethanol, add 50 µl of trypsin solution to the dried gel pieces and let the cubes swell. Add 25 µl of ABC buffer to avoid that the cubes dry out. Incubate the reaction tubes for digestion overnight at 37 °C [12].
5. Stop trypsin activity by adding 25 µl of 2 % TFA and transfer the supernatant to a new reaction tube. Wash the remaining peptides out of the cubes by two-time incubation in 100 µl ethanol on a shaker for 5 min and combine the supernatants of the respective slices.
6. Concentrate peptide solutions to less than 50 µl in a speedvac to remove ethanol.
7. Prepare StageTips for desalting of peptide solutions [13]. Punch two 0.5 mm discs out of a C18 Empore disc and pack them tightly into a 200 µl pipette tip. Wash the tip first with 50 µl buffer B to remove impurities followed by two washing steps with 50 µl buffer A for equilibration.
8. Add 200 µl of buffer A/A* (75 %/25 %) to the peptide solution and load the mixture onto the C18-tip. Wash the tip with 100 µl buffer A and elute the peptides in 50 µl buffer B into a new reaction tube.
9. Concentrate the sample to less than 5 µl to remove acetonitrile and add 10 µl of buffer A/A* (75 %/25 %).
10. The samples are ready to be injected into an HPLC connected to a mass spectrometer.

3.5 LC-MS

Fractionate the peptide mixture by a reversed-phase chromatography column (ID 75 μm) filled with C18 material, eluting the peptides directly into a mass spectrometer.

3.6 Data Analysis

Mass spectrometry data can be analyzed by specialized software for protein identification and quantitation [e.g., MaxQuant [14] or Proteome Discoverer (Thermo Fisher)]. With these tools, ratios of heavy to light peptides can be generated which are then combined to calculate respective protein ratios and standard deviations. All proteins with ratios in all six fractions can be used to generate respective distribution profiles over the gradient. These profiles can be clustered to identify proteins from specific compartments/organelles, e.g., by soft fuzzy c-means clustering [15] (*see* **Note 7**; Fig. 2). The number of clusters may vary depending on the experiment. As a *rule of thumb*, we increase cluster numbers as long as known organellar marker proteins stay together in one cluster (*see* **Note 8**). The remaining proteins in a cluster containing the autophagosomal marker proteins are new autophagosome candidate proteins.

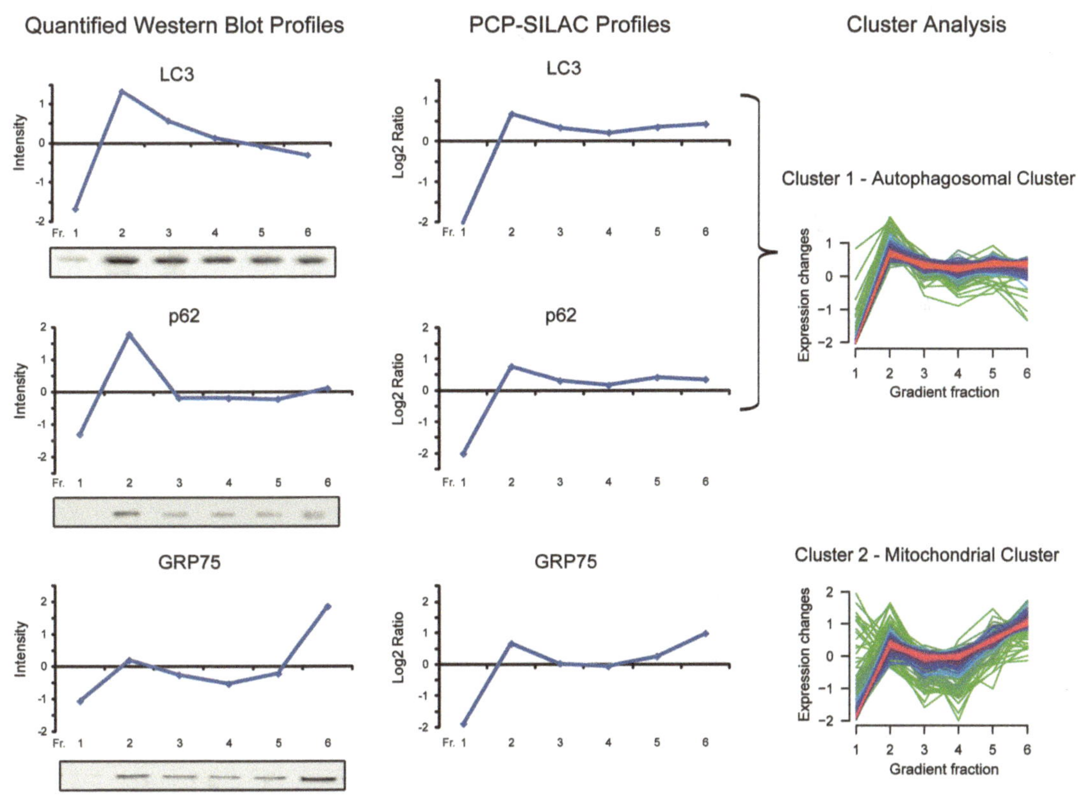

Fig. 2 Exemplified data of PCP-SILAC experiments. Cells were treated for 18 h with rapamycin and autophagosomes purified as described. Protein ratios are log_2 transformed and standardized. Profiles are recorded by western blot and MS. The MS profiles are clustered using GProX. Both methods highlight that the autophagosomal marker proteins LC3 and p62/SQSTM1 peak in fraction 2, and GRP75, a mitochondrial protein, in fraction 6, demonstrating resolution of different organelles by the gradient

4 Notes

1. Generally, we use approx. 5×10^7 cells per gradient. The content of arginine should be titrated to minimize the conversion to proline. Alternatively, unlabeled proline can be added to the medium. But again, its concentration should be titrated to avoid arginine conversion. Samples from the unmixed lysates can be used to check the level of incorporation of the heavy amino acids, in case of any troubleshooting.
2. Pellet can be stored at −80 °C for later use.
3. Detergent in the lysis buffer would destroy membranes and would lead to a loss of the organelles. Use protease inhibitor in the lysis buffer and perform all steps on ice to avoid degradation.
4. Resuspend the pellet properly by vortexing and pipetting to avoid aggregation. This could lead to a shift in density.
5. Fractionation by puncturing the bottom of the tube and sample collection from high to low density are supposed to be more accurate. However, we did not observe differences compared to fractionation from the top of the gradient.
6. Increase the volume of the mixed fractions with HM to ensure that 1 ml of the internal standard can be spiked to all collected fractions.
7. The clustering should be performed from log_2-transformed data. A useful analysis tool which can be used for soft clustering as well as other bioinformatics analyses is GProX [16].
8. Known organelles which can be used to check the quality of the clusters are the proteasome, mitochondrial proteins, large ribosomal subunits, and small ribosomal subunits. Autophagosomal marker proteins are SQSTM1/p62, MAP1LC3B (LC3), and GABARAPL (Fig. 2).

Acknowledgments

The research leading to these results has received funding from the Excellence Initiative of the German Federal and State Governments through FRIAS and BIOSS, from the Deutsche Forschungsgemeinschaft (GZ DE1757/2-1), and from the Federal Ministry of Education and Research through GerontoSys II—NephAge (031 5896 A).

References

1. Zimmermann AC, Zarei M, Eiselein S et al (2010) Quantitative proteomics for the analysis of spatio-temporal protein dynamics during autophagy. Autophagy 6:1009–1006
2. Engelke R, Becker AC, Dengjel J (2012) The degradative inventory of the cell: proteomic insights. Antioxid Redox Signal 17: 803–812
3. Dengjel J, Kratchmarova I, Blagoev B (2010) Mapping protein-protein interactions by quantitative proteomics. Methods Mol Biol 658: 267–278
4. Dengjel J, Kratchmarova I, Blagoev B (2009) Receptor tyrosine kinase signaling: a view from quantitative proteomics. Mol Biosyst 5: 1112–1121
5. Andersen JS, Wilkinson CJ, Mayor T et al (2003) Proteomic characterization of the human centrosome by protein correlation profiling. Nature 426:570–574
6. Dunkley TP, Hester S, Shadforth IP et al (2006) Mapping the Arabidopsis organelle proteome. Proc Natl Acad Sci U S A 103: 6518–6523
7. Dunkley TP, Watson R, Griffin JL et al (2004) Localization of organelle proteins by isotope tagging (LOPIT). Mol Cell Proteomics 3: 1128–1134
8. Sadowski PG, Groen AJ, Dupree P et al (2008) Subcellular localization of membrane proteins. Proteomics 8:3991–4011
9. Dengjel J, Jakobsen L, Andersen JS (2010) Organelle proteomics by label-free and SILAC-based protein correlation profiling. Methods Mol Biol 658:255–265
10. Jakobsen L, Vanselow K, Skogs M et al (2011) Novel asymmetrically localizing components of human centrosomes identified by complementary proteomics methods. EMBO J 30: 1520–1535
11. Dengjel J, Hoyer-Hansen M, Nielsen MO et al (2012) Identification of autophagosome-associated proteins and regulators by quantitative proteomic analysis and genetic screens. Mol Cell Proteomics 11:M111.014035
12. Shevchenko A, Tomas H, Havlis J et al (2006) In-gel digestion for mass spectrometric characterization of proteins and proteomes. Nat Protoc 1:2856–2860
13. Rappsilber J, Mann M, Ishihama Y (2007) Protocol for micro-purification, enrichment, pre-fractionation and storage of peptides for proteomics using StageTips. Nat Protoc 2: 1896–1906
14. Cox J, Mann M (2008) MaxQuant enables high peptide identification rates, individualized p.p.b.-range mass accuracies and proteome-wide protein quantification. Nat Biotechnol 26:1367–1372
15. Futschik ME, Carlisle B (2005) Noise-robust soft clustering of gene expression time-course data. J Bioinform Comput Biol 3: 965–988
16. Rigbolt KT, Vanselow JT, Blagoev B (2011) GProX, a user-friendly platform for bioinformatics analysis and visualization of quantitative proteomics data. Mol Cell Proteomics 10: O110.007450

Chapter 20

Design and Application of Super-SILAC for Proteome Quantification

Yair Pozniak and Tamar Geiger

Abstract

Stable isotope labeling with amino acids in cell culture (SILAC) is considered the most accurate method for proteome quantification by mass spectrometry. As it relies on active protein translation, it was traditionally limited to cells in culture and was not applicable to tissues. We have previously developed the super-SILAC mix, which is a mixture of several cell lines that serves as an internal spike-in standard for the study of human tumor tissue. The super-SILAC mix greatly improves the quantification accuracy while lowering error rates, and it is a simple, economic, and robust technique. Here we describe the design and application of super-SILAC to a broad range of biological systems, for basic biological research as well as clinical one.

Key words Mass spectrometry, Proteomics, Isotope labeling, Super-SILAC, FASP

1 Introduction

In the ever-progressing field of proteomics, mass spectrometry has become an indispensable tool due to its growing resolution, sensitivity, and accuracy of quantification [1–3]. Recent technological advances enable quantitative analyses of complex protein mixtures, including primary sequence identification, various posttranslational modifications, and protein-protein interactions [4–6]. Stable isotope labeling with amino acids in cell culture (SILAC) has emerged as an accurate, robust, and efficient technology for comparative analysis of different cell states [7]. However, since SILAC requires complete metabolic labeling of proteomes, it was applicable only to cultured cells and not to human tissue samples. Two related SILAC-based methodologies were developed in recent years to overcome this hurdle. The first introduced the use of SILAC-labeled cells as a spike-in standard [8]. The second broadened the applicability of this approach by combining an assortment of cell lines, a super-SILAC mix, to serve as the standard [9]. Super-SILAC is a mixture of several cell lines that differ in their origin, stage, and subtype. In the context of cancer studies, It was found

Bettina Warscheid (ed.), *Stable Isotope Labeling by Amino Acids in Cell Culture (SILAC): Methods and Protocols*, Methods in Molecular Biology, vol. 1188, DOI 10.1007/978-1-4939-1142-4_20, © Springer Science+Business Media New York 2014

to better represent the tumor the complexity and variability of tumors in comparison to a single cell line, as has been demonstrated for breast tumors, brain tumors, and lymphomas [9, 10]. Beyond these applications, spike-in SILAC standard in general and specifically super-SILAC can be applied to a large variety of samples, including various cells and tissues [11]. The super-SILAC method has several advantages even in cell culture experiments: In contrast to classical SILAC experiments, which routinely compare 2, 3, or 5 samples, here there is no limitation on the number of analyzed samples; the standard is common to all experimental samples and is prepared separately from—while being processed and analyzed together with—the "light"-labeled samples [11]; furthermore, the super-SILAC mix can be stored for years, and subsequent experiments can be analyzed and compared to previous ones using the same standard.

In this manuscript we emphasize the design of the super-SILAC mix, as it may have important implications on the outcome of the experiment. Two main parameters determine the quality of the spike-in standard: the ratios between the experimental samples and the standard, and the ability to quantify large proportions of the proteome. The design of the appropriate super-SILAC mix involves determination of the number and characteristics of cell lines for the mix. For the analysis of tumor samples, we recommend creating a mixture of 3–7 cell lines. Fewer cell lines might not represent the tumor adequately and too many will dilute each one and result in decreased representation of cell-specific proteins. The choice of cell lines to be used in the super-SILAC mix should be determined based on the experiment and the biological question. Design can start with a simple label-free evaluation of each candidate cell line and comparison to the target sample of interest. Statistical analysis can then show which cell line mixture would be most similar to the experimental sample and can further suggest the ratios in which these cells should be mixed. In the protocol below, we use the FASP digestion method [12] combined with strong anion exchange (SAX) fractionation [13] and data analysis with MaxQuant [14, 15]. In this type of experiment, one should expect to quantify >8,000 proteins with high accuracy. The super-SILAC method can also be used with alternative digestion methods, fractionations, and data analysis software. Moreover, it can be combined with enrichment steps for specific proteins of interest and posttranslational modifications.

2 Materials

All organic solvents should be of HPLC grade or higher. Buffers and solutions should be prepared with Milli-Q water. MS solutions should be prepared with HPLC-grade water.

2.1 Sample Preparation

1. Empore C_{18} disks for StageTips (3M).
2. Empore strong anion exchange (SAX) disks (3M).
3. FASP filters (30 kDa cutoff; Millipore or Sartorius) (*see* **Note 1**).
4. Proteomics-grade modified trypsin.
5. SDS lysis buffer: 4 % SDS, 100 mM Tris–HCl (pH 7.6).
6. Urea buffer: 8 M urea in 0.1 M Tris–HCl (pH 8.5) (*see* **Note 2**).
7. Dithiothreitol (DTT) solution: 1 M DTT in 50 mM ammonium bicarbonate.
8. Iodoacetamide (IAA) solution: 0.05 M IAA in urea buffer (*see* **Note 3**).
9. Digestion buffer: 10 % acetonitrile (ACN), 25 mM Tris-HCl (pH 8).
10. Strong anion exchange (SAX) buffers: Six buffers of different pH values, which are based on Britton-Robinson universal buffer (BRUB) [16]. To prepare 5× BRUB, mix 0.1 M acetic acid, 0.1 M phosphoric acid, and 0.1 M boric acid. Titrate this buffer with NaOH to the following pH values: 3, 4, 5, 6, 8, and 11. Prior to use, dilute each buffer fivefold in water and add NaCl to the pH 3 (1×) buffer to a concentration of 0.25 M (*see* **Note 4**).
11. MS buffer A: 0.1 % formic acid.
12. MS buffer B: 80 % ACN, 0.1 % formic acid.
13. MS buffer A*: 2 % ACN, 0.1 % trifluoroacetic acid.
14. Methanol.
15. Sodium hydroxide 1 M.
16. BCA Protein Assay Kit (Thermo Scientific Pierce).
17. Branson sonifier.
18. UV spectrophotometer.

2.2 Cell Culture and SILAC (see Note 5)

1. Cell lines of choice.
2. Appropriate cell culture medium without lysine and arginine (with glutamine).
3. Dialyzed serum.
4. Antibiotics.
5. SILAC amino acids: $^{13}C_6{}^{15}N_2$-L-lysine ("heavy lysine," Lys8) and $^{13}C_6{}^{15}N_4$-L-arginine ("heavy arginine," Arg10).
6. Lys8 and Arg10 stock solutions: Dissolve amino acids in sterile PBS to a concentration of 146 mg/ml for Lys8 and 84 mg/ml for Arg10.
7. Filter units for media (0.22 μm).

2.3 MS Acquisition

1. Nano-HPLC for online MS analysis
2. High-resolution mass spectrometer (preferentially Orbitrap based).

2.4 MS Data Analysis

1. MaxQuant software (or equivalent): Can be downloaded from www.maxquant.org.

3 Methods

Super-SILAC experiments consist of three basic steps: A label-free experiment for the selection of an appropriate mix, labeling of cells and incorporation testing, and the quantification of non-labeled sample using the super-SILAC mix as a standard (Fig. 1). Experimentally, these steps differ in cell culture conditions and peptide separation techniques, but share a common backbone of processing which includes sample digestion using the FASP protocol and LC-MS/MS analysis. The protocol below consists of the experimental steps from protein solubilization up to data analysis and interpretation.

3.1 Label-Free Evaluation of Cell Lines for the Super-SILAC Mix

1. Select candidate cell lines for the super-SILAC mix as described in Subheading 1.
2. Grow cells under standard conditions (without labeling). A total of 10^5 to 10^6 cells are sufficient for the preliminary tests.
3. Lyse cells with SDS lysis buffer and incubate the lysates at 95 °C for 5–10 min. Sonicate lysates for 20 s (2-s intervals, 40 % amplitude) to shear DNA and reduce sample viscosity

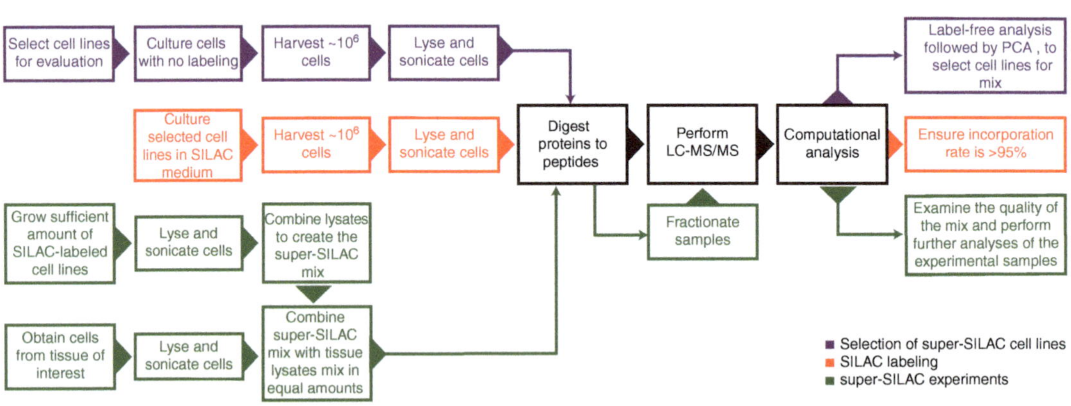

Fig. 1 A flow chart of a typical super-SILAC experiment. Procedures consist of three consecutive stages: Selection of candidate cell lines for the super-SILAC mix and label-free experiment for their evaluation (*purple*); SILAC labeling of the super-SILAC cell lines and testing of the labeling efficiency (*red*); quantification of the experimental sample proteome with super-SILAC as a spike-in standard (*green*). The common processing backbone is depicted in *black*

followed by centrifugation for 5 min at 16,000 × *g* at room temperature.

4. Transfer the supernatant to new test tubes.
5. Determine protein concentration using the bicinchoninic acid (BCA) method.
6. Add DTT to the samples to a final concentration of 100 mM. Incubate lysates with DTT for 20 min at room temperature.
7. Digest 50 μg of protein according to the FASP protocol as described below (**steps 8–21**) (*see* **Note 6**).
8. Dilute the sample 1:8 with urea buffer.
9. Place a FASP filter in a collection tube and load the diluted sample onto the filter.
10. Centrifuge at 8,000 × *g* for 10 min at room temperature.
11. Add 400 μl of urea buffer to the filter and centrifuge at 8,000 × *g* for 10 min at room temperature. Repeat this step twice. Discard flow-through from the collection tube when necessary (*see* **Note 7**).
12. Add 400 μl of IAA solution and mix at 600 rpm in a thermomixer for 1 min. Then incubate the sample in the dark for 20 min.
13. Centrifuge at 8,000 × *g* for 10 min at room temperature.
14. Wash the filter twice with urea buffer as indicated in **step 11**.
15. Wash the filter twice with 400 μl of digestion buffer.
16. Add 300 μl of digestion buffer to the filter, followed by 1 μg of trypsin (trypsin:protein ratio should be 1:50 to 1:100). Incubate the filter in a wet chamber at 37 °C for 4–18 h.
17. Transfer the filter to a new collection tube.
18. Centrifuge at 8,000 × *g* for 10 min at room temperature.
19. Add 200 μl of digestion buffer to the filter and repeat centrifugation.
20. Concentrate the sample by vacuum concentration.
21. Determine the peptide concentration with a UV spectrophotometer. Record the spectrum from 240 to 340 nm. It should have a distinct peak at 270–280 nm (*see* **Note 8**).
22. Purify the peptides on C_{18} StageTips [17] or equivalents according to published protocols (*see* **Note 9**).
23. Elute the samples from the StageTips with 60 μl of MS buffer B.
24. Remove ACN by vacuum concentration to reach a volume of 2–5 μl and dilute with MS buffer A* to a volume of 5–10 μl.

25. Inject 2–5 μl of each sample onto the column for LC-MS/MS analysis.
26. Perform LC-MS/MS analysis with a high-resolution MS instrument using a 4-h gradient or other methods that enable routine identification of more than 3,000 proteins.
27. Analyze the data with MaxQuant following the published protocol. After upload of the data, select the appropriate database file and use the following parameters: under "Group-specific parameters", select multiplicity 1 and "label free quantification". Under "Global parameters" select "Match between runs".
28. Perform principal component analysis (PCA) of the MaxQuant results (Fig. 2). Use the label-free quantification intensities (LFQ intensities) for the analysis. PCA will show the similarities of the samples and will enable selection of the appropriate super-SILAC mix. Furthermore, the analysis can highlight the optional ratios of various cell lines (*see* **Note 10**). The selected cell lines should resemble the experimental systems, but in order to represent a broad range of samples, the super-SILAC components should be as diverse as possible. Selection of samples with high similarity will not give any advantage in quantification.

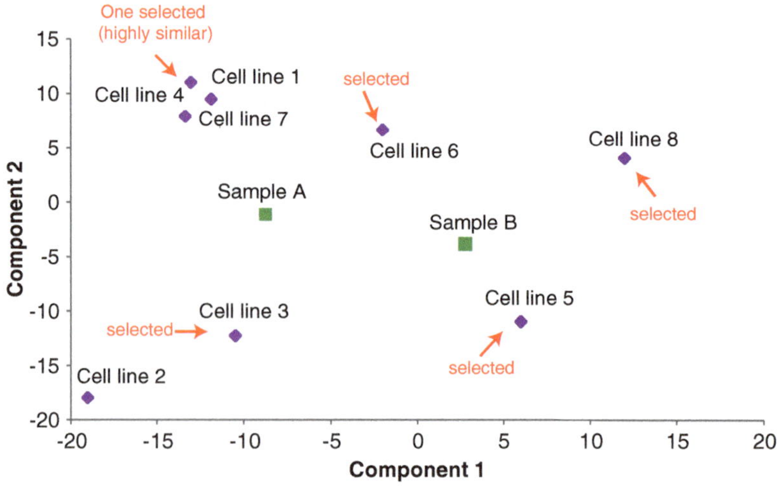

Fig. 2 Principal component analysis for the selection of super-SILAC cell lines. Candidate cell lines (*purple*) and representative experimental samples (*green*) are analyzed in a preliminary label-free experiment. Analysis shows the distribution of the samples and suggests the appropriate combination of cells. Cell lines should be selected according to their similarity to the experimental samples (cell line 2 is avoided), while cell lines with high similarity to each other will not benefit the mix and only one of them should be selected

3.2 SILAC Labeling of the Super-SILAC Cell Lines and Labeling Efficiency Testing

1. Dilute Lys8 and Arg10 1:2,000 into the appropriate cell culture medium deprived of those amino acids to a final concentration of 73 mg/l for lysine and 42 mg/l for arginine. These concentrations may need to be further titrated (*see* **Note 11**). Add dialyzed serum and antibiotics and filter the medium with a 0.22-μm filter.
2. Grow the cells in the SILAC medium for 5–10 doublings. If trypsin is used in passaging the cells, centrifuge the cells at 500 × g for 5 min after trypsinization to eliminate the trypsin; remains of trypsin may reduce labeling efficiency of the cells. Replace the medium every 2–3 days.
3. After five doublings, harvest a small amount of cells (~10^5 cells) and check labeling efficiency. For labeling check, lyse, digest, and perform LC-MS/MS analysis as described in Subheading 3.1 (**steps 3–26**).
4. In MaxQuant, under "Variable modifications", add Pro6 (proline 6 to determine arginine-to-proline conversion), select multiplicity 2, and mark the labels as Lys8 and Arg10. Do not select the "re-quantify" option. The labeling efficiency should be calculated based on the non-normalized SILAC ratios and should be calculated for lysine- and arginine-containing peptides separately. Determine the incorporation rate as [1 − (1/average ratio)]. Complete labeling is considered when the incorporation rate is higher than 95 %. Heavy proline should not exceed 1 % (*see* **Note 11**).

3.3 Performing the Super-SILAC Experiment

Grow a sufficient amount of SILAC-labeled cells for the complete experiment, including replicates. The total amount should be calculated by multiplying the number of samples by the amount of standard needed for each sample (for typical proteomics experiments, 50–100 μg of protein is sufficient).

1. Lyse the cells and determine the protein concentrations as described in Subheading 3.1 (**steps 3–5**).
2. To create the super-SILAC mix, combine the lysates of the labeled cell lines according to the ratios that were determined before.
3. Obtain cells from the sample of interest and lyse according to Subheading 3.1 (**steps 3–5**). Different tissues may require alternative methods.
4. Combine equal protein amounts of the super-SILAC mix with lysates from the sample of interest (50–100 μg of each).
5. Digest the combined samples according to the FASP protocol as described in Subheading 3.1 (**steps 7–21**). After determining the peptide concentrations, continue with the SAX fractionation protocol [13] as described below.

6. For each sample, prepare **one** SAX tip by stacking six layers of Empore anion exchange disks. Label the SAX tip with sample ID.
7. For each sample, prepare **six** C_{18} StageTips [17] by stacking three layers of Empore C_{18} disks. Label the StageTips with sample ID and pH value (11, 8, 6, 5, 4, and 3).
8. Activate the SAX tips using the following washes: first, 100 μl methanol wash; second, 100 μl 1 M NaOH wash; and third, two washes with 100 μl 1× SAX buffer pH 11.
9. Activate the C_{18} StageTips using the following washes: first, 50 μl methanol wash; second, 50 μl MS buffer B wash; and third, two 100 μl Milli-Q water washes.
10. Dilute 30–50 μg of the peptides in 1× SAX buffer pH 11 to a final volume of 200 μl. Adjust pH to 11 or higher with NaOH.
11. Assemble the SAX tips into the appropriate C_{18} StageTip (pH 11) and place both inside an adaptor and into a collection tube (*see* **Note 12**).
12. Load the diluted peptides onto the SAX tips and centrifuge at 1,500–3,000 × *g* for 3 min or until the entire sample has passed through both filters.
13. Add 100 μl of 1× SAX buffer pH 11 to the SAX StageTip. Repeat the centrifugation as described in **step 12**.
14. Transfer the SAX StageTip to the next C_{18} StageTip (pH 8) and repeat the previous steps with 1× SAX buffer pH 8.
15. Continue subsequently eluting with buffer pH 6, pH 5, pH 4, and pH 3, each into its corresponding StageTip.
16. Each sample should have six fractions on six C_{18} StageTips. Wash the StageTips with 50 μl of MS buffer A (*see* **Note 9**).
17. Perform the LC-MS/MS analysis as described in Subheading 3.1, **steps 23–26**.
18. Analyze the files using MaxQuant with the following parameters: Under the "Group-specific parameters" tab, select multiplicity 2 and mark the labels as Lys8 and Arg10; under the "Global parameters" tab, select "re-quantify" and "match between runs" if appropriate (*see* **Note 13**).
19. To examine the quality of the super-SILAC mix, analyze the proteinGroups.txt file. A spike-in standard is considered to be adequate when the protein ratios between the standard and the super-SILAC are low; typically more than 90 % of the proteins should have a ratio below fivefold. In addition, a histogram of these ratios should show a unimodal distribution. Further analysis of the results can include clustering of the data, PCA, network analysis, enrichment tests, and other bioinformatic analyses, which can eventually unravel the biological meaning of the experimental information.

4 Notes

1. There are two forms of Millipore filters: flat bottom and conical shaped. The protocol describes the use of the conical ones. For the use of the flat-bottom filters, the buffer volumes should be adjusted (decreased by two- to threefold). Our experience shows higher robustness but lower yield of the conical vs. the flat-bottom filters.
2. Urea buffer must be prepared fresh before use. Do not heat over 30 °C to avoid peptide and protein carbamylation.
3. IAA is light sensitive and must be kept in the dark.
4. The 5× BRUB stock solutions can be kept at room temperature for 1 year but the diluted solutions for the experiment must be prepared fresh.
5. This protocol describes the use of lysine and arginine in SILAC labeling, combined with trypsin. This ensures that every peptide, except for the C-terminal peptides of the proteins, would have at least one labeled amino acid. The user can select other amino acids and use appropriate medium and proteinase.
6. The FASP procedure described here is suitable for up to 500 μg of protein in 0.5-ml filters. When larger protein quantities are needed (i.e., prior to PTM enrichment steps), filters with higher capacity should be used and the volumes should be up-scaled accordingly.
7. Mix well in each step without touching the filter membrane. After centrifugation, ensure having sample retained on top of the filter. Transfer of the whole sample through the filter indicates that it is broken. In this case, the flow-through should be reloaded onto a new filter.
8. The yield of the FASP procedure should be 40–70 % of the original protein amount.
9. StageTips can be stored at 4 °C for several months.
10. It is possible to combine the cell lines in different ratios, according to the actual representation of different cells in the experimental tissue. For example, if a tumor tissue is known to contain a small fraction of fibroblasts, it is possible to combine a representative fibroblast cell line in an appropriate fraction of the super-SILAC mix.
11. In eukaryotic cells, the metabolic conversion of arginine to proline can hamper the accuracy of the quantitation due to incomplete labeling [18]. High proline-to-arginine conversion ratio would result in lower arginine incorporation rates in the labeling check experiment. Depending on the cell line of choice, arginine levels may have to be carefully titrated;

alternatively, proline levels can be titrated [18] or internal correction methods may be applied [19]. To calculate the arginine-to-proline conversion rate, refer to the "Pro6" column in the "evidence.txt" file. Count the incidence of nonzero values and divide by the total number of evidences. Conversion rate should not exceed 0.01.

12. We advise to cut the top (3–4 mm) of the SAX tip in order for the apparatus to fit conveniently in the microcentrifuge.
13. Human tissue lysates inevitably contain proteins that are not represented in the cell line-derived super-SILAC mix, such as extracellular matrix proteins. These proteins may have a high (>10-fold) ratio of quantification between the sample and the standard, which will lead to less accurate quantification. The "re-quantify" option in MaxQuant should estimate the ratio for these low-abundant SILAC proteins. Alternatively, perform an additional label-free analysis to quantify these proteins.

References

1. Gstaiger M, Aebersold R (2009) Applying mass spectrometry-based proteomics to genetics, genomics and network biology. Nat Rev Genet 10:617–627
2. Cravatt BF, Simon GM, Yates JR III (2007) The biological impact of mass-spectrometry-based proteomics. Nature 450:991–1000
3. Mann M, Kelleher NL (2008) Precision proteomics: the case for high resolution and high mass accuracy. Proc Natl Acad Sci U S A 105: 18132–18138
4. Cox J, Mann M (2011) Quantitative, high-resolution proteomics for data-driven systems biology. Annu Rev Biochem 80:273–299
5. Choudhary C, Mann M (2010) Decoding signalling networks by mass spectrometry-based proteomics. Nat Rev Mol Cell Biol 11: 427–439
6. Mallick P, Kuster B (2010) Proteomics: a pragmatic perspective. Nat Biotechnol 28:695–709
7. Ong SE, Mann M (2005) Mass spectrometry-based proteomics turns quantitative. Nat Chem Biol 1:252–262
8. Ishihama Y, Sato T, Tabata T et al (2005) Quantitative mouse brain proteomics using culture-derived isotope tags as internal standards. Nat Biotechnol 23:617–621
9. Geiger T, Cox J, Ostasiewicz P et al (2010) Super-SILAC mix for quantitative proteomics of human tumor tissue. Nat Methods 7:383–385
10. Deeb SJ, D'souza RC, Cox J et al (2012) Super-SILAC allows classification of diffuse large B-cell lymphoma subtypes by their protein expression profiles. Mol Cell Proteomics 11:77–89
11. Geiger T, Wisniewski JR, Cox J et al (2011) Use of stable isotope labeling by amino acids in cell culture as a spike-in standard in quantitative proteomics. Nat Protoc 6:147–157
12. Wisniewski JR, Zougman A, Nagaraj N et al (2009) Universal sample preparation method for proteome analysis. Nat Methods 6:359–362
13. Wisniewski JR, Zougman A, Mann M (2009) Combination of FASP and StageTip-based fractionation allows in-depth analysis of the hippocampal membrane proteome. J Proteome Res 8:5674–5678
14. Cox J, Matic I, Hilger M et al (2009) A practical guide to the MaxQuant computational platform for SILAC-based quantitative proteomics. Nat Protoc 4:698–705
15. Cox J, Mann M (2008) MaxQuant enables high peptide identification rates, individualized p.p.b.-range mass accuracies and proteome-wide protein quantification. Nat Biotechnol 26:1367–1372
16. Britton HTS, Robinson RA (1931) CXCVIII.-Universal buffer solutions and the dissociation constant of veronal. J Chem Soc (Resumed): 1456–1462

17. Rappsilber J, Mann M, Ishihama Y (2007) Protocol for micro-purification, enrichment, pre-fractionation and storage of peptides for proteomics using StageTips. Nat Protoc 2:1896–1906
18. Bendall SC, Hughes C, Stewart MH et al (2008) Prevention of amino acid conversion in SILAC experiments with embryonic stem cells. Mol Cell Proteomics 7:1587–1597
19. Van Hoof D, Pinkse MW, Oostwaard DW et al (2007) An experimental correction for arginine-to-proline conversion artifacts in SILAC-based quantitative proteomics. Nat Methods 4:677–678

Chapter 21

Proteomics Meets Genetics: SILAC Labeling of *Drosophila melanogaster* Larvae and Cells for In Vivo Functional Studies

Alessandro Cuomo, Roberta Sanfilippo, Thomas Vaccari, and Tiziana Bonaldi

Abstract

Stable isotope labeling by amino acids in cell culture (SILAC) is an established and potent method for quantitative proteomics. When combined with high-resolution mass spectrometry (MS) and efficient algorithms for the analysis of quantitative MS data, SILAC has proven to be the strategy of choice for the in-depth characterization of functional states at the protein level. The fruit fly *Drosophila melanogaster* is one of the most widely used model systems for studies of genetics and developmental biology. Despite this, a global proteomic approach in *Drosophila* is rarely considered. Here, we describe an adaptation of SILAC for functional investigation of fruit flies by proteomics: We illustrate how to perform efficient SILAC labeling of cells in culture and whole fly larvae. The combination of SILAC, a highly accurate global protein quantification method, and of the fruit fly, the prime genetics and developmental model, represents a unique opportunity for quantitative proteomic studies in vivo.

Key words Quantitative proteomics, *Drosophila melanogaster*, SILAC, Schneider cell, *Saccharomyces cerevisiae*, Intact organism labeling, Mass spectrometry

1 Introduction

The comprehensive study of proteomes has become an important part of the efforts aimed at unravelling the systemic properties of biological processes. In fact, the information provided by proteomics offers a close description of the phenotype in a specific state [1–5]. However, nowadays proteomics studies contribute successfully to "functional genomics" when they are able to provide quantitative information, as well as when they are rooted in the best discovery- and hypothesis-driven research. In recent years, stable isotope labeling by amino acids in cell culture (SILAC) has proved extremely successful for quantitative proteomics [6–9]. Such success would be enormously enhanced in combination with

Bettina Warscheid (ed.), *Stable Isotope Labeling by Amino Acids in Cell Culture (SILAC): Methods and Protocols*, Methods in Molecular Biology, vol. 1188, DOI 10.1007/978-1-4939-1142-4_21, © Springer Science+Business Media New York 2014

manipulation of model organisms, which provides the foundation of most modern discovery- and hypothesis-driven research.

Drosophila melanogaster represents an excellent model system to investigate genetic and developmental processes; hence, developing SILAC labeling of fruit fly cells and tissues has been an obligate step to apply quantitative proteomics for functional analyses in this model system. In the past, our group established the first protocol for the efficient labeling of *Drosophila* Schneider SL2 cells. In combination with RNA interference, our protocol allowed to analyze knockdown phenotypes on a global proteomic scale [10]. Cell culture models, however, do not always reflect accurately the multitude of regulatory mechanisms observed in vivo in multicellular eukaryotes. In particular, their applicability to the investigation of complex processes that involve interactions between different cell types, such as differentiation and development, is limited [11]. Krijgsveld and co-workers first reported the in vivo labeling of the fruit fly *D. melanogaster* for quantitation purposes, achieved by using the heavy stable isotope of nitrogen (^{15}N) [12]. This important proof of principle had however few applications in functional studies, due to the limited rate of protein identification and accuracy of protein quantification. The reason was that peptides were only partially labeled and produced complex isotope clusters, yielding varying mass shifts between the labeled and unlabeled peptides, and eventually making the identification by search algorithms very difficult.

Selbach and co-workers demonstrated the feasibility of in vivo SILAC labeling of *D. melanogaster* (SILAC flies) by feeding insects with SILAC-labeled yeast, which was grown in modified media to control the source of amino acids. With their method, they achieved almost full incorporation and proved applicability in vivo by analyzing the sexual dimorphism of protein abundance among female and male files [13]. More recently, Xu et al. published a slightly modified protocol, based on a series of assays to optimize the culture conditions for a more efficient SILAC labeling of flies; they then assessed the improved method for the proteomic characterization of a fly model for fragile X syndrome [14].

Elaborating on these studies, we discuss here a protocol with some adaptations from previously published ones for efficient in vivo labeling of *Drosophila*, using SILAC-labeled *Saccharomyces cerevisiae* strains as source of labeled amino acids. We also describe an improved method for culturing *Drosophila* Schneider SL2 cells in SILAC conditions (see the original description of the protocol with the full rationale in [15]). Our adaptations stem from direct experience in the setup and troubleshooting of existing methods and involve improved culturing of cells and flies and the ability to label efficiently larval stages, a workhorse of developmental and genetic studies. Our improved method is intended to facilitate successful application of quantitative strategies by the fly community and by researchers not currently considering *Drosophila* as a model system. We are confident that both will benefit from applying SILAC to the wealth of *Drosophila* mutants already available, to gain a proteomic perspective in vivo, on processes so far investigated only at the phenotypic and genomic level.

2 Materials

2.1 Cultivation of Drosophila SL2 Embryonic Cells

1. *Drosophila* SL2 cells.
2. Schneider's *Drosophila* medium.
3. Fetal bovine serum (heat inactivated, insect cell culture tested).
4. Penicillin, 50 units/ml.
5. Streptomycin, 50 mg/ml.
6. Bottles for growth of cells in suspension.
7. Cell culture dishes.
8. Trypan blue solution, 0.4 %.
9. Incubator at 26 °C, without CO_2 supply.
10. Acetone RPE ACS Reagent PEHD 1 l (Carlo Erba).

2.2 SILAC Labeling of SL2 Cells

2.2.1 Normal and Heavy Isotope-Enriched Amino Acids for SILAC Metabolic Labeling

1. L-lysine (light, Lys0), 100× stock solution: 165 g/l in PBS (2 g/l NaCl, 2 g/l KH_2PO_4, 11.5 g/l Na_2HPO_2, pH 7.4).
2. L-arginine (light, Arg0), 100× stock solution: 40 g/l in PBS.
3. $^{13}C_6{}^{15}N_2$-lysine (heavy, Lys8), 100× stock solution: 165 g/l in PBS.
4. $^{13}C_6{}^{15}N_4$-arginine (heavy, Arg10), 100× stock solution: 40 g/l in PBS.

2.2.2 Schneider's Culture Medium (See **Note 1**)

1. Solution A: 2.1 g NaCl, 0.43 g $Na_2HPO_4 \cdot 2H_2O$ (*see* **Note 2**), 0.68 g KH_2PO_4, 1.6 g KCl, 3.7 g $MgSO_4 \cdot 7H_2O$ (*see* **Note 3**), 0.2 g α-ketoglutaric acid, 0.10 g succinic acid, 0.10 g fumaric acid, and 0.10 g malic acid in 300 ml ddH_2O.
2. Solution B: 2.1 g glucose and 2.2 g trehalose in 50 ml ddH_2O.
3. Solution C: 0.55 g β-alanine, 0.04 g L-asparagine, 0.44 g L-aspartic acid, 0.07 g L-cysteine, 0.88 g L-glutamic acid, 0.27 g glycine, 0.44 g L-histidine, 0.16 g L-isoleucine, 0.16 g L-leucine, 0.88 g L-methionine, 0.16 g L-phenylalanine, 1.87 g L-proline, 0.27 g L-serine, 0.38 g L-threonine, 0.11 g L-tryptophane, 0.33 g L-valine, 0.44 g L-arginine (*see* **Note 4**), 0.10 g L-cystine (*see* **Note 5**), and 0.50 g L-tyrosine (*see* **Note 5**) in 450 ml ddH_2O.
4. Solution D: 2.0 g Yeastolate (*see* **Notes 6–8**) in 50 ml ddH_2O.
5. Solution E: 0.60 g $CaCl_2$ (*see* **Note 9**) in 50 ml ddH_2O.
6. 1 N KOH.
7. Stericup® Filter Units (Millipore).

2.3 In Vivo Labeling of Drosophila Melanogaster

2.3.1 Yeast Strain

S. cerevisiae strain BY4743 (MATa/α his3Δ1/his3Δ1 leu2Δ0/leu2Δ0 LYS2/lys2Δ0 met15Δ0/MET15 ura3Δ0/ura3Δ0) [16] (*see* **Note 10**).

2.3.2 Labeling Medium for Yeast

1. Yeast nitrogen base (YNB), 10× stock solution: 17 g YNB in 1 l ddH_2O.
2. D-Glucose, 25× stock solution: 50 g D-glucose in 100 ml ddH_2O.
3. Ammonium sulfate, 10× stock solution: 50 g ammonium sulfate in 1 l ddH_2O.
4. Amino acid drop-out solution, 25× stock solution: 5 g adenine, 0.5 g L-uracil, 2.5 g L-tyrosine, 2.5 g L-arginine, 1.5 g L-leucine, 1.5 g L-phenylalanine, 1 g L-tryptophane, 0.25 g L-histidine, and 0.25 g L-methionine in 1 l of 5 % EtOH.
5. Light and heavy L-lysine (Lys0 and Lys8), 30 mg each (*see* **Note 11**).

2.3.3 Minimal Fly Food for In Vivo Labeling

1. ddH_2O, 700 ml.
2. Sucrose, 150 g.
3. Agar, 7.5 g.
4. Labeled or unlabeled yeast pellet, 30 g.
5. Propionic acid, 4 ml.
6. Fly bottles and plugs.
7. Cheesecloth.
8. Spectrophotometer Ultraspec 2100 pro (GE Healthcare Life Science).

2.3.4 Embryo Collection

1. ddH_2O, 250 ml.
2. Molasses, 36 ml.
3. Agar, 8.8 g.
4. Tegosept (10 % in 95 % EtOH), 50 μl.
5. 100 mm plates.
6. Collection cages (Genesee, Cat # 59-101).
7. Small paintbrush.
8. Fly anesthesia apparatus.

2.4 Sample Preparation

2.4.1 Extraction and Fractionation of SL2 Cells

1. Swelling buffer: 20 mM HEPES-KOH pH 7.5, 250 mM sucrose, 0.5 mM EDTA, 1 mM DTT, protease inhibitors (Protease Inhibitor Cocktail, Roche Applied Science; *see* Subheading 2.4.2).
2. PBS.

3. RIPA buffer: 50 mM Tris–HCl pH 7.5, 150 mM NaCl, 1 % NP-40, 0.5 % deoxycholate, 0.1 % SDS, 0.4 mM EDTA, 10 % glycerol.
4. Trichloroacetic acid (TCA), 100 % solution.
5. Cell homogenizer (e.g., from Isobiotec, Heidelberg, Germany).
6. Vacufuge Concentrator 5301.

2.4.2 Preparation of Larvae/Embryo/Adult/Tissue Extracts

1. Wash solution: NaCl, 0.7 %.
2. Homogenization buffer:

8 M urea in 20 mM Hepes, pH 8	95 μl
PMSF, 100×	1 μl
25× solution (1 tablet in 2 ml ddH_2O) of Protease Inhibitor Cocktail	4 μl

3. A homogenization pestle for microcentrifuge tubes.

2.4.3 SDS-PAGE for Protein Separation

1. NuPAGE® Novex® 4–12 % Bis-Tris gels (Life Technologies).
2. NuPAGE® LDS sample buffer (4×; Life Technologies).
3. Colloidal Blue Staining kit (Life Technologies).

2.4.4 In-Gel Digestion

1. Acetonitrile (ACN) (HPLC grade).
2. Trifluoroacetic acid (TFA), 1 %.
3. Reduction buffer: 10 mM dithiothreitol (DTT) in 100 mM ammonium bicarbonate (ABC) in ultrapure water.
4. Alkylation buffer: 55 mM iodoacetamide (IAA) in 100 mM ABC in ultrapure water.
5. Digestion buffer: 50 mM ABC in ultrapure water (pH 8.0).
6. Destaining buffer: 25 mM ABC/50 % ACN.
7. Trypsin solution: 12.5 ng/μl sequencing-grade trypsin in 50 mM ABC.
8. Extraction buffer: 3 % trifluoroacetic acid (TFA)/30 % ACN.
9. Stage Tips [17]: Empore SPE C18 disks (3 M).

2.5 High-Performance Liquid Chromatography-Tandem Mass Spectrometry

1. HPLC solvent "A": 0.5 % acetic acid (AA) in ultrapure water.
2. HPLC solvent "B": 0.5 % AA/80 % ACN in ultrapure water.
3. Pico Tip Emitter 75 μm inner diameter (New Objective).
4. NANOBAUME Capillary Packing unit Western Fluids Engineering (http://www.westernfluids.net).
5. Reversed-phase material for nano-flow HPLC column: Reprosil-Pur C18-AQ, 3 μm (Dr. Maisch).
6. Online HPLC system: Easy-nLC (Proxeon Biosystems).

7. High-resolution mass spectrometer: Linear ion trap quadrupole (LTQ) Orbitrap hybrid instrument (LTQ-Orbitrap Classic, Xl or Velos, ThermoFisher Scientific).
8. Ion Source Kit for Thermo LTQ-FT (Proxeon Biosystems).

2.6 Data Acquisition and Analysis

1. MaxQuant software [18] (http://www.maxquant.org/).
2. PC 2 GB RAM minimum, with no upper limit on the number of cores, Microsoft Windows 7/Vista/2003/XP/2000; 64-bit version is recommended to enhance processing speed.
3. .NET Framework 3.5 from Microsoft.
4. MSFileReader; needed to access Thermo Fisher data.

3 Methods

3.1 Establishment of SILAC in Drosophila SL2 Cells

Drosophila cells are grown in Schneider's medium, which cannot be purchased in a formula adapted for SILAC labeling (e.g., specifically depleted of the standard SILAC amino acids arginine and lysine), and thus it must be assembled from individual components, based on published protocols [9, 10]. Furthermore, serum and total yeast extract, supplemented to the broth during culture, should be dialyzed in order to remove the unlabeled amino acids that may reduce labeling efficiency (*see* **Notes 6** and 7).

In order to detect possible alterations from normal physiology due to SILAC conditions, cell growth and viability should be carefully compared between cells cultivated in standard media and the same grown in SILAC media, as follows:

1. Inspect cells under the microscope to uncover any significant morphological alteration.
2. Count cells and plot growth curves of cells in the two conditions (*see* **Note 12**).
3. Estimate cell mortality by trypan blue staining.

3.1.1 Preparation of the "Minimal Schneider's Medium"

1. Combine sequentially solutions A–E (*see* **Notes 4** and **5**).
2. Adjust the pH slowly to 6.7 with 1 N KOH (~15 ml).
3. Bring the titrated A–E mix to a final volume of 1 l.
4. Sterilize by filtration using Stericup® and keep at 4 °C.

3.1.2 Preparation of the "Complete SILAC-Schneider's Medium"

1. Add the following components to 880 ml of "minimal Schneider's medium": 10 ml of arginine/lysine (either light or heavy) stock solution (100×), 6 ml of glutamine, 10 ml of penicillin/streptomycin (100×), 100 ml of serum (dialyzed against 0.9 % NaCl) (*see* **Note 13**).
2. Filter-sterilize the SILAC medium and store at 4 °C for up to 3 months.

3.2 Preparation of Yeast SILAC Labeling Medium

1. Prepare YNB (10×) solution and filter-sterilize; store at 4 °C until use.
2. Prepare D-glucose (25×) solution and autoclave; store at 4 °C until use.
3. Prepare ammonium sulfate (10×) solution and autoclave; store at 4 °C until use.
4. Prepare amino acid drop-out solution (25×) by combining all components, then add 50 ml of 100 % EtOH, and make sure that it covers the powder. Let it rest overnight, then add ddH_2O to 1 l, and store at 4 °C until use.
5. Just before use Combine in a bottle 100 ml of YNB (10×), 40 ml of D-glucose (25×), 100 ml of ammonium sulfate (10×), 40 ml of amino acid drop-out (25×), and 30 mg of heavy or light lysine; add ddH_2O to 1 l.

3.3 Preparation of Minimal Fly Food for In Vivo Labeling

1. Combine in a beaker ddH_2O, sucrose, agar, and light or heavy yeast pellet.
2. Bring to 1 l using ddH_2O and mix.
3. Stir and heat on the stirrer (or microwave) until boiling.
4. Cool down to below 50 °C.
5. Add propionic acid.
6. Aliquot in standard fly bottles using approximately 30 ml per bottle.
7. Allow bottled fly food to dry overnight. Cover bottles with the cheesecloth to prevent contaminations.
8. Plug the bottles.
9. Store at 4 °C for up to a few weeks or until use.

3.4 Preparation of Plates for Embryo Collection

1. Combine in a beaker ddH_2O, molasses, and agar.
2. Stir and heat on the stirrer (or microwave) until boiling.
3. Cool down to below 50 °C.
4. Add Tegosept.
5. Mix well; pour into 100-mm plates and cover.
6. Allow the agar to dry and store plates at 4 °C for up to a few weeks or until use.

3.5 Establishment of SILAC in Whole/Living Flies

Metabolic labeling of adult flies and larvae is achieved by feeding SILAC-labeled yeast as the sole source of amino acids. Such approach requires growth of a lysine auxotrophic yeast strain using yeast SILAC medium (*see* **Note 10**).

3.5.1 Preparation of Fly Food

1. The lysine auxotrophic BY4743 strain is inoculated from a plate or culture stub in 50 ml of heavy yeast SILAC medium

(prepared with heavy lysine, Lys8) and cultured overnight at 30 °C in a shaker set at 240 rpm.

2. The culture is diluted 1:5,000 in 50 ml of heavy yeast SILAC medium and allowed to grow overnight.
3. This second pre-culture is diluted 1:1,000 in heavy yeast labeling medium and grown overnight until the culture reaches an optical density of 4 (measured with a spectrophotometer set at 600 nm).
4. The yeast is pelleted by centrifugation at 3,220 × *g* and weighted, and 3 g aliquots are stored at −20 °C until use for preparation of fly food. A parallel set of light-labeled yeast is produced using yeast SILAC medium prepared with light lysine (Lys0) to finally produce bottles with heavy- and light-labeled food heavy- and light-labeled fly food (*see* **Note 14**).

3.5.2 Labeling of Animals (Fig. 1)

1. 150–200 adult flies (females and males in a 3:1 ratio) are anesthetized with CO_2 and transferred to a cage sitting on an embryo collection plate (*see* **Note 15**).
2. Fly cages are incubated at 25 °C overnight to allow flies to adapt to cages. Old plates are removed and discarded and new

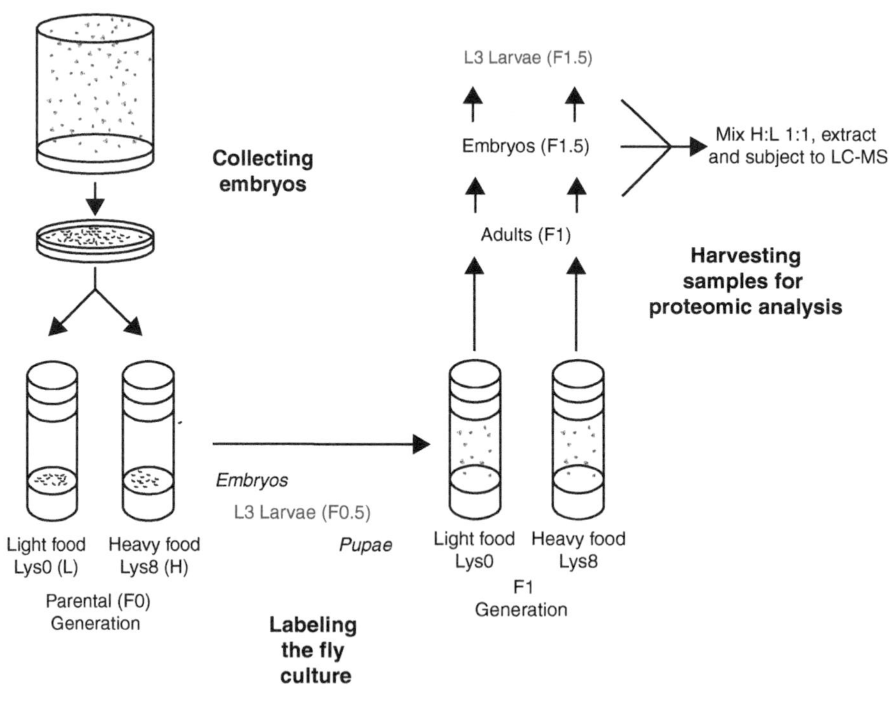

Fig. 1 Schematic representation of the protocol for in vivo labeling of *Drosophila*. Embryo collections are performed in cages. Recovered embryos are distributed on light- and heavy-labeled fly food in equal numbers. After 10–12 days of culturing, F1 adults are mixed 1:1 and processed for MS analysis. Alternatively, light- and heavy-labeled F1 adults are transferred to a new batch of light- and heavy-labeled fly food for F1.5 embryo or L3 larvae collection. Embryos or larvae (or their organs) are mixed 1:1 and processed for MS analysis

plates are added and the fly cages are moved back to 25 °C for egg deposition.

3. After 12 h, plates containing 0–12 h-old embryos are removed; embryos are counted, hand-picked with a moist paintbrush, and added to a set of bottles containing heavy-labeled and light-labeled fly food. Each bottle will sustain approximately 150 fly embryos (*see* **Notes 16** and **17**).
4. Incubate bottles for 10–12 days at 25 °C. If not collected, the F0.5 larvae will give rise to the F1 generation (*see* **Note 18**).
5. The hatching F1 flies are allowed to grow for a few more days before collection to set up labeling of F1.5 larvae (*see* **Note 19**).
6. For labeling of F1.5 larvae, 100–150 heavy-labeled and light-labeled F1 adults per bottle (females and males in a 3:1 ratio) are transferred to a new set of bottles containing fresh heavy-labeled and light-labeled fly food.
7. Bottles are moved back to 25 °C for 3 days for egg laying.
8. Adults are transferred to new bottles and larvae in old bottles are allowed to develop at 25 °C for a few more days until larval instar 3 (L3) larvae of the F1.5 generation are ready for harvesting (*see* **Note 20**).

3.6 Extraction of SL2 Cells and Subcellular Fractionation

Fractionating the total cellular extracts reduces sample complexity and thus improves the dynamic range of protein identification by LC-MS/MS (*see* **Note 21**).

1. Mix equal numbers of cells for each sample (heavy and light).
2. Wash cells with PBS.
3. Incubate cells in 4 ml of swelling buffer at 4 °C for 10 min.
4. Pass the cell suspension four times through a cell homogenizer containing a ball allowing for 10-μm clearance.
5. Centrifuge the cell lysate at 1,500 × *g* and 4 °C for 10 min.
6. Dissolve the pellet containing the nuclei (N) in 200 μl of RIPA buffer.
7. Centrifuge the supernatant at 120,000 × *g* and 4 °C for 15 min. Pellets contain membrane and cell debris. The supernatant is the cytosol.
8. Concentrate the supernatant (cytosolic fraction, C) by precipitation in 15 % TCA as follows:
   - Add the appropriate volume of 100 % TCA cold stock solution to the volume of the cytosolic fraction in order to reach a final concentration of 15 %.
   - Vortex for mixing and incubate for 30 min on ice.
   - Centrifuge at 120,000 × *g* and 4 °C for 15 min.

- Aspirate carefully the supernatant. The pellet contains the cytosolic proteins.
- Wash the pellet with 0.5 ml of ice-cold acetone.
- Discard supernatant and allow the pellet to air-dry at RT to completely remove acetone by volatilization.

9. Resuspend nuclear and cytosolic pellets in 4× LDS sample buffer for subsequent separation by SDS-PAGE.

3.7 Preparation of Total Extracts from Whole Organisms or Tissues

1. Samples are rinsed in wash solution to remove contaminations from fly food, mixed 1:1, and collected in tubes containing freshly prepared homogenization buffer (*see* **Notes 22** and **23**).
2. Samples are homogenized using a pestle.
3. Tubes are vortexed for 5–10 min to ensure complete dissolution.
4. Suspensions are centrifuged for 20 min at 16,000 × *g* at RT in a tabletop centrifuge.
5. Supernatants containing the solubilized proteins are transferred to a new tube, frozen in liquid nitrogen, and stored at −80 °C until further processing.

3.8 Sample Preparation Prior to MS: SDS-PAGE and In-Gel Digestion

1. Resolve the protein mixture by SDS-PAGE.
2. Cut the gel lanes in slices and each slice in small cubes, 1 mm3 in size.
3. Destain the gel slices three times for 20 min each in 500 μl destaining buffer and add 500 μl of absolute ACN for 20 min to dehydrate the gel slices. Repeat until the gel slices are completely distained.
4. Add 50 μl of reduction buffer to the gel pieces; incubate for 1 h at 56 °C.
5. Remove reduction buffer and add 50 μl of alkylation buffer; incubate for 45 min at RT in the dark.
6. Remove alkylation buffer and wash the gel pieces twice, as described in **step 3**.
7. Remove ACN by aspiration and dry the gel pieces in a vacuum centrifuge.
8. Rehydrate the gel pieces with sufficient volume of ice-cold trypsin solution in 50 mM ABC to cover completely the gel pieces.
9. Incubate on ice till the gel pieces are fully rehydrated, thus allowing for a maximum diffusion of trypsin into the gel pieces and reducing autolysis by low temperature.
10. After complete rehydration of the gel pieces, remove the trypsin solution in excess.

11. Add 50 mM of ABC to completely cover the gel pieces and incubate overnight at 37 °C.
12. On the next day spin down and collect liquid in a new tube.
13. Add the extraction buffer to the gel pieces; incubate in a thermomixer with strong agitation for 20 min at RT. Collect the supernatant and repeat the extraction twice.
14. Pool all supernatants. Dry the peptide mixture in a vacuum centrifuge.
15. Reconstitute dried peptides in 1 % TFA.
16. Desalt and concentrate peptides on a reversed-phase C18 micro-column (StageTips or equivalent), as previously described [17].
17. Elute peptides from the StageTips using HPLC solvent "B."
18. Remove the organic component by vacuum centrifugation and resuspend the peptides in a suitable injection volume (typically 5–10 μl) of 0.1 % AA.
19. Load approximately 1 up to 3 μg of the total peptide mixture onto the capillary column for the nano-flow LC-MS/MS analysis.

3.9 LC-MS Analysis

3.9.1 Liquid Chromatography

1. Tryptic peptides generated by in-gel digestion and desalted via StageTip are loaded onto the nano-flow RP-HPLC device directly connected with an electrospray source for subsequent MS analysis. Pack an analytical column in a 15 cm fused silica emitter with methanol slurry of reverse-phase C18 resin at a constant helium pressure (50 bar) using a bomb-loader device as described previously [19].
2. Connect the packed emitter (C18 RP HPLC column) directly to the outlet of the 6-port valve of the HPLC through a 20-cm-long (25 μm ID) fused silica, without using a precolumn or a split device (*see* **Note 24**).
3. Load the tryptic peptide mixtures onto the C18 column at a flow of 750 nl/min.
4. After sample loading, apply a gradient of 3–60 % HPLC solvent "B" at a flow of 250 nl/min over 120 min for peptide separation.

3.9.2 Mass Spectrometry Analysis

1. Mass spectrometry is performed on an LTQ Orbitrap instrument operating in the data-dependent acquisition (DDA) mode to automatically switch between MS and MS/MS experiments (*see* **Note 25**).
2. Use the following settings in the "Tune" acquisition file:
   (a) FT full scan: Accumulation target value 1×10^6; maximum filling time: 500 ms.
   (b) IT MSn: Accumulation target value 1×10^5; maximum filling time: 100 ms.

3. Standard acquisition method settings are as follows: electrospray voltage: 2.5 kV; no sheath and auxiliary gas flow; ion transfer (heated) capillary temperature: 250 °C; dynamic exclusion of up to 500 precursor ions for 45 s upon MS/MS; exclusion mass width of 10 ppm; normalized collision energy using wide-band activation mode: 35 %; ion selection threshold: 1,000 counts; activation $q = 0.25$; and activation time: 10 ms.

3.10 Data Analysis

All raw data files acquired are analyzed with the publicly available software MaxQuant (http://www.maxquant.org/) [18] (*see* **Note 26**).

1. Configure the built-in search engine "Andromeda" using the AndromedaConfig.exe [20]. This module allows configuration of all parameters related to the database search engine such as proteases and protein database modifications (*see* **Note 27**).
2. Define the enzyme that was used for digestion (i.e., trypsin/P) (*see* **Note 28**).
3. Missed cleavages: Up to 3.
4. Fixed modifications: Carbamidomethylation.
5. Variable modification: *N*-acetyl (Protein), Oxidation (M).
6. Upload the correct fasta file and the "contaminants" (*see* **Note 29**).
7. Specify the label parameter (Lys8).
8. Define "Max label amino acid": 3 for trypsin, 4 for LysC.
9. Mass accuracy of the parent ions in the initial "Andromeda" search: 7 ppm.
10. Mass accuracy for CID MS/MS: 0.5 Da (six top peaks per 100 Da).
11. Peptide false discovery rate (FDR) (*see* **Note 30**): 0.01.
12. Protein FDR (*see* **Note 30**): 0.01.
13. Maximum posterior error probability (PEP) (*see* **Note 31**): 1.
14. Minimum peptide length: 6.
15. Minimum number of peptides: 2.
16. Minimum number of unique peptides: 1.
17. Activate: Use only unmodified peptides and Oxidation (M)/ Acetyl (Protein N-term) (*see* **Note 32**).
18. Activate the function: Discard unmodified counterpart peptides (*see* **Note 33**).
19. Minimum score: 0.
20. Minimum ratio count: 2 (*see* **Note 34**).

3.11 Evaluation of Labeling Efficiency Prior to Large-Scale SILAC Experiment

Stable isotope amino acids available on the market have a reported purity of 99 % that already represents the upper limit for direct labeling in cell culture. When the labeling is performed indirectly by using labeled organic material (i.e., yeast), a reduced efficiency is expected. Moreover, in higher organisms, recycling of internal amino acid sources may also cause a reduction in incorporation rate. However, the label efficiency has to be monitored before starting a large-scale SILAC experiment. To this aim, the proportion of remnant light peptides found in the heavy-labeled sample measures the incorporation level. Ideally, peptides identified from this pool should contain only heavy amino acids without detectable signals at m/z values corresponding to the light peptide; however, in reality, light peptides are remaining and non-normalized ratio H/L is therefore calculated (*see* **Note 35**).

*3.11.1 Incorporation Test for SL2 Cells (See **Notes 35–38**)*

1. Resolve whole-cell extracts from heavy-labeled cells by SDS-PAGE.
2. Follow the protocol for in-gel digestion and subsequent MS analysis as described in Subheadings 3.6, 3.8, and 3.9.
3. Follow the protocol for data analysis as described in Subheading 3.10, disabling the re-quantify option in the MaxQuant software.

3.11.2 Incorporation Test for Extract from Whole Larvae (Fig. 2)

1. Grow F1.5 L3 fly larvae as described in Subheading 3.2 using heavy-labeled fly food only.
2. Extract and resolve protein extracts from heavy-labeled samples as outlined above in Subheadings 3.4 and 3.5.
4. Follow the protocol for in-gel digestion and subsequent MS analysis as described in Subheadings 3.7, 3.8, and 3.9.
5. Follow the protocol for data analysis as described in Subheading 3.10, disabling the re-quantify option in MaxQuant.

4 Notes

1. As a result from a continuous optimization of the protocol in a collaborative effort with Drosophilists, the recipe for SILAC-Schneider's medium described in this chapter contains slight modifications from the ones previously published by our group [10, 15]. We hence refer to these publications for the overall rationale of the protocol, but we advise to follow the refined recipe described here for the corrected quantities and concentrations of components, as they enable more efficient labeling.
2. Add 0.86 g of $Na_2HPO_4{\cdot}12H_2O$.
3. Add 1.8 g of anhydrous $MgSO_4$.
4. Add light arginine here if only lysine is used for SILAC.
5. L-Cystine is dissolved separately in 50 ml of hot acidified water (pH 2, adjust with HCl, i.e., 5 ml of 1 N HCl/45 ml ddH_2O).

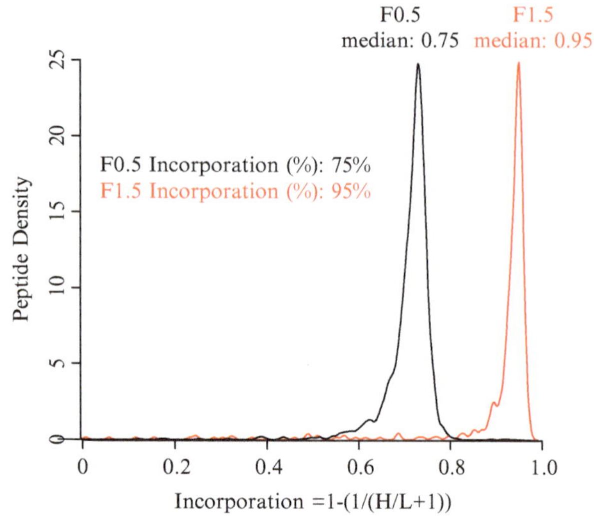

Fig. 2 Incorporation analysis of fly larvae F0.5 and F1.5. To assess the efficiency of SILAC labeling, the incorporation of Lys8 into proteins was monitored by mass spectrometry and MaxQuant analysis. The percentage of incorporation of Lys8 into all peptides was calculated as described in **Note 32**. The incorporation % values were 75 % (median of lysine peptide density distribution equal to 0.75, *black curve*) and 95 % (median of lysine peptide density distribution equal to 0.95, *red curve*) for F0.5 and F1.5, respectively. Thus, a sufficient degree of incorporation for a large-scale SILAC analysis was successfully achieved in F1.5 larvae

L-Tyrosine is dissolved separately in 50 ml of alkaline water (pH 9, adjust with NaOH, e.g., 5 ml of 1 N NaOH/45 ml ddH_2O). Cystine and tyrosine solutions are then added slowly (dropwise) to the general amino acid solution.

6. Dialyze overnight against 5 l of 0.9 % NaCl in 3,500 Da molecular weight cutoff (MCWO) tubes.
7. Different dialysis protocols need to be tested to find the optimal compromise between free amino acid removal and efficient cell growth. In our case, about 20 h of dialysis using an MWCO of 3,500 Da for serum and yeastolate guarantees both SL2 cell growth and efficient incorporation.
8. Rinse dialysis tubes well for 1 h in ddH_2O before use.
9. Add 0.80 g of $CaCl_2 \cdot 2H_2O$.
10. Lysine auxotrophic strains that bear point mutations in the Lys2 gene, rather than deletions, should be avoided as they can easily revert. To ensure the maintenance of auxotrophy, a test should be carried out by growing cells on plates and medium without lysine versus the complete medium.

11. Heavy arginine is to be avoided when lysine-only labeling is carried out.
12. A growth curve is a useful tool to evaluate the growth characteristics of a cell line. From a growth curve, the lag time, population doubling time, and saturation density can be determined. Plot the cell number on a log scale: the population-doubling time can be determined by identifying a cell number along the exponential phase of the curve, tracing the curve until the number has doubled, and calculating the time between the two.
13. In some cases, more sensitive cell types might suffer from the use of dialyzed serum or other components (e.g., yeastolate), due to the lack of small peptides functioning as growth factors. If so, supplementing the SILAC medium with single purified growth factors or with a small percentage (5–20 %) of normal serum might compensate this. During the optimization of the method, we observed that this second option could increase the growth of certain SL2 clones in SILAC medium for prolonged periods with no effect on labeling efficiency.
14. If frozen and thawed, before addition to fly food, yeast pellets should be centrifuged to remove excess medium.
15. Large 87.5 mm cages (Genesee Scientific 59-101) can sit on standard 100 mm plates. Dry baker's yeast freshly dissolved in water to form a paste and added as a small ball in the center of the plate will stimulate egg deposition and enhance embryo production. For general fly husbandry methods, please *see* [21].
16. One cage with 150–200 adults, if well fed, should yield 300–400 12-h embryos.
17. If food is not efficiently consumed after 5–6 days, the experiment can be repeated using more embryos or vials instead of bottles. Standard 28.5 mm × 95 mm plugged vials (vials: Genesee Scientific 32-114, plugs: Genesee Scientific 49-101) hold approximately 5 ml of food.
18. Labeled and unlabeled F1 adults can be collected and subjected to proteomic analysis.
19. Eggs laid from F1 parents can also be collected as explained in Subheading 3.5.2, **steps 1–3**, to be subjected to proteomic analysis. Embryos will hatch into larvae in 24 h at 25 °C. Collections at shorter times (6 or 12 h) allow to analyze early stages of development.
20. L3 larvae crawl out of the food and wander up the wall of the fly bottle in preparation for pupariation. Such larvae can be hand-collected from the wall of the bottles using tweezers.
21. Several techniques alternative to classical SDS-PAGE can be used to improve sample pre-fractionation such as filter-aided sample

preparation (FASP) [22] combined with anion-exchanger (SAX) filter plugs in pipet tips [23] prior to RPLC-MS analysis providing peptide separation complementary to C18-based column. FASP can also be used in combination with OFFGEL (Agilent) separation [24].

22. Mixing is based on the same number of flies, larvae, embryos, or organs. When using adults, the same male-to-female ratio for light and heavy must be maintained as the sexes differ in size. More in general, attention must be paid to avoid conditions leading to morphological changes between samples (i.e., wild-type animal versus mutant animals that might display changes in cell, tissue, organ, or organism size). In case of inhomogeneous samples, an internal reference, such as DNA content, can be used to ensure 1:1 mixing.
23. If only a certain tissue or organ is to be analyzed, animals can be dissected in wash solution prior to homogenization. For dissection of the larval organs used for genetics and development (such as imaginal discs, gut, salivary glands, fat tissue, central nervous system, lymph gland, or hemocytes) or of adult parts, please *see* [21].
24. It has been shown that packing long, narrow capillary RP columns greatly improves loading capacity, sensitivity, and dynamic range of the RPLC [25, 26]. Smith and co-workers have introduced long, small-particle-size (1.4 μm) RPLC columns with high peak capacity operated in an ultrahigh pressure regime [27]. This small particle size, operated at elevated temperature (65 °C) and under ultra-high-performance liquid chromatography (UHPLC) conditions, was further shown to improve the number of identified proteins as compared to standard HPLC [28]. Coupling an UHPLC system to a benchtop Orbitrap mass spectrometer (Q-Exactive) [29].
25. In a typical DDA experiment, the 5 (Orbitrap Classic and XL instruments) or 20 (Orbitrap Velos instrument) most intense peptide signals with a charge state >1 are isolated and then subjected to fragmentation.
26. The MaxQuant software package has been designed to perform highly automated protein identification and quantitation, controlling false-positive identifications and scoring posttranslational modifications. Using high-resolution data, it is able to enhance peptide identification rates and precision of quantitation for SILAC MS-based experiment. We recommend reading a MaxQuant dedicated publication [18, 20] for a detailed description about rationale and configuration of this software. Here, we describe a general workflow for data processing but remind readers that updated MaxQuant versions are released on a regular basis, since this software is under constant development. Hence, it is advisable to check the MaxQuant

homepage (http://www.maxquant.org) for the most recent versions and features made available.

27. MaxQuant already contains predefined modifications, enzymes, and protein databases that can be used by default. However, all those settings can be customized by including different number and types of modifications and/or different proteases, or also by configuring different protein databases. All settings can be changed and saved in the Andromeda configuration module.
28. Trypsin/P cleaves C-terminal of all K and R residues. In this case, the search is performed taking into account that the efficiency of enzyme cleavage is reduced when the next amino acid is P (/P).
29. The configuration folder contains a txt file (contaminants.fasta) with proteins—such as keratins—that contaminate large-scale proteomic experiment more frequently. The nature of these contaminants can however vary, depending on different conditions used in the experiments (i.e., cell types used, addition of bovine serum albumin in the experiment, the presence of immunoglobulin for affinity enrichment).
30. Peptide and protein FDR of 0.01 means that both peptides and proteins identified are expected to contain 1 % of false positives. This is estimated using a target-decoy database-based searching.
31. PEP is the probability that an individual peptide is a false-positive match. A PEP equal to 1 in your setting means that all peptides will be listed irrespective of the PEP; thus, filtering is based exclusively on the FDR.
32. Peptides with regulated modifications should generally not be counted for protein quantitation since their abundance may not reflect the ratio of the corresponding protein.
33. Enable the "Discard unmodified counterpart peptides" option in order to exclude both the modified and unmodified form of the same peptide from quantitation.
34. Increasing the minimum ratio count results in a more accurate quantitation but also in a reduced number of quantified proteins.
35. MaxQuant can be used to automatically estimate the degree of incorporation of heavy amino acids into proteins setting the parameters as described in Subheading 3.5 but disabling the re-quantify option. Incorporation percentage can then easily be calculated applying the following equation to non-redundant peptide ratios:

    $$\text{Incorporation }(\%) = \text{ratio } H/L/(\text{ratio } H/\text{L}+1)\times 100.$$

    However, considering factors that may potentially hamper the quantification such as overlapping isotope clusters, it is advisable to further check spectra from SILAC data manually.

36. Generally, after 7 days of SL2 growth, the heavy-to-light (H/L) non-normalized ratios are higher than 9 for both Arg10- and Lys8-containing peptides, corresponding to more than 90 % incorporation.
37. It is also possible to use an in-solution digestion of a whole-cell extract from heavy samples as previously described [9].
38. In some cell types grown in standard medium, the metabolic interconversion between arginine and proline can occur when arginine is provided to cells in excess. The reverse metabolic conversion of proline to arginine can also occur when cells are not provided with a sufficient amount of arginine. Thus, when assembling a new SILAC medium, the optimum concentration of arginine must be determined experimentally for the cell line under investigation. This is achieved by measuring the frequency of heavy proline in cells grown in heavy SILAC medium following a careful titration of arginine in SILAC-Schneider's medium.

Acknowledgments

Work in T.B.'s laboratory is supported by an Armenise-Harvard foundation career development program grant, a grant from the Associazione Italiana Ricerca sul Cancro (AIRC), a grant from the Association of International Cancer Research (AICR), and a Minister of Health grant (Giovani Ricercatori). T.V. lab is supported by a New Unit Start-up Grant from AIRC and a grant from Telethon. The authors wish to thank T. Straub for crucial suggestions on how to improve SL2 culturing in SILAC conditions and D. Corona, in which lab RS performed part of the work.

References

1. de Hoog CL, Mann M (2004) Proteomics. Annu Rev Genomics Hum Genet 5:267–293
2. Cox J, Mann M (2007) Is proteomics the new genomics? Cell 130:395–398
3. Cravatt BF, Simon GM, Yates JR 3rd (2007) The biological impact of mass-spectrometry-based proteomics. Nature 450:991–1000
4. Bachi A, Bonaldi T (2008) Quantitative proteomics as a new piece of the systems biology puzzle. J Proteomics 71:357–367
5. Gstaiger M, Aebersold R (2009) Applying mass spectrometry-based proteomics to genetics, genomics and network biology. Nat Rev Genet 10:617–627
6. Ong SE, Blagoev B, Kratchmarova I et al (2002) Stable isotope labeling by amino acids in cell culture, SILAC, as a simple and accurate approach to expression proteomics. Mol Cell Proteomics 1:376–386
7. Ong SE, Mann M (2005) Mass spectrometry-based proteomics turns quantitative. Nat Chem Biol 1:252–262
8. Ong SE, Mann M (2006) Identifying and quantifying sites of protein methylation by heavy methyl SILAC. *Curr Protoc Protein Sci* **14**, Unit 14 9
9. Ong SE, Mann M (2007) Stable isotope labeling by amino acids in cell culture for quantitative proteomics. Methods Mol Biol 359: 37–52
10. Bonaldi T, Straub T, Cox J et al (2008) Combined use of RNAi and quantitative proteomics to study gene function in Drosophila. Mol Cell 31:762–772

11. Vermeulen M, Selbach M (2009) Quantitative proteomics: a tool to assess cell differentiation. Curr Opin Cell Biol 21:761–766
12. Gouw JW, Tops BB, Mortensen P et al (2008) Optimizing identification and quantitation of 15N-labeled proteins in comparative proteomics. Anal Chem 80:7796–7803
13. Sury MD, Chen JX, Selbach M (2010) The SILAC fly allows for accurate protein quantification in vivo. Mol Cell Proteomics 9: 2173–2183
14. Xu P, Tan H, Duong DM et al (2012) Stable isotope labeling with amino acids in Drosophila for quantifying proteins and modifications. J Proteome Res 11:4403–44012
15. Cuomo A, Bonaldi T (2010) Systems biology "on-the-fly": SILAC-based quantitative proteomics and RNAi approach in Drosophila melanogaster. Methods Mol Biol 662:59–78
16. de Godoy LM, Olsen JV, Cox J et al (2008) Comprehensive mass-spectrometry-based proteome quantification of haploid versus diploid yeast. Nature 455:1251–1254
17. Rappsilber J, Mann M, Ishihama Y (2007) Protocol for micro-purification, enrichment, pre-fractionation and storage of peptides for proteomics using StageTips. Nat Protoc 2: 1896–1906
18. Cox J, Mann M (2008) MaxQuant enables high peptide identification rates, individualized p.p.b.-range mass accuracies and proteome-wide protein quantification. Nat Biotechnol 26:1367–1372
19. Olsen JV, Ong SE, Mann M (2004) Trypsin cleaves exclusively C-terminal to arginine and lysine residues. Mol Cell Proteomics 3:608–614
20. Cox J, Neuhauser N, Michalski A et al (2011) Andromeda: a peptide search engine integrated into the MaxQuant environment. J Proteome Res 10:1794–1805
21. Sullivan W, Ashburner M, Hawley SR (2000) Drosophila protocols. Cold Spring Harbor Laboratory Press, Cold Spring Harbor, NY
22. Wisniewski JR, Zougman A, Nagaraj N et al (2009) Universal sample preparation method for proteome analysis. Nat Methods 6:359–362
23. Wisniewski JR, Zougman A, Mann M (2009) Combination of FASP and StageTip-based fractionation allows in-depth analysis of the hippocampal membrane proteome. J Proteome Res 8:5674–5678
24. Hubner NC, Ren S, Mann M (2008) Peptide separation with immobilized pI strips is an attractive alternative to in-gel protein digestion for proteome analysis. Proteomics 8:4862–4872
25. Shen Y, Smith RD (2002) Proteomics based on high-efficiency capillary separations. Electrophoresis 23:3106–3124
26. Shen Y, Zhao R, Berger SJ et al (2002) High-efficiency nanoscale liquid chromatography coupled on-line with mass spectrometry using nanoelectrospray ionization for proteomics. Anal Chem 74:4235–4249
27. Shen Y, Zhang R, Moore RJ et al (2005) Automated 20 kpsi RPLC-MS and MS/MS with chromatographic peak capacities of 1000-1500 and capabilities in proteomics and metabolomics. Anal Chem 77:3090–3100
28. Thakur SS, Geiger T, Chatterjee B et al (2011) Deep and highly sensitive proteome coverage by LC-MS/MS without prefractionation. Mol Cell Proteomics 10(M110):003699
29. Nagaraj N, Wisniewski JR, Geiger T et al (2011) Deep proteome and transcriptome mapping of a human cancer cell line. Mol Syst Biol 7:548

Chapter 22

Analysis of Secreted Proteins Using SILAC

Jeanette Henningsen, Blagoy Blagoev, and Irina Kratchmarova

Abstract

Secreted proteins serve a crucial role in the communication between cells, tissues, and organs. Proteins released to the extracellular environment exert their function either locally or at distant points of the organism. Proteins are secreted in a highly dynamic fashion by cells and tissues in the body responding to the stimuli and requirements presented by the extracellular milieu. Characterization of secretomes derived from various cell types has been performed using different quantitative mass spectrometry-based proteomics strategies, several of them taking advantage of labeling with stable isotopes. Here, we describe the use of Stable Isotope Labeling by Amino acids in Cell culture (SILAC) for the quantitative analysis of the skeletal muscle secretome during myogenesis.

Key words Secreted proteins, Secretome, Differentiation, Cell culture, Mass spectrometry, Quantitative proteomics, SILAC

1 Introduction

The ability of the organism to react and adapt to changes in the environment is dependent on the induction of multiple coordinated events affecting virtually all cellular processes. The continuum of cellular communication and cross talk assures the correct transmission of information flow ultimately inducing the appropriate response. The cross talk between cells, tissues, and organs is orchestrated by secreted factors released into the extracellular space and exerting their actions by auto-, para-, and endocrine mechanisms. The secreted factors do not work in isolation and the outcome of their action is a result of the combinatorial response to various secreted molecules in space and time. The remarkable and still puzzling interplay of secreted factors that induces a multitude of cellular responses underlies the maintenance of total body homeostasis. The functional interaction map of the specific secretomes defines the physiological status of the organism, while dysfunction, mutations, or improper regulation of secreted proteins often leads to development of various types of diseases.

Bettina Warscheid (ed.), *Stable Isotope Labeling by Amino Acids in Cell Culture (SILAC): Methods and Protocols*, Methods in Molecular Biology, vol. 1188, DOI 10.1007/978-1-4939-1142-4_22, © Springer Science+Business Media New York 2014

Therefore, comprehensive investigation of the secretomes can have a great impact on improving quality of life and bears enormous clinical significance.

During the last 10 years, the improvements of the mass spectrometry (MS) instrumentation and introduction of different quantitative proteomics technologies have had a profound impact on proteomics research including several studies on cellular secretomes. The secretome represents a biologically defined subproteome consisting of a structurally and functionally diverse set of protein groups such as cytokines, growth factors, proteolytic enzymes, antibodies, and extracellular matrix components. Approximately 10 % of the genes encoded by the human genome are predicted to be secreted [1, 2]. Recently, there has been an intense focus to elucidate the identity of secreted proteins to define potential biomarkers for various clinical disorders including cancer and metabolic and inflammatory diseases [3–6]. In addition to understanding the interplay between distinct tissues and organs, it is essential to identify which proteins are being released to the extracellular milieu both under normal circumstances and in response to altered levels of cytokines, hormones, and different physiological states.

Generally, studies of secretomes are facilitated by using cells in culture representing various tissues and analyzing which proteins are being released under defined conditions. For example, murine 3T3-L1 and C2C12 cell lines have been used extensively as cellular models of adipose and muscle tissue, respectively, resulting in the identification of numerous adipokines and myokines involved in the maintenance of body homeostasis [7–10]. In addition, MS has also been employed to study the secretory profile of various cancer cell lines to unveil potential biomarkers [4, 11]. Quantitative proteomics analyses are typically performed in two different ways either relying on the incorporation of stable isotopes or via label-free approaches, retrieving quantitative information based on peak intensity, peptide counting, and/or spectral counting [12, 13]. The introduction of specific mass tags into cellular proteomes via stable isotopes is obtained by in vivo metabolic labeling or by in vitro labeling utilizing chemical methods. Quantitative secretome analysis has been successfully performed using chemical labeling strategies such as dimethyl labeling and in particular iTRAQ [14, 15], whereas SILAC is the method of choice using metabolic labeling [16]. Here we focus on the application of SILAC to study cellular secretomes. SILAC employs metabolic labeling of the entire proteome of cells in culture and is frequently used to facilitate quantitative MS-based proteomics [17–19]. The SILAC strategy has been successfully applied to study the dynamics of secreted proteins during various cellular processes or in response to specific stimuli [9, 10, 20–25]. Two major improvements in secretome studies is presented by the use of the SILAC protocol that facilitate (1) mixing of samples prior to enrichment of secreted proteins by ultrafiltration or precipitation reducing quantitation error generated by variation in sample-to-sample preparation and (2) the possibility to make a distinction between contaminating proteins and cell-derived proteins. In this chapter, we describe a protocol applying triple-encoding SILAC [26] combined with liquid chromatography tandem mass spectrometry (LC-MS/MS) for the identification and quantitation of secreted proteins (Fig. 1), discussing the advantages of employing the SILAC strategy in secretome studies.

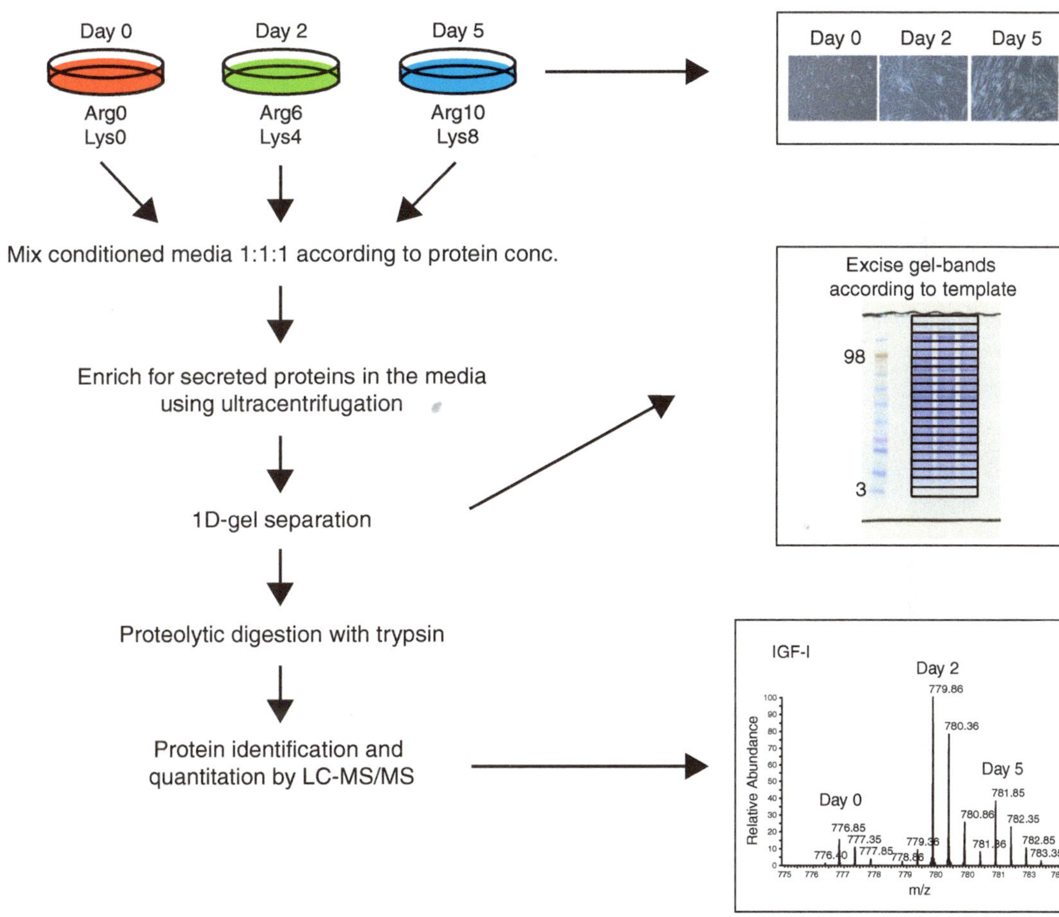

Fig. 1 Experimental workflow applied to quantitatively evaluate the secretome of skeletal myoblasts during differentiation. The outlined experimental approach is generally applicable to quantitatively identify any cellular secretome. Incorporation of isotopically labeled versions of specific amino acids introduces a mass shift that makes it possible to distinguish identical but isotopically distinct peptides within the same mass spectrum. CMs were collected on day 0, day 2, and day 5 of differentiation of C2C12 cells. "Day-0 cells" were cultured using unlabeled arginine (Arg0) and lysine (Lys0), "day-2 cells" were cultured using $^{13}C_6{}^{14}N_4$-arginine (Arg6) and $^{2}H_4$-lysine (Lys4), and finally "day-5 cells" were cultured using $^{13}C_6{}^{15}N_4$-arginine (Arg10) and $^{13}C_6$,$^{15}N_2$-lysine (Lys8). CMs collected from three time points of differentiation were combined in a 1:1:1 ratio according to protein determinations, concentrated using ultrafiltration column, and subsequently separated by 1D-gel electrophoresis. Gel bands were excised according to the demonstrated template, subjected to in-gel digestion, and analyzed by LC-MS/MS

2 Materials

2.1 Cell Culture

1. Skeletal muscle cell line: C2C12, murine myoblasts (American Type Culture Collection, ATCC).
2. Medium: Dulbecco's modified Eagle medium (DMEM), 4.5 g/l glucose, deficient in lysine (Lys) and arginine (Arg).
3. Stable isotope-labeled "heavy" amino acids: $^{13}C_6{}^{14}N_4$-L-arginine (Arg6), $^{13}C_6{}^{15}N_4$-L-arginine (Arg10), $^2H_4$-L-lysine (Lys4), and $^{13}C_6{}^{15}N_2$-L-lysine (Lys8).
4. Supplements: Dialyzed fetal bovine serum (dFBS); L-glutamine (200 mM), and penicillin-streptomycin (10,000 U/10,000 U).
5. Additional: Trypsin-EDTA solution (trypsin, 500 mg/l, and EDTA, 200 mg/l); sterile phosphate-buffered saline (PBS) without Ca^{2+} and Mg^{2+}.

2.2 Collection of Conditioned Media, Enrichment of Secreted Proteins, and Gel Electrophoresis

1. 0.2 μm filter units (Minisart® surfactant-free cellulose acetate (SFCA) membrane, Sartorius).
2. 50 ml polypropylene (PP) tubes (Sarstedt).
3. Protein determination: Coomassie Plus (Bradford) Assay Reagent (Pierce) and albumin standard ampules, 2 mg/ml (Pierce).
4. Ultrafiltration columns: Vivaspin 6 [3,000 molecular weight cutoff (MWCO), polyethersulfone (PES), and Vivaspin 20 (3,000 MWCO PES; Sartorius)].
5. NuPAGE® Novex® 4–12 % Bis-Tris gels (1.0 mm thick, 10-well), NuPAGE® LDS sample buffer (4×), NuPAGE® MES SDS running buffer (Life Technologies), and β-mercaptoethanol.
6. Colloidal Blue staining kit (Life Technologies) for visualization of proteins on gels.

2.3 In-Gel Digestion, StageTip Purification, and LC-MS/MS Analysis

1. Buffers for in-gel digestion [10, 27]: 50 mM ammonium bicarbonate (ABC buffer) and absolute ethanol.
2. 10 mM dithiothreitol (DTT) and 55 mM iodoacetamide (IAA), both dissolved in 50 mM ABC buffer for reduction and alkylation of proteins, respectively.
3. Sequencing-grade modified trypsin for the proteolytic digestion of proteins in gel slices.
4. Buffers for StageTip purification [28]:
   - Sample buffer: 5 % acetonitrile (ACN)/1 % trifluoroacetic acid (TFA).
   - Sample buffer A*: 1.66 % ACN/0.33 % TFA.
   - Buffer B: 0.5 % acetic acid/80 % ACN.
5. MaxQuant, a software program for protein analysis [29, 30].

3 Methods

Different labeling strategies have been explored to permit the unambiguous identification and functional analysis of the secretomes derived from various cell types taking advantage of both chemical and metabolic labeling [12, 13]. To study the dynamics of protein secretion during different cellular processes such as the development of specific cell types or in the response to any given stimulus or physiological state, the quantitative proteomics approach SILAC has become a popular choice (*see* **Note 1**) [21, 23, 31–34]. One of the major advantages of the SILAC strategy is the possibility to distinguish bona fide-secreted proteins from contaminating proteins such as sera-derived components and other growth media supplements (Fig. 2, *see* **Notes 2** and **3**).

In the following, we describe how to analyze the skeletal muscle secretome using triple-encoding SILAC combined with LC-MS/MS for protein identification and to quantitatively evaluate the dynamics of protein secretion. Application of this protocol leads to the identification of proteins known to be secreted from skeletal muscle cells in addition to novel proteins of which many are involved in different cellular processes such as cellular signaling, proliferation, differentiation, and extracellular matrix morphogenesis (*see* **Note 4**) [10, 20]. However, the protocol provided below is generally applicable to the study of any given cell secretome.

3.1 Cell Culture

1. Prepare SILAC-DMEM using custom-prepared DMEM 4.5 g/l glucose deficient in arginine and lysine adding different forms of Arg and Lys to final concentrations of 28 mg/l and 73 mg/l, respectively (Fig. 1; *see* **Notes 5** and **6**). Supplement the media with 10 % dFBS, 1 % L-glutamine, and 1 % penicillin-streptomycin.
2. Subculture the C2C12 cells by washing the cells with PBS without Ca^{2+} and Mg^{2+} and finally allowing cells to detach using a trypsin-EDTA solution. To ensure encoding of the entire proteome, cells should be cultured for at least five cell doublings [16, 35]. Complete incorporation of labeled amino acids can be analyzed by examining the cellular samples individually (*see* **Note 7**).
3. When cells reach confluence (day 0), myoblast differentiation is induced by reducing the amount of dFBS to 2 % with changes of the medium every 2 days. Conditioned media (CMs) are collected from three 10-cm dishes (*see* **Note 8**).

3.2 Collection of Conditioned Media

1. At the selected time points of differentiation, wash the cells six times with serum-free medium and finally starve the cells for 12 h in the respective serum-free SILAC medium (*see* **Note 9**).

a

```
mMMP2  1    APSPIIKFPGDVAPKTDKELAVQYLNTFYGCPKESCNLFVLKDTLKKMQKFFGLPQTGDL  60
bMMP2  31   APSPIIKFPGDVAPKTDKELAVQYLNTFYGCPKESCNLFVLKDTLKKMQKFFGLPQTGEL  90

mMMP2  61   DQNTIETMRKPRCGNPDVANYNFFPRKPKWDKNQITYRIIGYTPDLDPETVDDAFARALK  120
bMMP2  91   DQSTIETMRKPRCGNPDVANYNFFPRKPKWDKNQITYRIIGYTPDLDPQTVDDAFARAFQ  150

mMMP2  121  VWSDVTPLRFSRIHDGEADIMINFGRWEHGDGYPFDGKDGLLAHAFAPGTGVGGDSHFDD  180
bMMP2  151  VWSDVTPLRFSRIHDGEADIMINFGRWEHGDGYPFDGKDGLLAHAFAPGPGVGGDSHFDD  210

mMMP2  181  DELWTLGEGQVVRVKYGNADGEYCKFPFLFNGREYSSCTDTGRSDGFLWCSTTYNFEKDG  240
bMMP2  211  DELRTLGEGQVVRVKYGNADGEYCKFPFRFNGKEYTSCTDTGRSDGFLWCSTTYNFDKDG  270

mMMP2  241  KYGFCPHEALFTMGGNADGQPCKFPFRFQGTSYNSCTTEGRTDGYRWCGTTEDYDRDKKY  300
bMMP2  271  KYGFCPHEALFTMGGNADGQPCKFPFRFQGTSYDSCTTEGRTDGYRWCGTTEDYDRDKEY  330

mMMP2  301  GFCPETAMSTVGGNSEGAPCVFPFTFLGNKYESCTSAGRNDGKVWCATTTNYDDDRKWGF  360
bMMP2  331  GFCPETAMSTVGGNSEGAPCVLPFTFLGNKHESCTSAGRSDGKLWCATTSNYDDDRKWGF  390

mMMP2  361  CPDQGYSLFLVAAHEFGHAMGLEHSQDPGALMAPIYTYTKNFRLSHDDIKGIQELYGPSP  420
bMMP2  391  CPDQGYSLFLVAAHEFGHAMGLEHSQDPGALMAPIYTYTKNFRLSHDDIQGIQELYGASP  450

mMMP2  421  DADTDTGTGPTPTLGPVTPEICKQDIVFDGIAQIRGEIFFFKDRFIWRTVTPRDKPTGPL  480
bMMP2  451  --DIDTGTGPTPTLGPVTPELCKQDIVFDGISQIRGEIFFFKDRFIWRTVTPRDKPTGPL  508

mMMP2  481  LVATFWPELPEKIDAVYEAPQEEKAVFFAGNEYWVYSASTLERGYPKPLTSLGLPPDVQQ  540
bMMP2  509  LVATFWPELPEKIDAVYEDPQEEKAVFFAGNEYWVYSASTLERGYPKPLTSLGLPPGVQK  568

mMMP2  541  VDAAFNWSKNKKTYIFAGDKFWRYNEVKKKMDPGFPKLIADSWNAIPDNLDAVVDLQGGG  600
bMMP2  569  VDAAFNWSKNKKTYIFAGDKFWRYNEVKKKMDPGFPKLIADAWNAIPDNLDAVVDLQGGG  628

mMMP2  601  HSYFFKGAYYLKLENQSLKSVKFGSIKSDWLGC  633
bMMP2  629  HSYFFKGAYYLKLENQSLKSVKFGSIKSDWLGC  661
```

b

c

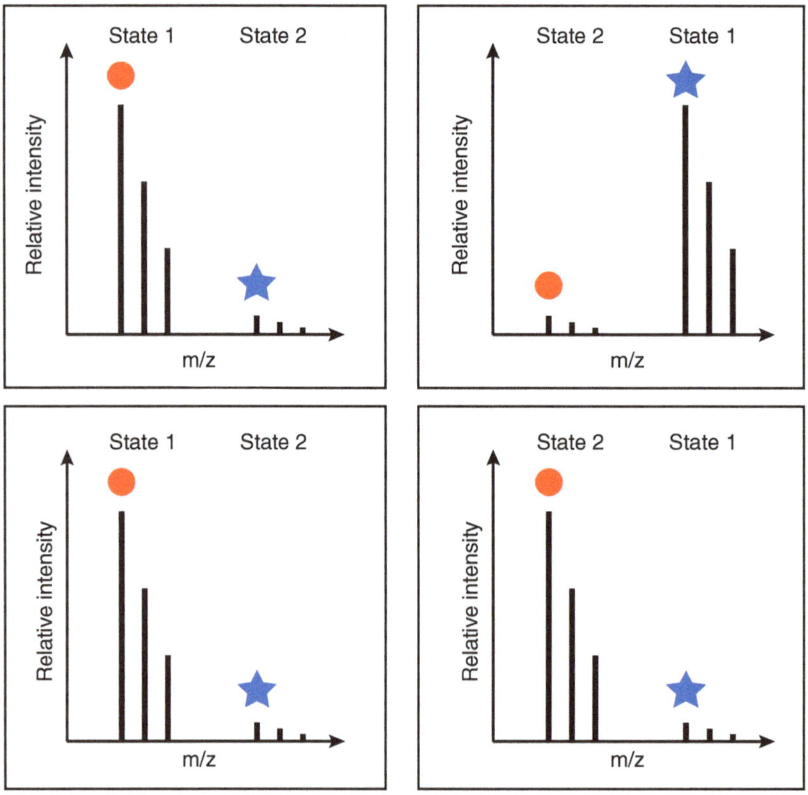

Fig. 2 Reverse SILAC labeling can be used to distinguish downregulated proteins from contaminating proteins such as keratins and serum-derived proteins. (**a**) High-sequence homology can be observed between secreted

2. Collect CM by centrifugation at 200 × *g* for 5 min followed by filtration using a 0.2 μm filter to ensure removal of any floating cells (*see* **Note 10**).
3. Keep CM at 4 °C for short-term storage to prevent precipitation of proteins by freezing (*see* **Note 11**).
4. Save a small aliquot of each CM as a future reference to visualize estimated sample concentrations by 1D-PAGE and Coomassie staining.

3.3 Concentration of Conditioned Media and Gel Electrophoresis

1. Measure the protein concentration of individual CMs using the Coomassie Plus Reagent (*see* **Note 12**). Mix CM and reagent 1:1, incubate at RT for 10 min, and then measure the absorbance at 595 nm. Use albumin as standard.
2. Mix the three collected CMs 1:1:1 according to determined protein concentrations (*see* **Note 13**).
3. Concentrate the pooled CM using the Vivaspin concentrator columns according to the manufacturer's recommendation (*see* **Note 14**).
4. Collect the concentrate (100–150 μl) and add 4× LDS sample buffer and 3 % β-mercaptoethanol (*see* **Note 15**).
5. Load the sample in 1–5 lanes on a 4–12 % Bis-Tris gel (*see* **Note 16**) and separate the proteins by SDS-PAGE.
6. Visualize the proteins using Coomassie Blue and cut the gel lane(s) into slices of equal size. The size of slices below 8 kDa can be doubled (*see* Fig. 1). To improve protein recovery, minimize the size of gel piece per sample. Therefore increase the total number of slices with the number of loaded gel lanes (for example cut 1 gel lane into 12 slices or 3–5 gel lanes into 20–24 slices).

3.4 In-Gel Digestion, LC-MS/MS, and Data Analysis

1. Excise whole gel lanes of proteins into separate slices, which are cut into small 1 mm cubes.
2. Wash the gel pieces 2–4 times to remove residual protein stain using 1:1 (v/v) 50 mM ABC buffer and 50 % ethanol followed by dehydration of gel pieces after final wash using absolute ethanol.

Fig. 2 (continued) proteins and proteins derived from serum supplement as demonstrated for matrix metalloproteinase-2 (MMP-2); mMMP2: MMP2_mouse, P33434 and bMMP2: MMP2_bovin, Q9GLE5. Applying both forward and reverse labeling in your proteomics experiments helps to identify downregulated cell-derived secreted proteins from residual proteins originating from the serum. (**b**) The spectra representing the reverse labeling of a downregulated protein will exhibit an inverted pattern of intensities. (**c**) In contrast, contaminating proteins will give rise to the same pattern in a mass spectrum representing reverse labeling since these proteins are not labeled (*red-filled circle* and *blue-filled star* represent the light and heavy SILAC label, respectively)

3. Reduce proteins incubating gel pieces in 10 mM DTT for 45 min at 56 °C followed by alkylation using 55 mM IAA for 30 min in the dark.
4. After alkylation wash gel pieces with 50 mM ABC and dehydrate using absolute ethanol. Repeat one time. Rehydrate gel pieces with 12.5 ng/ml trypsin in 50 mM ABC. Perform trypsin digestion overnight at 37 °C.
5. Extract peptides using two times 30 % acetonitrile (ACN) and 3 % trifluoroacetic acid (TFA) followed by dehydration using 100 % ACN. Reduce the volume of extracted peptides via a vacuum centrifuge to ensure complete removal of ACN.
6. Acidify peptides adding 1/3 volume of sample buffer 5 % ACN + 1 % TFA before enrichment and desalting by StageTips.
7. Bind peptides to disks of C_{18} reverse-phase material, wash using 0.33 % TFA + 1.66 % ACN, and finally elute with 80 % ACN containing 0.5 % acetic acid.
8. Reduce the volume of eluted peptide mixtures to near dryness via a vacuum centrifuge and resuspend peptides using 0.33 % TFA and 1.66 % ACN.
9. Perform mass spectrometry as previously described here [10, 27, 28].
10. Process the acquired raw data using programs such as MaxQuant, a software package for protein identification and quantitation [29, 30]. Triple-encoding SILAC allows for the comparison of three different cellular states. In this study, the ratios given by the analysis describe the dynamics of the level of secreted proteins during myogenesis.
11. Use software tools to isolate secreted proteins from the total list of identified proteins (*see* **Note 17**).

4 Notes

1. SILAC [16] is a powerful tool commonly used in MS-based quantitative proteomics. The principle of SILAC involves growing cell populations in media containing isotopically distinct amino acids (Fig. 1). The mass differences introduced by the use of different isotope forms of amino acids enable a direct comparison within the same mass spectrum of isotopically labeled peptides obtained from identical proteins but at different cellular states.
2. A fundamental challenge in the analysis of cell culture-derived secretomes is the capability to discriminate between actually secreted proteins and contaminating proteins originating from serum supplements like FBS, which is an essential supplement in many growth media. Despite extensive washes (*see also* **Note 9**),

traces of serum proteins such as albumin can still be detected by MS. SILAC as a metabolic labeling strategy offers a great advantage compared to other known labeling strategies using chemical tags since the isotopic labels introduced via the SILAC approach are only incorporated into cellular proteins.

3. Reverse SILAC labeling is beneficial when investigating secreted proteins to distinguish bona fide-secreted proteins from contaminating background proteins originating from the serum or other growth supplements (Fig. 2). Reversing the SILAC label is in particular useful when determining and analyzing the downregulated secreted proteins in the samples.
4. Recent literature has demonstrated that the skeletal muscle serves an important function as an endocrine organ releasing proteins conceptualized as myokines which can influence whole-body metabolism and cytokine production [36, 37]. Different quantitative proteomics strategies have been employed to study the secretome of the skeletal muscle including both isotope labeling and label-free protocols [10, 20, 31, 38, 39]. Application of SILAC to analyze the dynamics of the skeletal muscle secretome during myogenesis resulted in the quantitative identification of 624 secreted proteins; 188 of these were found to be differentially secreted during myoblast differentiation [10].
5. Combining both Arg and Lys in SILAC experiments improves the probability of positive protein quantitation by increasing the number of labeled peptides. Trypsin is one of the most frequently used proteolytic enzymes in MS studies due to its high efficiency and specificity [40]. It cleaves solely C-terminal to Arg and Lys residues; thereby the use of labeled versions of both Arg and Lys ensures that all tryptic peptides (except for the most C-terminal peptide) contain one labeled residue and can therefore potentially be used for quantitation.
6. The SILAC protocol requires the use of dialyzed sera (dFBS) to prevent the presence of non-labeled amino acids from sera, which will otherwise obstruct the accuracy of quantitation. Commercially available dFBS are dialyzed using 10,000 MWCO filters to remove any traces of amino acids. Unfortunately, this can also result in the removal of low-molecular-weight proteins including certain growth factors and cytokines. Therefore, dFBS is not compatible with all cell types and slower growth rates can be observed in rare cases. Dialysis with an MWCO of 1,000 Da would be adequate to remove amino acids, but it is costly.
7. Titration of a favorable Arg concentration can be necessary since conversion of Arg to proline (Pro) has been observed in certain cell lines [35, 41]. This would affect quantitation accuracy due to the presence of a labeled Pro giving rise to an additional

mass shift of Pro-containing peptides. Supplementing additional proline to cell cultures can help to circumvent this problem. However, it should also be noted that reverse metabolic conversion of Pro to Arg can occur if the level of Arg is limited. Despite these minor obstacles, successful SILAC labeling using Arg has been reported in numerous cases. Complete incorporation of labeled amino acids and Arg-to-Pro conversion can be evaluated analyzing the collected CM or, more feasibly, the corresponding cell lysates separately by LC-MS/MS.

8. The number of cells and the volumes of collected CM depend on the cellular system. Some cell types such as metabolically active adipocytes and muscle cells have higher secretion profiles than other cells of mesenchymal origin. Therefore, we recommend performing an initial screen to determine the general secretion pattern and concentration for any given cell type. In addition, since different cells are phenotypically different, the number of cell culture dishes used varies accordingly. In general, we recommend using three to eight 10-cm dishes per condition.
9. Washing and starvation are required to minimize the presence of serum proteins. High levels of sera-derived proteins will interfere with the subsequent concentration of proteins by blocking the membranes of the concentrator columns. In addition, and more importantly, traces of serum proteins will interfere with the MS analysis by masking the presence of bona fide-secreted proteins either due to sequence homology or low levels of specific proteins. The time of starvation needs to be evaluated for any given cell line prior to a SILAC experiment to assure cell viability.
10. CMs are filtrated using 0.2-μm filters to remove any whole cells or cell debris, to minimize the risk of contaminating the CM with proteins of intracellular origin.
11. Storage at freezing temperatures can result in loss of protein caused by precipitation.
12. It is not possible to use protein measurements of individual lysates as a reference for the level of secreted proteins [10]. Investigating the dynamic regulation of secretion of a specific group of proteins (semaphorins) did not reflect their intracellular level. One should therefore be cautious to extrapolate secretion profiles based on intracellular protein levels.
13. Another major advantage of the SILAC protocol is the possibility to combine samples obtained from the different labeled cell populations prior to any further sample preparation such as ultrafiltration, enrichment by different fractionation protocols, or antibody-directed immunoprecipitation. The combined workflow minimizes differences in sample handling, which will translate to higher quantitative accuracy of the experiment.

14. The pool of CM is concentrated by ultrafiltration using Vivaspin columns with an MWCO of 3,000 to ensure that virtually all proteins are retained in the concentrate. Add the maximum volume of CM to the Vivaspin columns (Vivaspin 6: 6 ml; Vivaspin 20: 14 ml when using a fixed rotor and 20 ml when using a swing-bucket rotor). Using an MWCO of 3,000 Da in contrast to 10,000 Da does enhance the concentration time significantly. However, to ensure that we retain all secreted proteins including the low-molecular-weight proteins such as the insulin-like growth factors and small chemokines, we utilized an MWCO of 3,000 [10, 20]. Reduce the time of concentration using additional concentration columns and replace columns if needed (when reduction of media becomes extremely slow). However, it should be noted that there are small losses in terms of total protein amount due to proteins attaching to the membrane of the Vivaspin columns.
15. During ultrafiltration make sure that the volume of the concentrate does not get too small to confer solubility of the secreted proteins.
16. The concentrated CM is size-separated by 1D-gel electrophoresis to reduce sample complexity, thereby effectively increasing the dynamic range of the subsequent MS analysis. It also provides useful information for the approximate molecular mass of the secreted form of the protein.
17. One of the great challenges facing secretome studies is the isolation of truly secreted proteins from the total number of identified proteins in a proteomics study. The increased sensitivity of mass spectrometers does improve not only the identification of secreted proteins but also the identification of intracellular proteins which are released to the extracellular space due to apoptosis and cell leakage or via exosomes carrying intracellular cargo. Different tools are being used to classify the extracellular compartment; most common is classification by literature mining, Gene Ontology (GO) term "extracellular," and/or algorithms predicting secretion by classical (SignalP) or nonclassical (SecretomeP) mechanisms. Many of these applications are available as open source.

    GO annotations can be retrieved using databases provided by the GO consortium (http://www.geneontology.org/) or QuickGO provided by the Gene Ontology Annotation (GOA) group (http://www.ebi.ac.uk/GOA), which is a bioinformatics resource integrating various databases to assign subcellular localization and functional annotation according to GO terms [42, 43].

    SignalP (http://www.cbs.dtu.dk/services/SignalP) and SecretomeP (http://www.cbs.dtu.dk/services/SecretomeP/) prediction servers are both provided by the Center for

Biological Sequence analysis (CBS) as open software programs. SignalP is used to predict the presence of a signal peptide, suggesting that proteins are secreted by the classical secretory pathway [44–46], whereas SecretomeP has been designed to predict nonclassical secreted proteins [47]. Finally, it should also be mentioned that there are commercially available databases such as the ProteinCenter (http://www.proxeon.com), which combines the information provided by several databases to facilitate analyses of large-scale proteomics experiments according to different criteria including GO annotations such as cellular component and prediction of specific sequence motifs including signal peptides using the freely available PrediSi algorithm [48].

Acknowledgments

We thank all CEBI group members for useful discussions. This work was supported by funding obtained from the Novo Nordisk Foundation, the Lundbeck Foundation, and the Augustinus Foundation.

References

1. Clamp M, Fry B, Kamal M et al (2007) Distinguishing protein-coding and noncoding genes in the human genome. Proc Natl Acad Sci U S A 104:19428–19433
2. Skalnikova H, Motlik J, Gadher SJ et al (2011) Mapping of the secretome of primary isolates of mammalian cells, stem cells and derived cell lines. Proteomics 11:691–708
3. Lehr S, Hartwig S, Sell H (2012) Adipokines: a treasure trove for the discovery of biomarkers for metabolic disorders. Proteomics Clin Appl 6:91–101
4. Pavlou MP, Diamandis EP (2010) The cancer cell secretome: a good source for discovering biomarkers? J Proteomics 73:1896–1906
5. Herder C, Karakas M, Koenig W (2011) Biomarkers for the prediction of type 2 diabetes and cardiovascular disease. Clin Pharmacol Ther 90:52–66
6. Brown KJ, Formolo CA, Seol H et al (2012) Advances in the proteomic investigation of the cell secretome. Expert Rev Proteomics 9: 337–345
7. Kratchmarova I, Kalume DE, Blagoev B et al (2002) A proteomic approach for identification of secreted proteins during the differentiation of 3T3-L1 preadipocytes to adipocytes. Mol Cell Proteomics 1:213–222
8. Chan XC, McDermott JC, Siu KW (2007) Identification of secreted proteins during skeletal muscle development. J Proteome Res 6:698–710
9. Molina H, Yang Y, Ruch T et al (2009) Temporal profiling of the adipocyte proteome during differentiation using a five-plex SILAC based strategy. J Proteome Res 8: 48–58
10. Henningsen J, Rigbolt KT, Blagoev B et al (2010) Dynamics of the skeletal muscle secretome during myoblast differentiation. Mol Cell Proteomics 9:2482–2496
11. Makridakis M, Vlahou A (2010) Secretome proteomics for discovery of cancer biomarkers. J Proteomics 73:2291–2305
12. Walther TC, Mann M (2010) Mass spectrometry-based proteomics in cell biology. J Cell Biol 190:491–500
13. Schulze WX, Usadel B (2010) Quantitation in mass-spectrometry-based proteomics. Annu Rev Plant Biol 61:491–516
14. Zhong J, Krawczyk SA, Chaerkady R et al (2010) Temporal profiling of the secretome during adipogenesis in humans. J Proteome Res 9:5228–5238
15. Kovanich D, Cappadona S, Raijmakers R et al (2012) Applications of stable isotope dimethyl

labeling in quantitative proteomics. Anal Bioanal Chem 404(4):991–1009
16. Ong SE, Blagoev B, Kratchmarova I et al (2002) Stable isotope labeling by amino acids in cell culture, SILAC, as a simple and accurate approach to expression proteomics. Mol Cell Proteomics 1:376–386
17. Ong SE (2012) The expanding field of SILAC. Anal Bioanal Chem 404(4):967–976
18. Rigbolt KT, Blagoev B (2012) Quantitative phosphoproteomics to characterize signaling networks. Sem Cell Dev Biol 23(8):863–871
19. Kirchner M, Selbach M (2012) In vivo quantitative proteome profiling: planning and evaluation of SILAC experiments. Methods Mol Biol 893:175–199
20. Henningsen J, Pedersen BK, Kratchmarova I (2011) Quantitative analysis of the secretion of the MCP family of chemokines by muscle cells. Mol Biosyst 7:311–321
21. Chen CY, Chi LM, Chi HC et al (2012) Stable isotope labeling with amino acids in cell culture (SILAC)-based quantitative proteomics study of a thyroid hormone-regulated secretome in human hepatoma cells. Mol Cell Proteomics 11(M111):011270
22. Kashyap MK, Harsha HC, Renuse S et al (2010) SILAC-based quantitative proteomic approach to identify potential biomarkers from the esophageal squamous cell carcinoma secretome. Cancer Biol Ther 10:796–810
23. Greco TM, Seeholzer SH, Mak A et al (2010) Quantitative mass spectrometry-based proteomics reveals the dynamic range of primary mouse astrocyte protein secretion. J Proteome Res 9:2764–2774
24. Lamoureux F, Gastinel LN, Mestre E et al (2012) Mapping cyclosporine-induced changes in protein secretion by renal cells using stable isotope labeling with amino acids in cell culture (SILAC). J Proteomics 75:3674–3687
25. An E, Gordish-Dressman H, Hathout Y (2008) Effect of TNF-alpha on human ARPE-19-secreted proteins. Mol Vis 14:2292–22303
26. Blagoev B, Ong SE, Kratchmarova I et al (2004) Temporal analysis of phosphotyrosine-dependent signaling networks by quantitative proteomics. Nat Biotechnol 22:1139–1145
27. Lundby A, Olsen JV (2011) GeLCMS for in-depth protein characterization and advanced analysis of proteomes. Methods Mol Biol 753: 143–155
28. Rappsilber J, Mann M, Ishihama Y (2007) Protocol for micro-purification, enrichment, pre-fractionation and storage of peptides for proteomics using StageTips. Nat Protoc 2: 1896–1906
29. Cox J, Mann M (2008) MaxQuant enables high peptide identification rates, individualized p.p.b.-range mass accuracies and proteome-wide protein quantification. Nat Biotechnol 26: 1367–1372
30. Cox J, Neuhauser N, Michalski A et al (2011) Andromeda: a peptide search engine integrated into the MaxQuant environment. J Proteome Res 10:1794–1805
31. Hittel DS, Berggren JR, Shearer J et al (2009) Increased secretion and expression of myostatin in skeletal muscle from extremely obese women. Diabetes 58:30–38
32. Pelosi L, Giacinti C, Nardis C et al (2007) Local expression of IGF-1 accelerates muscle regeneration by rapidly modulating inflammatory cytokines and chemokines. FASEB J 21: 1393–1402
33. Kristensen LP, Chen L, Nielsen MO et al (2012) Temporal profiling and pulsed SILAC labeling identify novel secreted proteins during ex vivo osteoblast differentiation of human stromal stem cells. Mol Cell Proteomics 11(10): 989–1007
34. Henningsen J, Blagoev B, Kratchmarova I (2012) Quantitative proteomics for investigation of secreted factors: focus on muscle secretome. In: Flores RJ, Man TK (eds) Proteomics - human diseases and protein functions. InTech, Rijeka, Croatia, pp 417–438
35. Ong SE, Mann M (2006) A practical recipe for stable isotope labeling by amino acids in cell culture (SILAC). Nat Protoc 1:2650–2660
36. Pedersen BK, Febbraio MA (2008) Muscle as an endocrine organ: focus on muscle-derived interleukin-6. Physiol Rev 88:1379–1406
37. Pedersen BK, Febbraio MA (2012) Muscles, exercise and obesity: skeletal muscle as a secretory organ. Nat Rev Endocrinol 8:457–465
38. Yoon JH, Yea K, Kim J et al (2009) Comparative proteomic analysis of the insulin-induced L6 myotube secretome. Proteomics 9:51–60
39. Yoon JH, Song P, Jang JH et al (2011) Proteomic analysis of tumor necrosis factor-alpha (TNF-alpha)-induced L6 myotube secretome reveals novel TNF-alpha-dependent myokines in diabetic skeletal muscle. J Proteome Res 10:5315–5325
40. Olsen JV, Ong SE, Mann M (2004) Trypsin cleaves exclusively C-terminal to arginine and lysine residues. Mol Cell Proteomics 3:608–614
41. Ong SE, Kratchmarova I, Mann M (2003) Properties of 13C-substituted arginine in stable

isotope labeling by amino acids in cell culture (SILAC). J Proteome Res 2:173–181

42. Barrell D, Dimmer E, Huntley RP et al (2009) The GOA database in 2009 – an integrated Gene Ontology Annotation resource. Nucleic Acids Res 37:D396–D403
43. Binns D, Dimmer E, Huntley R et al (2009) QuickGO: a web-based tool for Gene Ontology searching. Bioinformatics 25:3045–3046
44. Bendtsen JD, Nielsen H, von Heijne G et al (2004) Improved prediction of signal peptides: SignalP 3.0. J Mol Biol 340:783–795
45. Emanuelsson O, Brunak S, von Heijne G et al (2007) Locating proteins in the cell using TargetP, SignalP and related tools. Nat Protoc 2:953–971
46. Petersen TN, Brunak S, von Heijne G et al (2011) SignalP 4.0: discriminating signal peptides from transmembrane regions. Nat Methods 8:785–786
47. Bendtsen JD, Jensen LJ, Blom N et al (2004) Feature-based prediction of non-classical and leaderless protein secretion. Protein Eng Des Sel 17:349–356
48. Hiller K, Grote A, Scheer M et al (2004) PrediSi: prediction of signal peptides and their cleavage positions. Nucleic Acids Res 32: W375–W379

Chapter 23

Identification of MicroRNA Targets by Pulsed SILAC

Markus Kaller, Silke Oeljeklaus, Bettina Warscheid, and Heiko Hermeking

Abstract

Pulsed stable isotope labeling by amino acids in cell culture (pulsed SILAC or pSILAC) allows to monitor and quantify the de novo synthesis of proteins in an unbiased fashion on a proteome-wide scale. The high applicability of this metabolic labeling technique has been demonstrated for the identification of posttranscriptional changes in gene expression on the proteome level, in particular those caused by microRNAs. The application of pSILAC allows the selective quantification of newly synthesized proteins and thus the detection of differences in protein translation. This is of particular interest in the case of microRNA-mediated regulations, which characteristically cause rather modest decreases in protein amounts that may be difficult to detect by other proteomic methods. Here, we describe a detailed protocol for using pSILAC to track miRNA-mediated changes in protein expression, using the p53-induced *miR-34a* microRNA as a prototypic example of microRNA-mediated regulations.

Key words Pulsed SILAC, Quantitative mass spectrometry, microRNA targets, De novo protein synthesis, miR-34a, p53

1 Introduction

Pulsed stable isotope labeling by amino acids in cell culture (pulsed SILAC or pSILAC) is currently the most elegant, quantitative mass spectrometry (MS)-based technique that allows for monitoring and quantifying the de novo synthesis of proteins in an unbiased fashion on a proteome-wide scale. pSILAC has been employed to study the global protein turnover in yeast or cultured cells [1] and the impact of inhibitors on protein synthesis [2, 3]. Furthermore, pSILAC was used to globally profile posttranslational modifications [4, 5] and assess protein dynamics during cell differentiation [6]. The high applicability of this metabolic labeling technique for the identification of posttranscriptional changes in gene expression on the proteome level, in particular those induced by microRNAs, has further been demonstrated [7–11]. During such pSILAC analyses, induction or inhibition of microRNA expression was followed

Bettina Warscheid (ed.), *Stable Isotope Labeling by Amino Acids in Cell Culture (SILAC): Methods and Protocols*, Methods in Molecular Biology, vol. 1188, DOI 10.1007/978-1-4939-1142-4_23, © Springer Science+Business Media New York 2014

by a pulse of isotope-labeled amino acids, which are incorporated into all newly synthesized proteins. Subsequent MS analysis of the proteome eventually allows to detect changes in de novo protein translation.

MicroRNAs, also referred to as miRNAs, are a class of small, ~21 nucleotide-long noncoding RNAs derived from RNA hairpin precursors (pri- and pre-mRNAs) that act as posttranscriptional repressors of gene expression by binding to the cognate mRNAs, usually in their 3′-untranslated regions (3′-UTRs) [12]. Based on computational predictions of target mRNAs, the expression of more than 60 % of human protein coding genes is regulated by miRNAs indicating the importance of miRNAs for modulating gene expression at a posttranscriptional level [13]. The primary determinant of target recognition is the so-called "seed" sequence, a short, ~7 nucleotide-long stretch in the 5′ region of the miRNA that pairs to the largely complementary target sequence in the target mRNA. Additional base pairing between the remainder of the miRNA and its target mRNA may occur, but in most cases this involves only a lower degree of complementarity. Binding of the miRNA to its target mRNA causes inhibition of translation initiation, with or without destabilization and degradation of the mRNA, in sum leading to a repression of the protein production from the respective target gene/mRNA. Individual miRNAs generally regulate dozens to probably hundreds of target mRNAs. In most cases, however, they only cause a modest decrease in protein translation, consistent with their assumed primary role as rheostats buffering transcriptional noise and reinforcing pre-patterned transcriptional programs [14]. This general feature makes target identification for a given miRNA challenging. The "traditional" SILAC approach [15] has been shown to be quite suitable for the global analysis of miRNA-mediated changes in protein abundance and for identifying distinct miRNA targets [16, 17]. However, the detection of miRNA-mediated regulation of certain proteins, in particular of those with long half-lives, may fail when measuring steady-state protein levels only. This shortcoming can be resolved by using the pSILAC strategy, which facilitates the selective quantification of newly synthesized proteins and, thus, the detection of differences in protein translation [11].

In this chapter, we describe a detailed protocol for using pSILAC to track miRNA-mediated changes in protein expression in the colorectal SW480 cancer cell line containing an episomal vector, which allows conditional expression under control of the tet-repressor system [18]. We have used the system to express *miR-34a*, which is directly regulated by the p53 tumor suppressor [7, 19]. However, the vectors are suitable for expression of any other pri-microRNA of interest. Moreover, numerous variants of the experimental setup suggested here can be envisioned. For examples, a cell line of choice can be transfected with synthetic miRNA mimics or miRNA inhibitors (antagomiRs) or protein expression in cell lines carrying a targeted deletion of the respective miRNA can be compared with the wild-type counterpart.

Figure 1 illustrates the outline of a pSILAC analysis. Cells treated with doxycycline and non-treated control cells are cultured in medium containing "light" (L) arginine and lysine. 16 h after addition of doxycycline, cells are shifted to "heavy" (*H*) medium containing $^{13}C^{15}N$-labeled variants of arginine (Arg10) and lysine (Lys8) while non-treated control cells are shifted to "medium-heavy" (*M*) medium supplemented with $^{13}C^{14}N$-coded arginine (Arg6) and deuterated lysine (Lys4). Upon this pulse, newly synthesized proteins will incorporate the isotope-coded variants of arginine and lysine. Cells are grown in "heavy" or "medium-heavy" medium for further 24 h. Proteins are extracted from each cell population, combined, separated by SDS-PAGE and subjected to enzymatic digestion using trypsin followed by high performance liquid chromatography tandem mass spectrometry (HPLC/MS/MS) analysis. Differences in de novo protein translation resulting from the expression of the miRNA are reflected by the ratio of MS signal intensities of "heavy" versus "medium-heavy" peptides (H/M) of the respective protein. "Light" peptide ion signals are derived from preexisting proteins and can be ignored for the analysis of changes in protein biosynthesis.

The results obtained in a pSILAC analysis allow to identify proteins whose translation rates are affected by induction of a given microRNA. However, many of these effects may be secondary and not directly linked to the regulation by the microRNA of interest. Figure 2 shows a suggested flow-chart for the identification of microRNA targets from the pSILAC-derived MS data set after quantitative analysis with the software MaxQuant.

Initially, a threshold for considering protein biosynthesis (ratio H/M) to be downregulated by microRNA expression should be defined. It should be noted that too stringent thresholds reduce the number of potential microRNA targets since most mammalian targets only show modest decreases in protein translation. Conversely, if the threshold is set too low, the number of putative microRNA targets increases but the weak regulation of individual targets may be difficult to validate by independent methods such as Western blot analysis. Others and we have defined a "strong" translational downregulation of targets as a $\log_2$-fold change of ≤ -0.3, corresponding to a ~1.23-fold reduction of translation rates after induction of the microRNA [7, 11].

Next, the set of downregulated proteins is screened for the presence of predicted microRNA binding sites. This can be achieved by screening for the presence and distribution of the hexameric sequence representing the seed-matching sequence of the microRNA in the 3′-UTR of the mRNAs corresponding to the proteins present in the pSILAC results. However, not every hexameric sequence

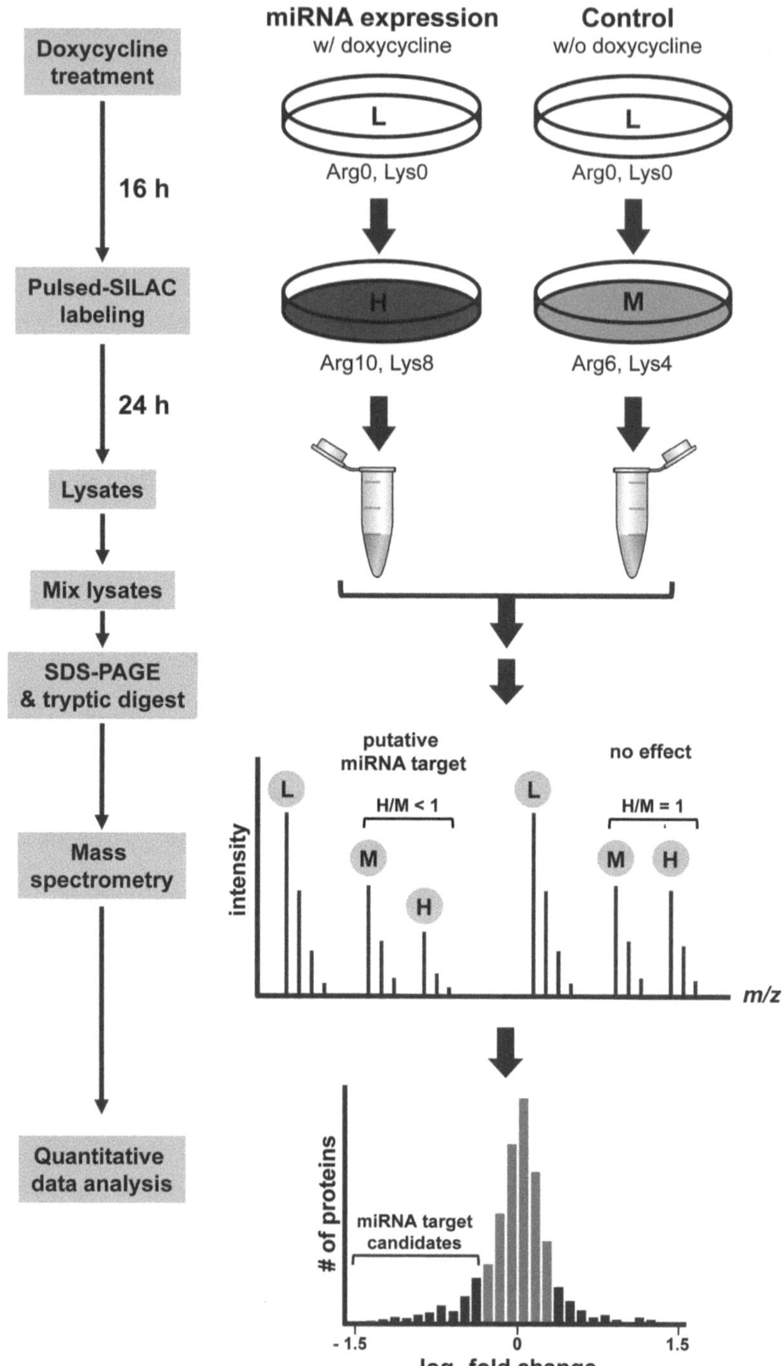

Fig. 1 Identification of potential miRNA targets using pSILAC. Cells harboring an episomal pRTS-*miR-34a* plasmid are grown in "light" medium (*L*) containing unlabeled arginine (Arg0) and lysine (Lys0). To induce expression of the miRNA, cells are treated with doxycycline for 16 h. The medium of doxycycline-treated cells is then changed to "heavy" SILAC medium (*H*) containing $^{13}C^{15}N$-labeled variants of arginine and lysine (Arg10, Lys8) while non-treated control cells are

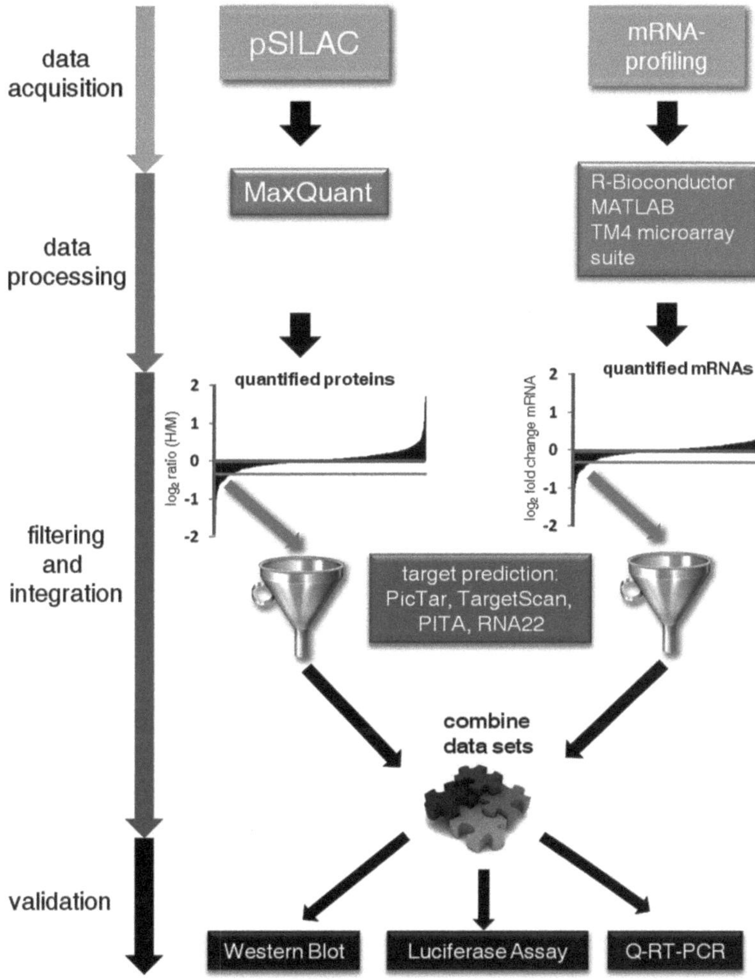

Fig. 2 Flowchart for the analysis and integration of data derived from a pSILAC experiment and mRNA profiling after microRNA expression as well as subsequent validation of candidate microRNA target genes. For detailed explanations, *see* Subheading 1

Fig. 1 (continued) shifted to "medium-heavy" SILAC medium (*M*) supplemented with $^{13}C_6$-arginine (Arg6) and $^{2}H_4$-lysine (Lys4). 24 h after the pulse, cells are lysed, lysates are mixed, and proteins are separated by SDS-PAGE. Following tryptic in-gel digestion, the resulting peptide mixture is analyzed by high performance liquid chromatography–mass spectrometry for protein identification and relative protein quantification. Proteins newly synthesized after the SILAC-pulse are labeled with either "medium-heavy" or "heavy" amino acids. Potential miRNA targets, i.e., newly synthesized proteins with significantly reduced abundance following miRNA expression, are identified by *H*/*M* ratios lower than one. Proteins containing the "light" versions of arginine and lysine are preexisting proteins that are not relevant for this analysis

matching the microRNA seed represents a functional microRNA binding site. Therefore, the use of bioinformatic microRNA target site prediction tools is recommended to cross-compare with the pSILAC-derived data sets. Several bioinformatic algorithms, which make use of different parameters to weigh features involved in the miRNA/mRNA interaction, have been developed for the prediction of miRNA targets. The data sets of predicted microRNA targets are available online and can be downloaded from the associated Web resources. The algorithms TargetScan and PicTar [13, 20] highlight the presence of perfect, evolutionarily conserved seed matches, whereas PITA and RNA22 [21, 22] prioritize the ΔG of the entire miRNA/mRNA duplex and the accessibility of the site within the mRNA. TargetScan and PicTar provide high predictive power for experimentally obtained proteomic data [11, 16, 23], however, they may be less useful in the prediction of miRNA target sites that lack a perfect seed-sequence, are not evolutionarily conserved, or lie outside the 3′-UTR of the target gene. Therefore, the combined use of several different algorithms may be helpful to identify target mRNAs of a given miRNA.

In order to determine whether the translationally downregulated proteins are indeed *bona fide* targets of a given microRNA, data derived from pSILAC experiments can be compared to other genome-wide data sets derived from microarray or RNAseq experiments (Fig. 2). The raw data from these experiments can be processed with software packages such as Bioconductor [24], the TM4 microarray suite [25] or MATLAB® to obtain fold changes in mRNA expression levels after miRNA expression when compared to a control. These data sets can further be filtered for differential expression and the presence of predicted microRNA binding sites analogous to the pSILAC-derived data. The two data sets can then be combined into a composite set of predicted microRNA targets that ideally pass the applied threshold and filtering in both experiments, i.e., show a regulation both at the mRNA and at the protein level.

The regulation of translation initiation and mRNA stability/degradation are tightly coupled processes during miRNA-mediated repression [26]. In fact, mammalian miRNAs have been shown to reduce the mRNA levels of the majority of their targets [27]. Hence, in most cases the inhibition of translation of a miRNA-regulated protein correlates with downregulation of mRNA expression. We have also found that changes in de novo protein synthesis in general correlate with changes in mRNA abundance for predicted miRNA targets, and accordingly, miRNA targets with a strong translational regulation also displayed a strong regulation at the level of mRNA stability [7]. However, it should be noted that weakly regulated targets, which do not pass a defined threshold of differential regulation in one or the other assay, may still be considered targets of the miRNA of interest and subjected to further experimental validation.

In addition, pSILAC data can be complemented and cross-compared with data derived from transcriptome-wide miRNA binding methods such as *hi*gh-*th*roughput *se*quencing of RNAs isolated by *c*ross*li*nking and *i*mmuno*p*recipitation (HITS-CLIP) or the related *p*hoto*a*ctivatable-*r*ibonucleoside-enhanced *c*ross*li*nking and *i*mmuno*p*recipitation (PAR-CLIP) [28, 29], which would provide more direct, physical evidence for binding of a microRNA to a given mRNA. However, since both methods essentially rely on the immunoprecipitation of the RISC/miRNA/mRNA ternary complex, they do not distinguish between binding sites of individual, selected miRNAs of choice. Hence, a more direct, alternative approach would involve the use of biotinylated miRNAs of choice, which can be purified together with RISC in a tandem affinity purification approach [30, 31].

Finally, identified target genes should be validated by additional biochemical or molecular biological methods such as luciferase 3′-UTR reporter assays, Western blot analysis, and quantitative real-time (qRT)-PCR. We recommend the use of luciferase 3′-UTR reporter assays as the method of choice to determine if a given mRNA is directly regulated by a microRNA. It allows the targeted mutagenesis of the microRNA binding site in the context of the 3′-UTR, which should disrupt the regulation by the microRNA.

Subsequently, a Western blot analysis should be performed to verify whether the microRNA reduces the steady-state protein levels of its target(s) to a physiologically relevant extent. Since microRNAs often regulate their targets both at the protein and at the mRNA level, the effect of a microRNA on its target mRNA should also be analyzed by qRT-PCR. A detailed protocol for the design of a qRT-PCR analysis can be found elsewhere [32]. Furthermore, the combination of different genome-wide approaches to achieve an integrated analysis of microRNA-mediated regulations, which occur after activation of the p53 transcription factor, has been recently described [33].

2 Materials

2.1 Pulsed SILAC Labeling and Cell Culture

2.1.1 SILAC Medium

1. DMEM lacking arginine and lysine (*see* **Note 1**).
2. Amino acids (*see* **Note 2**): $^{2}H_{4}$-L-lysine (Lys4), $^{13}C_{6}{}^{15}N_{2}$-L-lysine (Lys8), $^{13}C_{6}{}^{14}N_{4}$-L-arginine (Arg6), $^{13}C_{6}{}^{15}N_{4}$-L-arginine (Arg10) as well as unlabeled L-lysine and L-arginine (Lys0 and Arg0).
3. Syringe and sterile filter (0.22 μm).
4. Dialyzed fetal bovine serum (FBS; *see* **Note 3**); filter-sterilized.
5. Penicillin (100×, i.e., 10,000 units/ml) and streptomycin (100×, i.e., 10 mg/ml).

2.1.2 Cell Culture and Cell Lysis

1. SW480 cells (or other cell line of choice) containing the pRTS-miR-34a vector (*see* **Notes 4** and **5**).

2. 10-cm cell culture dishes, pipettes.
3. Hemocytometer.
4. Doxycycline (1,000× stock solution: 100 μg/ml in water).
5. Cell scraper.
6. Phosphate-buffered saline (PBS).
7. RIPA buffer: 50 mM Tris–HCl (pH 7.4), 150 mM NaCl, 1 mM EDTA, 1 % (v/v) NP-40, 0.1 % (w/v) SDS, 0.5 % (w/v) sodium deoxycholate.
8. Roche Complete Protease Inhibitor tablets.
9. Bradford reagent, BSA protein standards (0.25–2.0 mg/ml), and photometer or alternative method to determine protein concentrations.

2.2 Gel Electrophoresis, Protein Staining, and Tryptic In-Gel Digestion of Proteins

2.2.1 Sample Preparation and SDS Polyacrylamide Gel Electrophoresis

1. 5× SDS sample buffer: 0.25 M Tris–HCl (pH 6.8), 10 % (w/v) SDS, 50 % (w/v) glycerol, 0.05 % (w/v) bromophenol blue, 25 % (v/v) β-mercaptoethanol.
2. 4–12 % NuPAGE™ Bis-Tris gradient gels (Life Technologies; *see* **Note 6**).
3. XCellSureLock™ Mini-Cell Electrophoresis System (Life Technologies; *see* **Note 6**).
4. MOPS running buffer: 50 mM MOPS, 50 mM Tris, 1 mM EDTA, 0.1 % (w/v) SDS.
5. Prestained molecular weight marker (e.g., PageRuler™, Prestained Protein Ladder, Thermo Scientific).

2.2.2 Visualization of Proteins Using Colloidal Coomassie Brilliant Blue

1. Fixing solution: 50 % (v/v) methanol, 2 % (v/v) phosphoric acid.
2. Incubation solution: 34 % (v/v) methanol, 2 % (v/v) phosphoric acid, 17 % (w/v) ammonium sulfate.
3. Colloidal Coomassie Brilliant Blue G-250.
4. Scanner.

2.2.3 Tryptic In-Gel Digestion

1. Scalpel.
2. Microcentrifuge tubes or mini glass tubes with caps.
3. Glass vials for liquid chromatography (LC) analysis.
4. Solution A: 10 mM ammonium bicarbonate (NH_4HCO_3), pH 7.8 (*see* **Note 7**).
5. Solution B: 5 mM NH_4HCO_3 in 50 % (v/v) acetonitrile.
6. 100 % ethanol.
7. 10 mM dithiothreitol (DTT) in 5 mM NH_4HCO_3.
8. 55 mM iodoacetamide in 5 mM NH_4HCO_3.
9. SpeedVac.

10. Trypsin (modified sequencing grade), dissolved in solution A, used at a final concentration of 0.033 μg trypsin/μl.
11. Solution C: 2.5 % (v/v) formic acid in 50 % (v/v) acetonitrile.
12. Solution D: 0.1 % (v/v) trifluoroacetic acid in 5 % (v/v) acetonitrile.

2.3 Nano-HPLC/ESI-MS/MS Analysis and Mass Spectrometric Data Processing

1. Solvent A: 0.1 % (v/v) formic acid (*see* **Note 7**).
2. Solvent B: 0.1 % (v/v) formic acid in 84 % acetonitrile.
3. Nano-flow HPLC system (e.g., UltiMate™3000 or RSLC, Thermo Fisher Scientific/Dionex, Idstein, Germany) equipped with a C18 μ-precolumn and a C18 reversed-phase nano-LC column.
4. High resolution ESI-MS instrument such as the LTQ-Orbitrap XL, Velos, or Elite (Thermo Fisher Scientific, Bremen, Germany).
5. Software application(s) for protein identification and quantification such as MaxQuant (version 1.1.1.25 or higher, http://www.maxquant.org; [34, 35]).
6. Protein sequence database of the organism used (*see* **Note 8**).

2.4 Suggestions for the Analysis of Pulsed SILAC Data

The calculations described in these sections can be performed in MATLAB, GraphPad, and Excel.

1. MATLAB software package.
2. Microsoft Excel.
3. GraphPad prism software.

2.5 Validation of MicroRNA Targets Using the Luciferase Assay

We recommend the use of the Dual reporter luciferase assay kit (Promega); however, other commercially available kits may be used.

2.5.1 Transfection of Cells with Firefly and Renilla Luciferase Reporter Vectors

1. Cell line of choice (Hela, HEK293 or H1299) (*see* **Note 9**).
2. High-glucose DMEM supplemented with 10 % FBS and penicillin/streptomycin.
3. Opti-MEM (reduced serum medium).
4. Reagent for transfection of eukaryotic cells with microRNA, e.g., HiPerfect (QIAGEN).
5. Synthetic microRNA precursors (e.g., from Ambion) for both the negative control and the microRNA of choice (*see* **Note 10**).
6. Firefly luciferase vectors (pGL3) containing the 3′-UTR of interest downstream of the luciferase gene and *Renilla* luciferase vector (pRL) for normalization (Promega) (*see* **Note 11**).
7. 12-well cell culture plates, pipettes.

2.5.2 Luciferase Assay

1. PBS.
2. White 96-well microtiter plates (Nunc).
3. PLB buffer, LAR II buffer, and Stop & Glo buffer provided with the Dual reporter luciferase assay kit (Promega).
4. Luminometer, e.g., the Orion II microplate reader (Berthold Detection Systems) and the associated *Simplicity 4.1* software for data acquisition.

2.6 Western Blot Analysis for Validation of MicroRNA Targets

2.6.1 Transfection of Cells with MicroRNA Precursors (See Note 4)

1. SW480 cells (or other cell line of choice).
2. Media, transfection reagent and synthetic microRNA precursors as described in Subheading 2.5.1.
3. 10-cm cell culture dishes, pipettes.

2.6.2 Sample Preparation and SDS-PAGE

1. RIPA buffer supplemented with protease inhibitors (*see* Subheading 2.1.2).
2. Reagents and equipment for the determination of protein concentrations (*see* Subheading 2.1.2).
3. Reagents and equipment required for SDS-PAGE (*see* Subheading 2.2.1).

2.6.3 Protein Transfer and Antibody-Based Detection of Proteins

1. Whatman paper.
2. PVDF membrane, e.g., Immobilon™-P transfer membrane (Millipore).
3. Western blotting device of choice (semi-dry or tank blotting). We use a PerfectBlue™ semi-dry blotting device (Peqlab) for most applications.
4. 25× transfer buffer stock solution: 300 mM Tris–HCl (pH 8.3), 2.4 M glycine.
5. Methanol.
6. TBST buffer: 10 mM Tris–HCl (pH 7.4), 15 mM NaCl, 0.1 % (v/v) Tween 20.
7. Blocking solution: TBST containing 10 % (w/v) skim milk powder.
8. Specific primary antibodies against proteins of interest and a suitable loading control (e.g., α-tubulin).
9. Horseradish peroxidase (HRP)-conjugated secondary antibodies with specificities for the species the primary antibodies were developed in (e.g., goat anti-rabbit IgG or goat anti-mouse IgG).
10. ECL detection reagent, e.g., Immobilon™ Western HRP substrate (Millipore) or Western Lightning Chemoluminescence Reagent (PerkinElmer).

11. Transparent overhead foil.
12. Chemoluminescence imager for the detection and documentation of chemoluminescence, for example, the Kodak 440CF digital imaging system and the associated Kodak Molecular Imaging Software for image acquisition, image processing and quantification of signal intensities. However, X-ray films may also be used.

3 Methods

3.1 Pulsed SILAC Labeling

3.1.1 Preparation of Media

1. The final concentration of L-arginine (Arg) and L-lysine (Lys) in DMEM used for culturing SW480 cells is 84 mg/l and 40 mg/l, respectively. Prepare 100× stock solutions of all variants of L-arginine (84 mg/10 ml) and L-lysine (40 mg/10 ml) in Milli-Q water.
2. Filter-sterilize amino acid stock solutions (*see* **Note 12**).
3. Prepare "light", "medium-heavy", and "heavy" SILAC medium by supplementing DMEM deficient in Arg and Lys with dialyzed FBS (10 % final concentration), penicillin, streptomycin, and the combination of Arg0/Lys0 for "light", Arg6/Lys4 for "medium-heavy", and Arg10/Lys8 for "heavy" SILAC medium.

3.1.2 Cell Culture and Cell Lysis

1. Seed 5×10^5 SW480 cells harboring the pRTS-*miR-34a* vector onto 10-cm cell culture dishes and grow them in "light" medium for 24 h. Per experiment, use one dish for *pri-miR-34a*-induction and one dish as a control (i.e., non-induced cells). To generate biologically significant data, perform at least three independent biological replicates.
2. Induce expression of *pri-miR-34a* by adding doxycycline (100 ng/ml final concentration) to the cells.
3. 16 h after induction, metabolically label cells by shifting the doxycycline-induced cells to "heavy" and the non-induced control cells to "medium-heavy" SILAC medium (*see* **Note 13**).
4. Incubate the cells for further 24 h.
5. Place dishes on ice and wash the cells twice with 6 ml of ice-cold PBS. Remove PBS completely.
6. Cover the cells with 500 μl of ice-cold RIPA buffer supplemented with protease inhibitors.
7. Scrape the cells off the plate and transfer them to a 1.5-ml reaction tube. The samples can be stored at −80 °C at this step.
8. Sonicate the samples 5× for 1 s to shear the DNA. Keep samples on ice during sonification and keep intervals between sonification pulses long enough to avoid heating of the samples.

9. Centrifuge the samples for 20 min at 16.000 × *g* and 4 °C to remove insoluble debris.
10. Take off the supernatants, i.e., the lysates.
11. Combine lysates of induced and control cells of each replicate and determine the protein content using, for example, the Bradford assay [36] according to the standard protocol. Lysates can be stored at −80 °C.

3.2 Gel Electrophoresis, Protein Staining, and Tryptic Digestion of Proteins

3.2.1 Sample Preparation and SDS–Polyacrylamide Gel Electrophoresis

1. Add SDS sample buffer to an aliquot of the combined lysates that corresponds to 20 μg of protein and heat the sample at 95 °C for approx. 5 min.
2. Assemble the gel unit according to the manufacturer's instructions.
3. Fill the buffer tank with MOPS running buffer until gel cassettes are covered and carefully remove the comb. Load samples and 5 μl of the molecular weight marker onto the gel.
4. Run the gel with a constant voltage of 150 V. Stop electrophoresis just before the dye front reaches the end of the gel.
5. Disassemble the gel unit and carefully open the gel cassette.

3.2.2 Visualization of Proteins Using Colloidal Coomassie Brilliant Blue

All incubation and washing steps are performed at room temperature (RT) with gentle agitation.

1. Briefly wash the gel with deionized water and transfer it to fixing solution. Incubate for at least 30 min.
2. Remove fixing solution and wash the gel three times for 10 min with deionized water.
3. Incubate the gel for 30 min in incubation solution.
4. Add colloidal Coomassie Brilliant Blue G-250 to reach a final concentration of approx. 0.1 % (w/v). Incubate the gel overnight or until bands are visible.
5. Wash the gel several times with deionized water to remove residual Coomassie particles and background staining.
6. Scan the gel for documentation.

3.2.3 Tryptic In-Gel Digestion

1. Cut lanes of replicate experiments into 20 equal slices each using a clean scalpel and place them into separate microcentrifuge or glass tubes. Cut each slice into smaller pieces.
2. Wash the gel pieces alternating with 20–50 μl of solution A and solution B for 10 min each. Gel pieces should be covered with liquid. Discard the supernatants after each washing step. Incubate three times with each solution.
3. Dehydrate the gel pieces by incubating them for 10 min in 50 μl ethanol.

4. Remove ethanol and add 50 µl of the DTT solution. Incubate for 30 min at 65 °C (*see* **Note 14**).
5. Remove DTT, add 50 µl of iodoacetamide and incubate for 30 min at RT *in the dark* (*see* **Note 15**).
6. Remove iodoacetamide and wash the gel pieces for 10 min in 50 µl solution A.
7. Dry the gel pieces in a SpeedVac. Dried samples can be stored at −80 °C.
8. Digest proteins in-gel with 2–3 µl of trypsin solution overnight at 37 °C.
9. For extraction of the tryptic peptides from the gel, add 20 µl of solution C. Following incubation (10 min) in a cooled sonicator bath, transfer the supernatants to LC vials.
10. Repeat the extraction step and combine the peptide extracts.
11. Dry the peptides in a SpeedVac and reconstitute them by adding 15–20 µl of solution D and sonicating for 10 min.
12. Prior to LC/MS analysis, remove remaining particles by centrifugation (5 min at 16,000 × *g*). If necessary, transfer supernatants to a fresh glass vial.

3.3 Nano-HPLC/ESI-MS/MS Analysis and Data Processing

1. Analyze tryptic peptide mixtures by nano-HPLC/ESI-MS/MS using a nano-flow HPLC system directly coupled to an LTQ-Orbitrap XL instrument, for example [7].
2. Use MaxQuant or suitable alternative software applications and search algorithms for peptide identification, protein assembly, and relative protein quantification.

3.4 Suggestions for the Analysis of Pulsed SILAC Data

When analyzing the pSILAC data set, it should be determined first if the miRNA expression had an effect, i.e., if the changes in protein translation rates can be assigned to the action of the expressed miRNA. For visualization, fold changes in the abundance of newly synthesized proteins are plotted on a $\log_2$ basis.

1. Determine the set of those quantified proteins that contain ≥one predicted binding site in the 3′-UTRs of their corresponding mRNAs.
2. Determine the set of those quantified proteins that do not contain a predicted binding site in the 3′-UTRs of their corresponding mRNAs.
3. On the *X*-axis, plot both sets of proteins in ascending order based on the $\log_2$ basis of their ratios (*H*/*M*).
4. On the *Y*-axis, plot the fraction (in %) of quantified proteins from both sets of proteins that display, at any given ratio (*H*/*M*) on the *X*-axis, a ratio (*H*/*M*) that is ≤ the given ratio (*H*/*M*) on the *X*-axis.

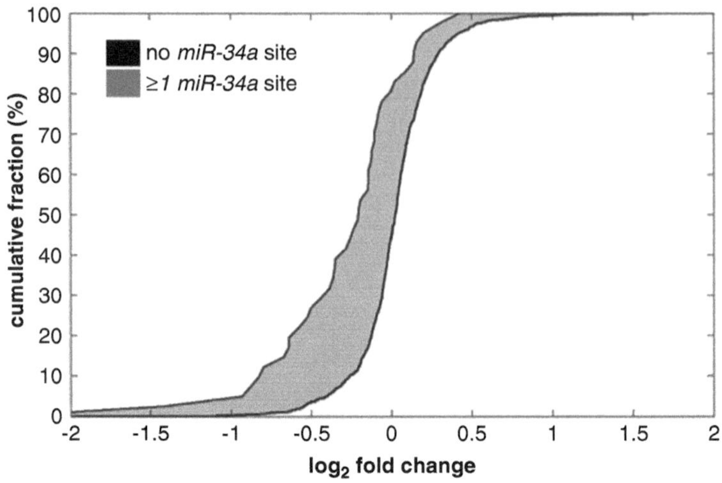

Fig. 3 *miR-34a* expression results in specific changes in protein expression. The cumulative distribution of *miR-34a* targets with ≥1 seed-matching sequence in the 3′-UTR (*grey line*) among proteins detected by pSILAC is compared to proteins without the respective seed-matching sequence in the 3′-UTR of their mRNAs (*black line*). Proteins with ≥1 seed-matching sequence in the 3′-UTR of their mRNAs show a clear shift toward lower fold changes in translation, i.e., reduced expression levels. The figure was originally published in Molecular & Cellular Proteomics (Kaller, M., Liffers, S. T., Oeljeklaus, S., Kuhlmann, K., Roh, S., Hoffmann, R., Warscheid, B., Hermeking, H. Genome-wide characterization of miR-34a induced changes in protein and mRNA expression by a combined pulsed SILAC and microarray analysis. *Molecular & Cellular Proteomics* 2011, 10, M111.010462. © the American Society for Biochemistry and Molecular Biology)

The cumulative distribution of predicted miRNA targets with ≥1 predicted binding site should show a clear bias to negative fold changes compared to proteins without a predicted binding site (*see* Fig. 3). This type of analysis has the advantage to circumvent the definition of a threshold of regulation but is able to visualize differences in translation rates over the entire range of regulations including minor changes in protein translation.

Alternatively, when applying a threshold of differential regulation, the predicted targets of the microRNA should show a bias toward the downregulated proteins (Fig. 4).

To further substantiate if this bias is specific for the ectopically expressed microRNA of interest, the predicted target sets of all known microRNAs (which can be obtained, for example, from TargetScan) can be cross-compared to the set of downregulated proteins derived from the pSILAC data.

1. Determine the number of all quantified proteins with *no* predicted binding site that display a differential regulation with a $\log_2$-fold change of ≤ -0.3 and the number of quantified proteins with $\log_2$-fold change of > -0.3.

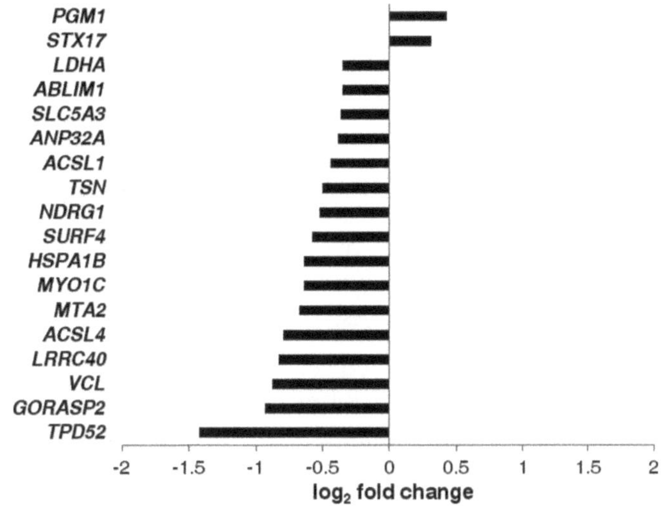

Fig. 4 Distribution of changes in protein synthesis of differentially regulated candidate proteins detected by pSILAC with *miR-34a*. Proteins detected by pSILAC with ≥1 seed-matching sequence in the 3′-UTR of their mRNAs, i.e., predicted *miR-34a* targets, were defined as differentially regulated with a $\log_2$-fold change in protein translation ≤–0.3 or ≥0.3. The distribution of predicted *miR-34a* targets is shifted toward translationally downregulated proteins. The figure was originally published in Molecular & Cellular Proteomics (Kaller, M., Liffers, S. T., Oeljeklaus, S., Kuhlmann, K., Roh, S., Hoffmann, R., Warscheid, B., Hermeking, H. Genome-wide characterization of miR-34a induced changes in protein and mRNA expression by a combined pulsed SILAC and microarray analysis. *Molecular & Cellular Proteomics* 2011, 10, M111.010462. © the American Society for Biochemistry and Molecular Biology)

2. Similarly, determine the number of all quantified proteins with ≥1 predicted binding site in the 3′-UTRs of their corresponding mRNAs site that display a differential regulation with a $\log_2$-fold change of ≤–0.3 and the number of quantified proteins with a $\log_2$-fold change of >–0.3.
3. Use a 2×2 contingency table (Fisher's Exact test), e.g., with the GraphPad software package, to determine if the predicted microRNA targets are significantly overrepresented among the number of proteins with a $\log_2$-fold change of ≤–0.3.

Using this approach, we found that the ectopically expressed microRNA was the *only* microRNA whose predicted targets showed statistically significant overrepresentation among the downregulated proteins, indicating a highly specific effect of the microRNA on protein translation rates [7].

3.5 Dual Reporter Luciferase Assay

In a dual reporter luciferase assay, the activities from firefly (*Photinus pyralis*) and Renilla (*Renilla reniformis*) luciferases are measured simultaneously in one assay. To determine whether a given mRNA is a target for direct regulation by a miRNA, the 3′-UTR of the putative target mRNA is placed downstream of a *Firefly* luciferase reporter gene. This reporter construct is co-transfected either with miRNA mimics or miRNA inhibitors and a *Renilla* luciferase vector for standardization (*see* **Note 16**). In case of specific, direct regulation, the 3′-UTR reporter is repressed by ~20–80 %. In order to map and validate the seed-matching sequences, these should be mutated in the context of the 3′-UTR sequence. The resulting constructs should ideally show resistance toward the respective miRNAs.

1. Seed cells the day before transfection in a 12-well plate in 1 ml medium. Cells should have a confluency of 50–70 % at the time of transfection. Cells should be seeded in triplicates for each transfection to assess the reproducibility of the results.
2. On the day of transfection, prepare the transfection mix. We commonly use 100 ng of *Firefly* reporter plasmid (or molar equivalents of the reporters bearing the 3′-UTRs of interest), 20 ng of *Renilla* reporter, and miRNA mimics at a final concentration of 25 nM in a total volume of 100 μl per single transfection. For triplicate transfections, the volumes of Opti-MEM, luciferase vectors, and miRNA mimics have to be adjusted accordingly.
3. Prepare a master mix consisting of the *Firefly* and *Renilla* reporter plasmids in a 1.5-ml reaction tube. Mix thoroughly (*see* **Note 17**).
4. Divide the master mix in two parts.
5. Add the negative control oligo and the miRNA mimic to the respective reaction tube. Mix thoroughly.
6. Add 5 μl of transfection reagent per single transfection (15 μl per triplicate transfection). Mix by vortexing at low intensity or tapping the tube and incubate for 10 min at RT.
7. Add the transfection mix drop-wise to the cells. Place the cells in an incubator.
8. Harvest the cells 48 h after transfection. Wash the cells twice with PBS, completely remove PBS and add 250 μl of passive lysis buffer (PLB) per well. Incubate for 20 min on a horizontal shaker for complete cell lysis (*see* **Note 18**).
9. Transfer 50 μl of the lysate from each transfection into a 96-well white microtiter plate. This step does not require clearing of the lysate (e.g., by centrifugation), but avoid the transfer of large clumps of cellular debris.
10. We use the following settings to measure the luciferase activities in each 96-well plate: inject 50 μl of LARII, 3 s delay, 5 s measurement, inject 50 μl of Stop & Glo, 3 s delay, 5 s measurement.

11. The effect of the microRNA on the corresponding 3′-UTR luciferase reporter is determined as follows:
    (a) Divide the *Firefly* relative light units (RLUs) by the *Renilla* RLUs for the triplicate control miRNA transfection as well as for the triplicate transfection for your miRNA of interest to determine the *Renilla*-normalized luciferase activity.
    (b) Determine the mean ratio of each triplicate. Normalize each *Firefly–Renilla* ratio to the mean ratio of the control transfection by dividing each *Firefly–Renilla* ratio by the mean ratio of the control transfection.
    (c) Determine the standard deviation for both triplicate transfections as well as the significance using the Student's *t*-test.
12. In case of a reduction in luciferase activity, the microRNA binding site(s) in the 3′-UTR of interest can be mutated to prove that the predicted binding site(s) is/are indeed required for the observed reduction in luciferase activity. For this, one may use the QuickChange™ Site-Directed Mutagenesis kit (Stratagene) following the manufacturer's instructions. In our lab, microRNA binding sites are usually inactivated by four nucleotide substitutions in the seed-matching sequence which disrupts the pairing between the microRNA and the mRNA.

3.6 Western Blot Analysis for the Validation of MicroRNA Targets

When designing the analysis, it should be taken into consideration that the regulation of steady-state protein levels may be only moderate for many microRNA targets, e.g., with a 20–30 % reduction in protein amounts. On the other hand, changes in expression levels of a given protein can also be an indirect consequence of miRNA expression. Hence, a compromise has to be made regarding the duration of miRNA expression to achieve a maximum effect on protein levels but, at the same time, avoiding secondary, indirect effects on the expression of a given protein due to global changes in cellular physiology after miRNA expression.

3.6.1 Transfection of Cells with MicroRNA Precursors

1. Seed 5×10^5 SW480 cells in 10-cm dishes the day before transfection (*see* **Note 19**). You need one dish per condition, e.g., negative control oligo and miRNA of choice.
2. The following day, change the medium before transfection.
3. Prepare the transfection mix:
    (a) Add 500 μl of Opti-MEM in a 1.5-ml reaction tube per condition (negative control oligo and miRNA of choice).
    (b) Add 25 μl of the 10 μM stock solution of pre-miRNA mimic. The final concentration of pre-miRNA mimics will be 25 nM. Mix by tapping.
    (c) Add 25 μl of the HiPerFect reagent. Mix by tapping or gentle, brief vortexing.

4. Incubate for 10 min at RT.
5. Add the transfection mix drop-wise to cells, swirl gently, and place cells in an incubator.
6. Incubate the cells for an appropriate length of time (*see* **Note 20**).

3.6.2 Sample Preparation and SDS-PAGE

1. Prepare protein lysates as described in Subheading 3.1.2 but without mixing lysates of control oligo and miRNA-of-choice transfected cells!
2. Determine the protein concentration (*see* **Note 21**).
3. Mix the amount of lysates needed with SDS sample buffer and heat for 5 min at 95 °C. Start out with a protein amount of 40–50 μg per sample (*see* **Note 22**).
4. Perform SDS-PAGE as described in Subheading 3.2.1 (*see* **Note 23**).

3.6.3 Protein Transfer and Antibody-Based Detection of Proteins

1. Soak Whatman paper in 1× transfer buffer freshly prepared from a 25× stock solution with 20 % (v/v) methanol.
2. Activate the PVDF membrane in methanol for 5 min. Then transfer the membrane into 1× transfer buffer.
3. Assemble the array of Whatman paper, membrane, and SDS gel.
4. Perform the transfer of the proteins onto the PVDF membrane using settings specifically determined for your protein(s) of interest (*see* **Note 24**).
5. Disassemble Whatman paper, membrane, and SDS gel and block the membrane for 1 h at RT in blocking solution on a rocking platform.
6. Wash the membrane three times for 10 min with TBST.
7. Incubate the membrane with primary antibody diluted in blocking solution (*see* **Note 25**) and wash again three times for 10 min with TBST.
8. Add secondary HRP-coupled antibody diluted in blocking solution according to the manufacturer's recommendations and incubate the membrane for 1 h at RT on a rocking platform.
9. Wash the membrane three times for 10 min with TBST.
10. Remove TBST.
11. Prepare the ECL reagent by mixing the luminol reagent and peroxide solution in a 1:1 ratio in a 1.5-ml reaction tube.
12. Place the membrane on a sheet of transparent overhead foil with the protein side facing upwards. Add ECL reagent. Cover with a second sheet of overhead foil and wipe away superfluous ECL reagent.
13. Place the membrane between the two overhead foils on the chemoluminescence imager with the protein side facing downwards.

14. Acquire an image of the chemoluminescence signal (*see* **Note 26**).
15. The quantification of signal intensities can be performed with the Kodak Molecular Imaging software. For this, the signal intensities for both the loading control (e.g., α-tubulin) and the protein of interest are normalized to cells transfected with a control oligonucleotide.

4 Notes

1. Cell culture media lacking the amino acids chosen for SILAC are commercially available from several companies such as Life Technologies, PAN Biotech, or Sigma-Aldrich. Alternatively, media can be prepared from individual components.
2. SILAC amino acids are available from Cambridge Isotope Laboratories, Eurisotop, Sigma-Aldrich, or Silantes.
3. Non-dialyzed serum contains growth factors which represent a source for unlabeled amino acids that may be incorporated into the proteome of the cells. Therefore, in order to guarantee accurate protein quantification, dialyzed serum should be used. However, it is recommended to ensure that growth and proliferation of the cell line of choice is not affected in the presence of dialyzed serum.
4. For both the pSILAC and Western blot experiments, we use stable polyclonal cell pools bearing episomal Pol-II-driven, doxycycline-inducible pri-miRNA expression vectors suitable for inducible miRNA expression [7, 37]. The generation of stable cell pools is time-consuming (although single-cell cloning is not necessary) and sometimes not possible in the cell line of choice but has the advantage of avoiding the transfection step and giving highly reproducible results between independent experiments. In many cases, transfection of microRNA mimics or microRNA inhibitors (antagomiRs) is a suitable alternative for the modulation of miRNA expression in the cell line of choice.
5. The cell line used for SILAC experiments should be checked for arginine-to-proline conversion and the ability of the cells to completely incorporate the stable isotope-coded amino acids (even if the pSILAC analysis is not aiming at complete incorporation). In case arginine-to-proline conversion is observed, the concentration of arginine in the medium needs to be decreased and/or unlabeled proline may be added.
6. The use of NuPAGE™ Bis-Tris gradient gels requires the XCell Sure Lock™ system from Life Technologies. Any other SDS-PAGE system may be used as well. However, we recommend

the use of gradient gels since they assure equal separation of proteins across a large molecular weight range.

7. All solvents and reagents used for LC/MS sample preparation and analysis should be of HPLC grade or higher purity; water should be of Milli-Q purity.
8. When MaxQuant is used for mass spectrometric data processing, databases of most common model species are provided with the download.
9. We use the H1299 lung cancer cell line which is seeded at 3×10^4 per well of a 12-well plate. The following protocol is adjusted to this particular format. H1299 cells display high transfection efficiencies and thus robust luciferase activities. However, SW480 cells can also be used for the assay, albeit with slightly lower transfection efficiencies.
10. The microRNAs should be reconstituted as a 10 μM stock solution and stored in aliquots at −80 °C. Thaw immediately before use and avoid repeated freeze–thaw cycles.
11. We use a modified version of the pGL3-control vector that has a multi-cloning site (MCS) downstream of the luciferase gene for insertion of 3′-UTRs [38]. However, there are alternative luciferase expression vectors available, e.g., the Ambion® pMIR-REPORT™ miRNA Expression Reporter Vector System (Life Technologies).
12. Stock solutions of amino acids solved in water can be stored at −20 °C.
13. In order to account for potential artifacts based on the use of different stable isotope-coded amino acids, it is recommended to switch the "medium-heavy" and "heavy" label between doxycycline-induced and non-induced cells in at least one replicate.
14. Reduction of cysteine residues with DTT and subsequent irreversible alkylation of the SH groups using iodoacetamide, for example, improves the MS-based identification of cysteine-containing peptides. Instead of DTT, tris(2-carboxyethyl) phosphine (TCEP) may be used as well.
15. Iodoacetamide is light-sensitive; therefore, the incubation needs to be performed in the dark. Instead of iodoacetamide, the light-insensitive chloroacetamide may also be used as alkylating agent.
16. Alternatively, the 3′-UTR can also be cloned downstream of a *Renilla* reporter gene with the *Firefly* luciferase vector as standardization control.
17. It is important to make one plasmid mastermix for both transfections to ensure equal amounts of reporter plasmid in each transfection.

18. We only use freshly prepared lysates for the assay and do not store lysates for subsequent use.
19. The amount of cells to be used varies between cell lines but should be adjusted to reach a confluency of max. 50–70 % (depending on the duration of miRNA expression) on the day of transfection. The given protocol is adjusted for the SW480 cell line and a 10-cm dish format.
20. Ideally, a time course experiment should be performed to determine the kinetics of target protein regulation and, thus, the optimal time point at which a significant reduction in protein levels can be seen. This can vary from protein to protein, with proteins of high abundance and long half-lives showing a visible reduction in steady-state levels only after 72 h or more.
21. It is of paramount importance to properly determine the protein concentration and adjust the amount of proteins between different samples because the reduction of the protein of interest after miRNA expression often is only moderate.
22. The amount of protein lysate required depends on the abundance of the protein of interest. Moreover, the amount of lysate needed may have to be adjusted in order to stay within the linear range of the subsequent interaction of the antibody with the protein epitope.
23. The polyacrylamide concentration to be used depends on the molecular weight of the protein to be analyzed. For proteins >100 kDa, a concentration of 7.5 % is recommended, whereas small proteins (<20 kDa) may require a concentration of up to 15 % polyacrylamide.
24. The choice of the device as well as the blotting time largely depends on the protein of interest (molecular weight) and has to be determined empirically. For example, a 50-kDa protein blotted onto a membrane of ~47 cm^2 may be blotted at 100 mA for 45 min at RT using a PerfectBlue™ semi-dry blotting device (Peqlab).
25. The dilution of the primary antibody as well as the conditions for the incubation depend on the protein of interest. It is important to ensure that the antibody detection reaction is within the linear range to avoid signal saturation. Incubation with primary antibody solution may be performed overnight at 4 °C or for 1 h at RT on a rocking platform.
26. The exposure time depends on the abundance of the protein of interest and the specificity of the antibody–protein interaction. Make sure to stay within the linear range of the detection reaction to avoid signal saturation. If the signal is too weak, the amounts of protein loaded onto the gel and/or the concentration of the primary antibody need to be adjusted.

References

1. Boisvert FM, Ahmad Y, Gierlinski M et al (2012) A quantitative spatial proteomics analysis of proteome turnover in human cells. Mol Cell Proteomics 11(M111):011429
2. Huo Y, Iadevaia V, Yao Z et al (2012) Stable isotope-labelling analysis of the impact of inhibition of the mammalian target of rapamycin on protein synthesis. Biochem J 444:141–151
3. Zhang L, Zhao H, Blagg BS, Dobrowsky RT (2012) C-terminal heat shock protein 90 inhibitor decreases hyperglycemia-induced oxidative stress and improves mitochondrial bioenergetics in sensory neurons. J Proteome Res 11:2581–2593
4. Martin BR, Wang C, Adibekian A et al (2012) Global profiling of dynamic protein palmitoylation. Nat Methods 9:84–89
5. Zee BM, Levin RS, Dimaggio PA, Garcia BA (2010) Global turnover of histone post-translational modifications and variants in human cells. Epigenetics Chromatin 3:22–31
6. Kraft-Terry SD, Gendelman HE (2011) Proteomic biosignatures for monocyte-macrophage differentiation. Cell Immunol 271:239–255
7. Kaller M, Liffers ST, Oeljeklaus S et al (2011) Genome-wide characterization of miR-34a induced changes in protein and mRNA expression by a combined pulsed SILAC and microarray analysis. Mol Cell Proteomics 10(M111):010462
8. Maragkakis M, Alexiou P, Papadopoulos GL et al (2009) Accurate microRNA target prediction correlates with protein repression levels. BMC Bioinform 10:295
9. Patron JP, Fendler A, Bild M et al (2012) MiR-133b targets antiapoptotic genes and enhances death receptor-induced apoptosis. PLoS One 7:e35345
10. Schwanhausser B, Gossen M, Dittmar G et al (2009) Global analysis of cellular protein translation by pulsed SILAC. Proteomics 9: 205–209
11. Selbach M, Schwanhausser B, Thierfelder N et al (2008) Widespread changes in protein synthesis induced by microRNAs. Nature 455:58–63
12. Fabian MR, Sonenberg N, Filipowicz W (2010) Regulation of mRNA translation and stability by microRNAs. Annu Rev Biochem 79:351–379
13. Friedman RC, Farh KK, Burge CB et al (2009) Most mammalian mRNAs are conserved targets of microRNAs. Genome Res 19:92–105
14. Ebert MS, Sharp PA (2012) Roles for microRNAs in conferring robustness to biological processes. Cell 149:515–524
15. Ong SE, Blagoev B, Kratchmarova I et al (2002) Stable isotope labeling by amino acids in cell culture, SILAC, as a simple and accurate approach to expression proteomics. Mol Cell Proteomics 1:376–386
16. Baek D, Villen J, Shin C et al (2008) The impact of microRNAs on protein output. Nature 455:64–71
17. Vinther J, Hedegaard MM, Gardner PP et al (2006) Identification of miRNA targets with stable isotope labeling by amino acids in cell culture. Nucleic Acids Res 34:e107
18. Epanchintsev A, Jung P, Menssen A et al (2006) Inducible microRNA expression by an all-in-one episomal vector system. Nucleic Acids Res 34:e119
19. Hermeking H (2012) MicroRNAs in the p53 network: micromanagement of tumour suppression. Nat Rev Cancer 12:613–626
20. Krek A, Grun D, Poy MN et al (2005) Combinatorial microRNA target predictions. Nat Genet 37:495–500
21. Kertesz M, Iovino N, Unnerstall U et al (2007) The role of site accessibility in microRNA target recognition. Nat Genet 39:1278–1284
22. Miranda KC, Huynh T, Tay Y et al (2006) A pattern-based method for the identification of MicroRNA binding sites and their corresponding heteroduplexes. Cell 126:1203–1217
23. Alexiou P, Maragkakis M, Papadopoulos GL et al (2009) Lost in translation: an assessment and perspective for computational microRNA target identification. Bioinformatics 25: 3049–3055
24. Reimers M, Carey VJ (2006) Bioconductor: an open source framework for bioinformatics and computational biology. Methods Enzymol 411:119–134
25. Saeed AI, Bhagabati NK, Braisted JC et al (2006) TM4 microarray software suite. Methods Enzymol 411:134–193
26. Djuranovic S, Nahvi A, Green R (2011) A parsimonious model for gene regulation by miRNAs. Science 331:550–553
27. Guo H, Ingolia NT, Weissman JS et al (2010) Mammalian microRNAs predominantly act to decrease target mRNA levels. Nature 466: 835–840
28. Chi SW, Zang JB, Mele A et al (2009) Argonaute HITS-CLIP decodes microRNA-

mRNA interaction maps. Nature 460: 479–486

29. Hafner M, Landthaler M, Burger L et al (2010) Transcriptome-wide identification of RNA-binding protein and microRNA target sites by PAR-CLIP. Cell 141:129–141
30. Nonne N, Ameyar-Zazoua M, Souidi M et al (2010) Tandem affinity purification of miRNA target mRNAs (TAP-Tar). Nucleic Acids Res 38:e20
31. Orom UA, Lund AH (2007) Isolation of microRNA targets using biotinylated synthetic microRNAs. Methods 43:162–165
32. Jackstadt R, Menssen A, Hermeking H (2013) Genome-wide analysis of c-MYC-regulated mRNAs and miRNAs, c-MYC DNA-binding by next generation sequencing. Methods Mol Biol 1012:145–185
33. Hünten S, Siemens H, Kaller M et al (2013) The p53/microRNA network in cancer: experimental and bioinformatics approaches. In: Schmitz U, Wolkenhauser O, Julio V (eds) miRNA cancer regulation: advanced concepts, bioinformatics and systems biology tools. Springer, New York, NY, pp 77–102
34. Cox J, Matic I, Hilger M et al (2009) A practical guide to the MaxQuant computational platform for SILAC-based quantitative proteomics. Nat Protoc 4:698–705
35. Cox J, Neuhauser N, Michalski A et al (2011) Andromeda: a peptide search engine integrated into the MaxQuant environment. J Proteome Res 10:1794–1805
36. Bradford MM (1976) A rapid and sensitive method for the quantitation of microgram quantities of protein utilizing the principle of protein-dye binding. Anal Biochem 72: 248–254
37. Bornkamm GW, Berens C, Kuklik-Roos C et al (2005) Stringent doxycycline-dependent control of gene activities using an episomal one-vector system. Nucleic Acids Res 33:e137
38. Welch C, Chen Y, Stallings RL (2007) MicroRNA-34a functions as a potential tumor suppressor by inducing apoptosis in neuroblastoma cells. Oncogene 26:5017–5022

Chapter 24

MaxQuant for In-Depth Analysis of Large SILAC Datasets

Stefka Tyanova, Matthias Mann, and Jürgen Cox

Abstract

Proteomics experiments can generate very large volumes of data, in particular in situations where within one experimental design many samples are compared to each other, possibly in combination with pre-fractionation of samples prior to LC-MS analysis. Here we provide a step-by-step protocol explaining how the current MaxQuant version can be used to analyze large SILAC-labeling datasets in an efficient way.

Key words Computational proteomics, Protein quantification, Peptide quantification, Experimental design, Large-scale data analysis

1 Introduction

Proteomics has recently been catching up with other omics types—in particular with transcriptomics—in terms of coverage of the expressed gene products and feasibility [1–3]. While relative label-free quantification is a convenient and increasingly accurate method for quantitative large-scale comparisons of multiple samples, labeling techniques provide the highest precision of quantification. In particular, the SILAC technology [4–6] is still considered the "gold standard" of quantitative proteomics. One specific approach that combines the quantitative precision of metabolic labeling with ease of applicability to experimental designs with many samples is the use of SILAC as an internal or a "spike-in" standard [7, 8]. In this case, the "heavy" SILAC partner is a common reference sample which is solely used to make the comparison between different mass spectrometric measurements more precise. This strategy can be applied to a multitude of situations since the samples of interest do not need to be labeled at all and it easily allows the comparison of large numbers of samples to each other.

MaxQuant is a computational shotgun proteomics platform providing a complete data analysis workflow from raw MS files to finished output tables, with detailed information about identified (modified) peptides and proteins, relative changes in abundance,

Bettina Warscheid (ed.), *Stable Isotope Labeling by Amino Acids in Cell Culture (SILAC): Methods and Protocols*, Methods in Molecular Biology, vol. 1188, DOI 10.1007/978-1-4939-1142-4_24, © Springer Science+Business Media New York 2014

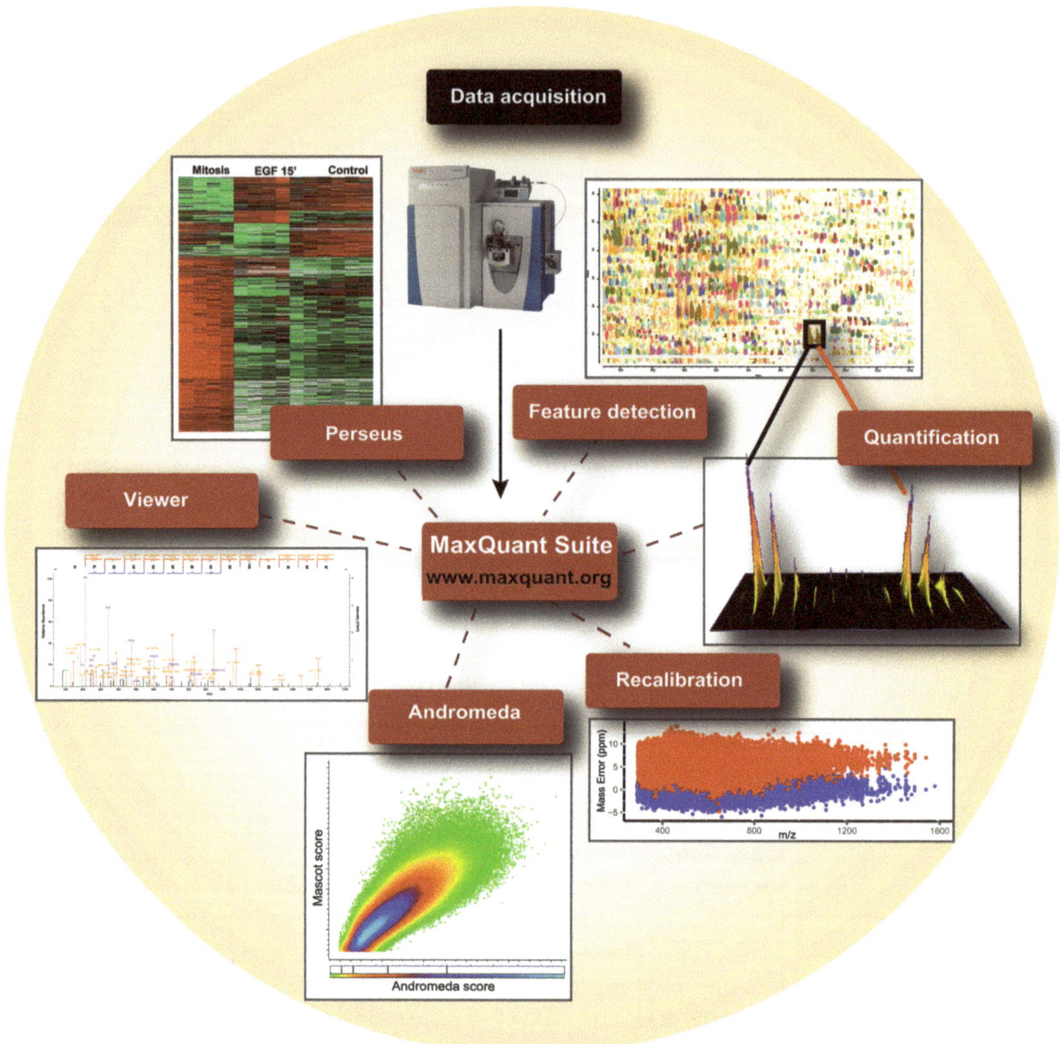

Fig. 1 Overview of the independent modules of the MaxQuant suite. The MaxQuant framework consists of several independent modules that together enable the complete processing of raw files. It supports protein identification, quantification, recalibration, and quality control of the raw and annotated spectra. Andromeda is the search engine used to match peptide fragmentation spectra to a sequence database. The main module of MaxQuant unites the processes of feature identification, quantification, and recalibration. To account for non-linear dependencies of the peptide mass error on the *m/z* and the retention time MaxQuant applies a re-calibration algorithm. Visualization and quality control of both raw file spectra and annotated spectra is possible with the Viewer module. Data normalization and transformation and downstream analysis of the detected and quantified features can be performed in the statistical module of MaxQuant—Perseus

and estimation of absolute abundance [9, 10] (Fig. 1). It employs its own peptide database search engine called Andromeda [11]. Downstream analysis of MaxQuant output tables can conveniently and comprehensively be performed with the extensible Perseus package [12]. While historically MaxQuant was first developed for

SILAC data analysis, by now most standard labeling techniques are supported, including iTRAQ, TMT, di-methyl, and many other MS1- or MS2-based labels. Furthermore a workflow for robust relative label-free protein quantification is included [13], which is in widespread use.

Here we focus on the analysis of SILAC data in the core part of the MaxQuant software. The reader will learn how to process the mass spectrometric raw data to obtain manageable output tables containing the results for identification and quantification of peptides, proteins, and posttranslational modification sites. For the downstream bioinformatics analysis—which is not covered here—we recommend the abovementioned Perseus software which can be downloaded from www.perseus-framework.org. MaxQuant also contains a powerful data visualization tool called "Viewer," which is available with the MaxQuant executable.

2 Materials

2.1 System Requirements

A personal computer with at least 2 GB RAM and a 64-bit Windows operating system is required. Supported versions are Windows Vista SP2, Windows 7, Windows 8, Windows Server 2008, and Windows Server 2012. Make sure that the "Regional and Language Options" are set to English. We recommend at least a dual-core processor. To take full advantage of the parallelization abilities of the software multi-core processors operating on shared memory should be used, as the speed of the processing scales with the number of cores. For the multithreading mode we recommend to have 2 GB RAM for each thread that is executed in parallel.

The raw data can be stored on a local or an external drive. It is important that enough space is available on this device to store all intermediate and output files which use approximately half of the space needed for the raw files.

2.2 Software Dependencies

If not already installed, the .NET Framework 4.5 from Microsoft should be downloaded. Usually it is already pre-installed on newer operating systems. It is required to install locally the Thermo Fisher Scientific MSFileReader software (downloadable from the manufacturer's website). No external 3D rendering library is needed for the Viewer (*see* **Note 1**).

2.3 Software Installation

To obtain the freely available MaxQuant software go to the MaxQuant home page (www.maxquant.org) and from there navigate to "Downloads." We require that all users have read and agreed to the "MaxQuant Freeware Software License Agreement" before acquiring or using the software. Download the compressed file containing the MaxQuant executables, and save and decompress the file at a suitable location on a local computer.

No installation is required and MaxQuant can readily be used. The explanations described in Subheading 3 refer to MaxQuant version 1.4.1.2.

3 Methods

3.1 Andromeda Configuration: Modifications Tab

1. Open the AndromedaConfig.exe file.
2. In the General tab, press the plus sign button to add a new modification to be used during the MaxQuant search.
3. Specify the "Title" (a short name for the modification) and the "Full name" (a longer description).
4. Set the elemental composition of the desired modification by clicking on the "Change" button. A pop-up will allow the selection of the correct composition and the "Monoisotopic mass" parameter will be updated automatically.
5. Set the position and type of the modification. Isotopic and isobaric labels (e.g., Arg10, Lys8, iTRAQ) are also specified in the Modifications tab by setting "Type" to *label* or *isobaricLabel*, respectively.
6. In the Specificity tab, specify the amino acids carrying the modification. Click on the plus sign button to add an empty site and select the amino acid from the drop-down "Site" menu.
7. If applicable, add "Neutral Loss" and "Diagnostic peak" parameters associated with the specified amino acid by pressing the plus button and specifying the "Name" and chemical "Composition" (pressing the composition cell will open a composition pop-up window).
8. In the Correction factors tab, in case you are defining an isobaric label, you can specify the correction factors in the provided fields.

3.2 Andromeda Configuration: Enzymes Tab

1. Examine the list of digestion enzymes and their specificities. To add a new enzyme, press the plus sign button. A blank entry will appear (Fig. 2a).
2. Specify the enzyme "Title" and "Description" in the provided fields.
3. Use the amino acid buttons marked with plus and minus signs to update the specificity matrix. For example, to define an enzyme that cleaves C-terminal to lysine but not before proline, first select all possible cleavage sites C-terminal to K (the "K+" row button) and then deselect the case preceding P ("P-" column button).

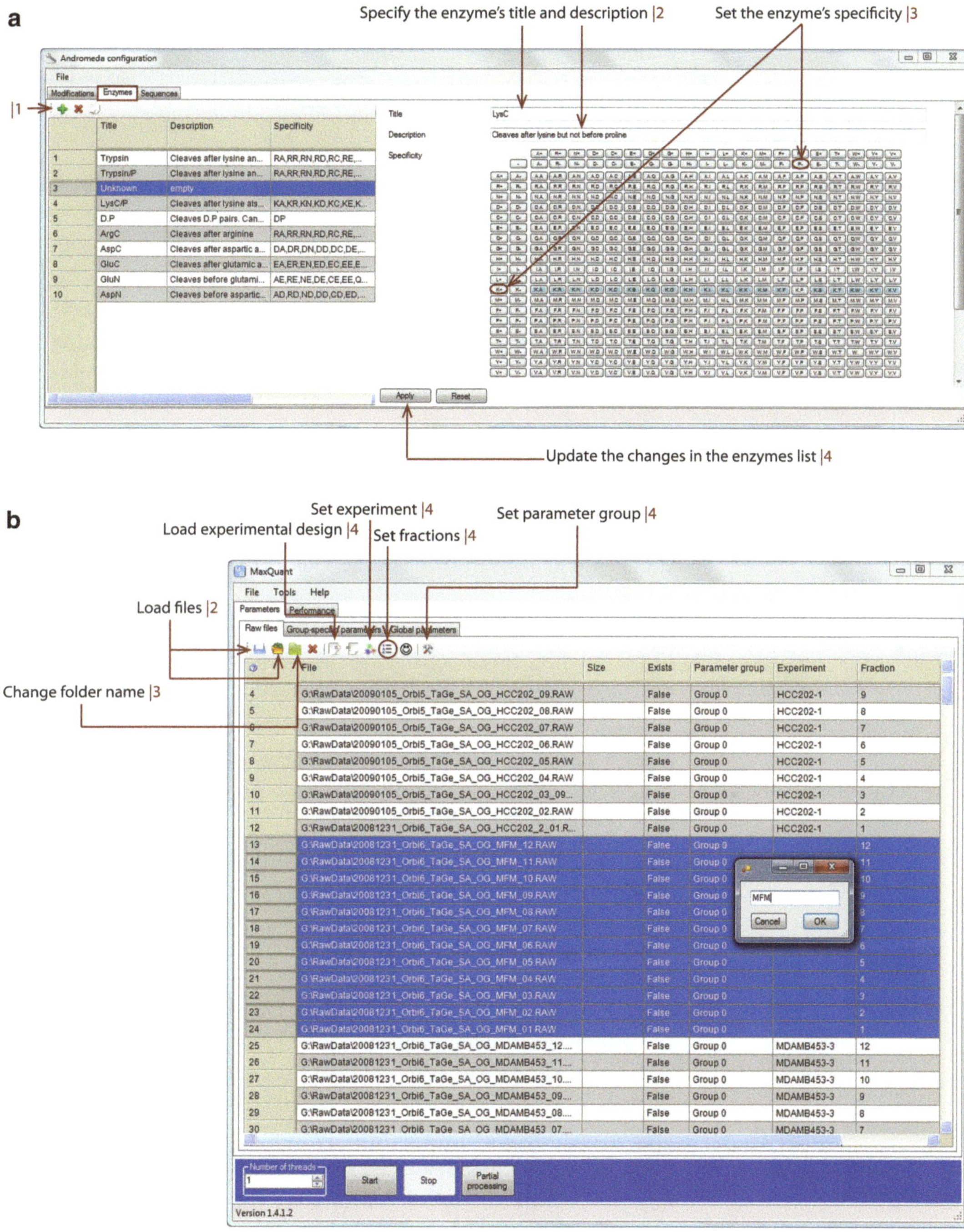

Fig. 2 Andromeda configuration and loading raw files. The workflow is outlined with *red arrows* and the numbered steps are described in detail in the text. Panel **a** shows the "Enzymes" tab of the AndromedaConfiguration software. An example of adding an enzyme that cleaves C-terminal to a lysine residue, but not before a proline residue, is presented. The steps required to add the new digestion enzyme and to set its specificity are marked by the *red arrows*. The selected specificity properties are highlighted in *light blue*. Panel **b** displays the "Raw files" tab of MaxQuant. An example of setting the experiment property as part of the experimental design for a group of raw data files can be seen. The selected group of files is highlighted in *dark blue* and the pop-up window allowing the specification of the desired group is shown

4. Press the "Apply" button to update the enzymes list in the left panel.
5. Save your changes to ensure that the added enzymes will be visible by MaxQuant.

3.3 Andromeda Configuration: Sequences Tab

1. Select a protein sequence fasta file from the available list or load a new fasta file.
2. Specify the "Select rule" according to which the fasta file will be read and the protein names and the fasta headers will be extracted. Choose an already predefined rule from the "Select Rule" panel or specify a new rule by pressing the plus sign button and using regular expressions (http://msdn.microsoft.com/en-us/library/az24scfc.aspx). Press the update button to select the desired rule from the rules list. Click on the "Test Rule" tab to check if Andromeda is extracting the information correctly.
3. Make sure to always save changes and to re-launch MaxQuant every time configuration changes are introduced in Andromeda.

3.4 MaxQuant: Raw Files Tab

1. Open the MaxQuant.exe file (*see* **Note 2**).
2. Load raw files by selecting a group of files from a folder or load the complete folder using the load folder button.
3. In case previously processed raw files are loaded using the mqpar file and the location of these raw files has changed, modifying the folder name adjusts the file paths to the correct new location.
4. Set experimental design by (1) loading an already available experimental design file ("Read from file" button); (2) creating an experimental design file template ("Write template") and modifying the fraction and experiment information (the template will be generated in a folder named "combined" in the same location as the raw files, unless specified differently); and (3) interactively selecting files and setting the information about fraction, experiment, or group by pressing the corresponding button (e.g., Set experiment, Fig. 2b) (*see* **Notes 3** and **4**).

3.5 MaxQuant: Group-Specific Parameters Tab

1. Set the "Type" according to the measurement machine.
2. Set the "Multiplicity" according to the number of labels in your experiment. Select 1 if no isotopic labeling was used; 2 for SILAC labeling with light and heavy labels (where the light state often corresponds to unlabeled samples); and 3 for triple SILAC labeling with light, medium, and heavy labels (Fig. 3a).
3. Set the maximum number of labeled amino acids per peptide.
4. Select the appropriate amino acids that carry the above-specified labels (e.g., in a typical SILAC experiment Arg10 and

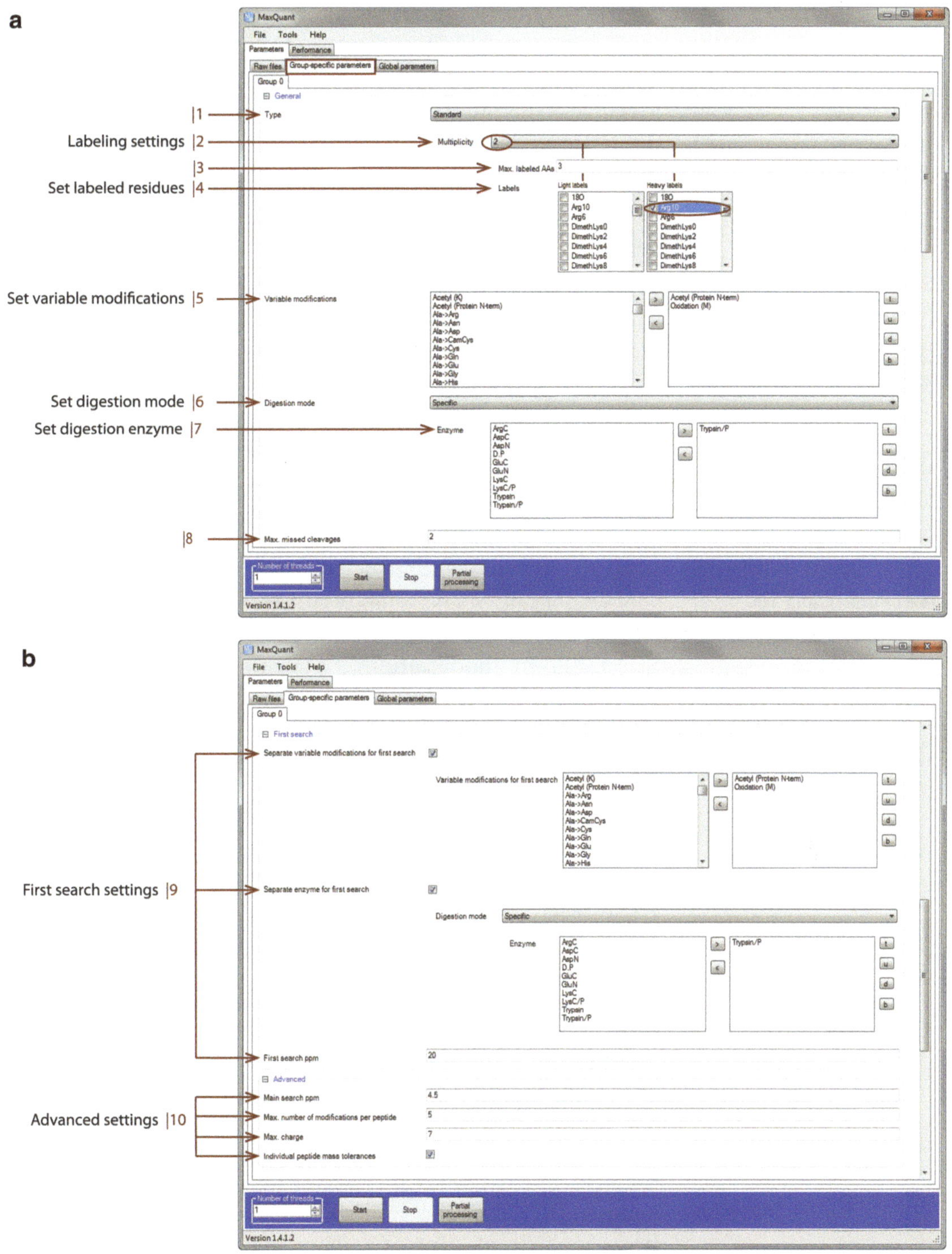

Fig. 3 Group-specific parameters. The workflow is outlined with *red arrows* and the numbered steps are described in detail in the text. Panel **a** shows the "Group-specific parameters" tab of MaxQuant. An example of setting the correct "Multiplicity" parameter in case of double-labeling experiment is shown. When the multiplicity is set to 2, two "Labels" panels appear, corresponding to "Light" and "Heavy" labels, respectively. Arginine10 is selected as a heavy-labeled amino acid. Panel **b** displays the "First search" options, which may be needed to increase the processing speed, and the default parameters of the advanced settings

Lys8). In case the light version corresponds to amino acids that are not labeled, only the heavy labels need to be set.

5. Select "Variable modifications" that may or may not be present by moving modifications from the left to the right panel using the arrows. By default methionine oxidation and protein N-terminal acetylation are selected.
6. Select enzyme specificity mode.
7. Select digestion enzyme.
8. Set maximum number of missed cleavage sites per peptide.
9. Set variable modifications, digestion enzyme, and ppm mass range to be used in the first search (Fig. 3b) (*see* **Note 5**).
10. The default settings of the advanced options (main search ppm, maximum number of modifications, maximum charge per peptide, and individual peptide mass tolerance) are suitable for achieving optimal results; however, if desired, they can be modified (Fig. 3b).

3.6 MaxQuant: Global Parameters Tab

1. Load a protein sequence fasta file to be used as a search database. The file should be pre-configured with the AndromedaConfig.exe. Multiple fasta files can be loaded simultaneously (Fig. 4a).
2. Select "Fixed modifications" analogously to the previously defined variable modifications. Fixed modifications are deliberately introduced modifications during the sample preparation and are considered to be present for the selected residues at any time during the analysis [e.g., carbamidomethyl (C)].
3. Select the "Re-quantify" option to enable quantification of isotopic patterns that are not assembled as SILAC pairs prior to protein identification. The missing isotope pattern will be reconstructed based on the shape of the found isotope pattern, the expected shift among the m/z retention time plane, and the intensities integrated over these regions. The option is highly recommended, especially in the presence of large or small ratios, but should be avoided in rare cases of extreme ratios (e.g., incorporation studies).
4. Check the "Match between runs" option to transfer MS/MS identifications between different LC-MS/MS runs based on the exact mass and the retention time of the peptides. Set the "Match time window" to account for possible retention time fluctuations remaining after the retention time alignment has been applied (*see* **Note 6**). The "Alignment time window" specifies the size of the time window in which a solution for the retention time alignment is searched for.
5. If desired, unidentified features (without MS/MS or with unidentified MS/MS) can also be matched. The result of the matched unidentified features is stored in a separate output table called "matchedFeatures.txt."

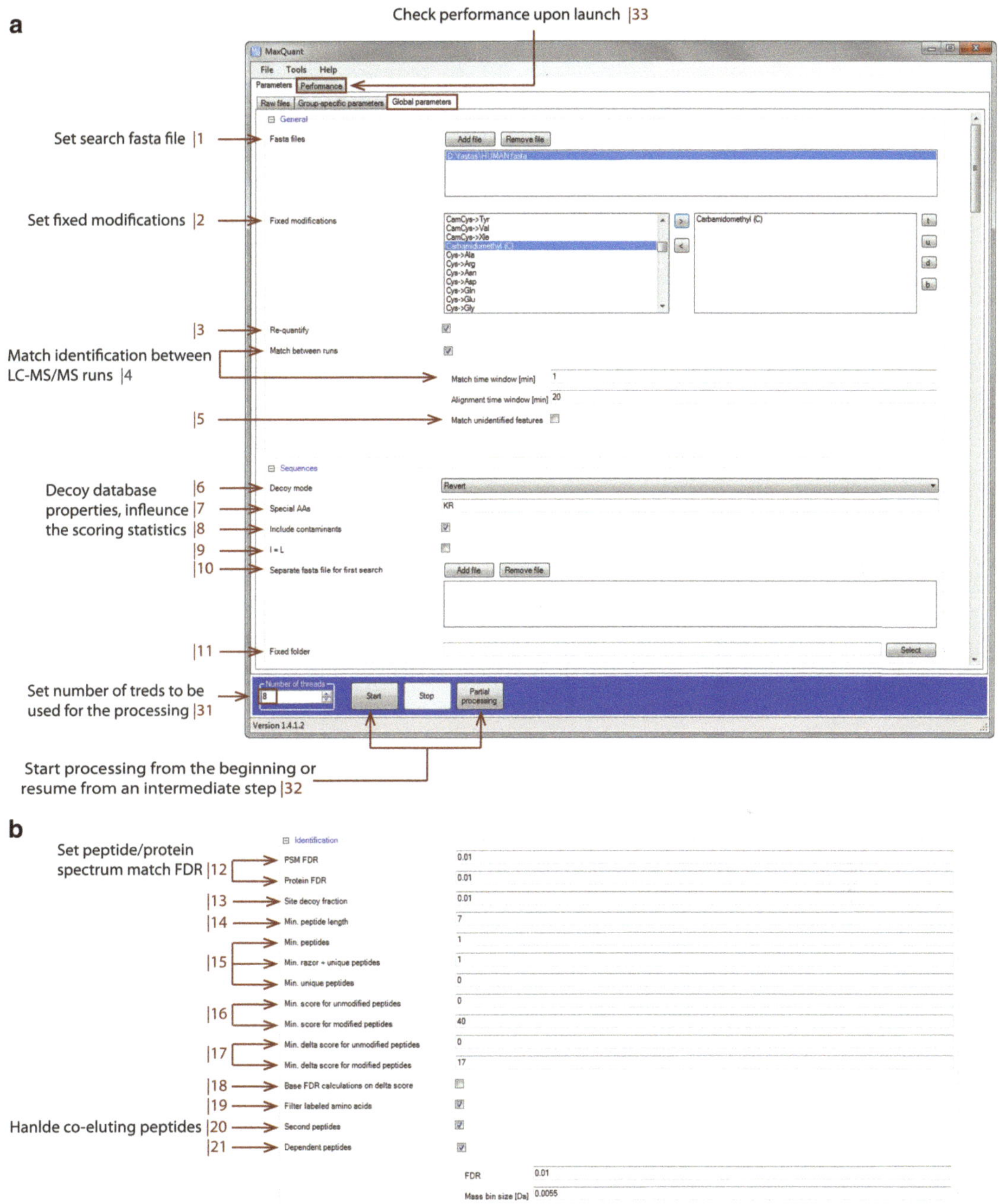

Fig. 4 Global parameters. The workflow is outlined with *red arrows* and the numbered steps are described in detail in the text. Panel **a** shows the "Global parameters" tab of MaxQuant. A protein sequence fasta file is selected to be used as a search database. The default parameters are kept. The number of threads to be used in the processing is specified to speed the computation time. Panel **b** displays the default settings of the peptide identification parameters. Additional options for handling co-eluting peptides are also included in the Identification tab

6. Select the decoy database mode to be used by MaxQuant for scoring statistics validation. Three options are available: (1) "Revert"—generating reversed fasta sequences, (2) "Randomize"—generating randomized fasta sequences, and (3) "Reward"—peptide-based decoy rule suitable for peptides with posttranslational modifications (*see* **Note 7**).
7. Selected "Special" amino acids are treated differently in the generation of the decoy database in order to avoid the situation that reverse peptides have exactly the same mass as forward peptides. The selected residues are swapped with the residue preceding them in the generation of the reverse database. In case of trypsin digestion these are usually R and K.
8. Check the "Include contaminants" box to include known laboratory-originating contaminants in the search database file. If features are match to any of the entries in this list, they will be marked in the output table in the column with header Contaminants.
9. Check the "I = L" box to use isoleucine and leucine as indistinguishable during the MaxQuant processing.
10. If applicable, add a second smaller fasta sequence database search file to be used only in the first search (*see* **Note 5**). This option may reduce the overall processing time and is recommended in cases where the main fasta file is extremely large, e.g., six-frame translations of whole genomes.
11. Specify a "Fixed folder" to store the index files for the search fasta database.
12. Set the desired Peptide Spectrum Match (PSM) and Protein FDRs. The default values are set to 1 %. To relax the search stringency the values can be increased.
13. Set the fraction of allowed decoy hits in the sites table.
14. Set the minimum peptide length. Shorter peptides are not considered for identification or quantification.
15. If desired, modify the minimum number of peptides, unique and razor peptides, and only unique peptides that a protein group should have to be considered and reported as identified in the final results tables.
16. If desired, set minimum scores for the unmodified and the modified peptides.
17. If desired, set minimum delta scores for the unmodified and the modified peptides. The delta score is the difference between the actual peptide score and the closest match to any other sequence.
18. Select the "Base FDR calculations on delta score" box to use the delta score in the calculation of the posterior error probabilities.

19. Filter labeled amino acids applies to MS1 labels that allow for the determination of the number of certain amino acids in the peptide. Only peptide candidates are admissible that have the correct number of amino acid consistent with the mass difference between the MS1 features.
20. Select the "Second peptides" option to enable identification and quantification of co-eluting peptides that have been co-fragmented in the same MS/MS spectrum.
21. Dependent peptides comprise peptides with unknown modifications and mutations or resulting from unknown proteases that usually remain unidentified. Switching this option enables their characterization based on the idea that such peptides create MS/MS spectra that are related to spectra from identified peptides in the same LC/MS run (Fig. 4b).
22. Set the minimum number of peptide feature ratios needed for a protein to be quantified.
23. Specify which peptides should be used for the protein ratio calculations. The "All" option picks all peptides to be used for protein quantification. Select the "Unique" option to use only peptides that are unique for a given protein group. If "Razor" is selected, only razor peptides are used for protein quantification (razor peptides are not unique for a protein group, but are assigned to the group with the maximum number of peptides).
24. If needed, select modifications, which will be used for protein quantification. These are usually posttranslational modifications without independent biological regulation. Phosphorylation is an example of a modification that should NOT be specified in this field when phospho-proteomes are studied.
25. Select "Normalized ratios" to use the normalized peptide ratios (centered on zero) for the site quantification (*see* **Note 8**). Normalized ratios should be used when the distributions are well behaved (e.g., not bimodal, not characterized by strong asymmetry or skewness).
26. Set the location of the temporary folder in order to specify where temporary files that are needed and produced during the analysis will be written out. As these files are frequently accessed by MaxQuant during the analysis, it is recommended that the temporary folder is located on a fast drive.
27. Set the location of the "combined" folder, which stores all output tables by MaxQuant. By default the "combined" folder is written in the same directory as the raw files.
28. When "Calculate peak properties" is checked, advanced properties of 3D peaks and isotope patterns are calculated and reported in the output tables. This includes the precursor intensity fraction (PIF) indicating the percentage of MS1-level

intensity that originates from the targeted precursor for an MS/MS spectrum.

29. Mass difference search: This option allows for listing all MS1 features that have a suitable mass difference to an identified peptide in order to be a modified form of the same peptide using one of the specified modifications. Results are listed in the allPeptides.txt table.
30. The default settings are well suited for common experimental situations.
31. Select the number of parallel threads that will be used for the processing (Fig. 4a).
32. Press the "Start" button for the processing to begin (*see* **Note 9**).
33. Go to the "Performance" tab to examine the state and the speed of the processing. A notification window will pop up once the analysis has finished.
34. Navigate to the combined folder → txt. The results are stored in tab-delimited .txt files suitable for further downstream analysis (*see* **Note 10**).

4 Notes

1. Viewer as a quality control tool: Although it is completely independent from the rest of the Max Quant suite, the Viewer module allows for visual inspection of the quality of both the raw files and the already processed with MaxQuant annotated data. It can also be used to generate figures for publications.
2. Organization of files: Relocating the data to a faster disk can make a big difference in processing time. External drives, such as the ones using USB2, are the slowest option. Going towards RAID systems and solid-state disks can speed up the data analysis tremendously. We recommend all raw files that need to be analyzed to be organized in one folder for convenience.
3. Experimental design configuration: The "Experiment" specifications allow for different experiments to be analyzed together without losing the information about the individual ratio values for each sample. In case of fractionation the "Fractions" can be set accordingly and used to enhance the identification and quantification with the "match between runs option."
4. The definition of groups is important: Parameter groups enable the specification of different values of parameters for different sets of raw files. In case all raw files should be analyzed with the same parameter settings, you do not need to consider groups. A prominent example in which multiple parameter groups are recommended is the joint analysis of the proteome and the

phosphoproteome of corresponding samples. In that case the raw files would be divided into two groups and only one of the groups would contain the variable modification for phosphorylation of serine, threonine, and tyrosine.

5. First search options are useful in case of large peptide search space, for example due to unspecific enzyme search or large number of variable modifications. In these cases it can be more time efficient to use a smaller database for the first search as this search is performed only to define the mass- and time-dependent mass recalibration curves [14]. To reduce the search space and the analysis time only some of the modifications and the cleavage enzymes used can be specified. The larger ppm range compared to the main search is needed for detecting and solving mass calibration problems.
6. Match time window length: Experience in our laboratory indicates that 1 min is a good choice for optimal identification and minimum error rates. Larger windows can improve identification in cases where the chromatography is less reproducible, but should be used with caution as they may also increase the false discovery rate.
7. Reward database: (REverse + forWARD) consists of peptides generated by taking the first half of a reversed peptide and then completing it by adding the second half of the forward peptide. This decoy database form is particularly suitable for modified peptide searches.
8. Identification and quantification of modification sites associated with multiply modified peptides: A multiply modified peptide is almost always represented by several forms that have a different number of modification sites. The site abundance of a modification site located on a multiply modified peptide varies depending on the modified peptide form used for the quantification. In order to retain maximum information MaxQuant quantifies three levels of site abundance: _1 from the singly modified peptide, _2 from the doubly modified peptide, and _3 from the peptide with more than two modifications. Additional column is provided that combines the three values taking the least modified peptide. We recommend entering the downstream data analysis with the quantities ending on _1 _2, and _3 instead of the ratio of the least modified peptide since the information content is higher.
9. Partial processing: To resume a MaxQuant analysis from an intermediate processing step go to "File" → "Load parameters" and navigate to the MaxQuant output folder where all files needed during the processing are stored. Select the mqpar.xml file. Press the Partial processing button and select the desired step from which the processing should continue by typing the corresponding number in the provided field.

Make sure that the same MaxQuant version is used for all processing steps.

10. Brief output description: The "proteinGroups.txt" table contains a comprehensive list of the identified and quantified proteins. In case of SILAC experiments the quantitative information that can directly be used for analysis in Perseus is stored in the Ratio H/L (*H*: heavy; *M*: medium; *L*: light labels) and Ratio H/L Normalized columns. Additionally, the table contains references to the other output tables. In experiments involving modifications, e.g., phosphorylation, the site-specific identifications and quantifications are available in the "Phospho (STY) Sites.txt" table. MaxQuant search specifications such as digestion enzyme, labels, identification and quantification success rates, and others are recorded in the "summary.txt" table. The "parameters.txt" table contains information about various parameters used during the computations, e.g., the software version, group-specific and general identification and quantification parameters, thresholds, and others. Detailed information on the meaning of tables is written with every MaxQuant analysis into the file \combined\txt\tables.pdf.

References

1. Cox J, Mann M (2007) Is proteomics the new genomics? Cell 130:395–398
2. Cox J, Mann M (2011) Quantitative, high-resolution proteomics for data-driven systems biology. Annu Rev Biochem 80:273–299
3. de Godoy LM, Olsen JV, Cox J et al (2008) Comprehensive mass-spectrometry-based proteome quantification of haploid versus diploid yeast. Nature 455:1251–1254
4. Ong SE, Mann M (2006) A practical recipe for stable isotope labeling by amino acids in cell culture (SILAC). Nat Protoc 1:2650–2660
5. Ong SE, Blagoev B, Kratchmarova I et al (2002) Stable isotope labeling by amino acids in cell culture, SILAC, as a simple and accurate approach to expression proteomics. Mol Cell Proteomics 1:376–386
6. Mann M (2006) Functional and quantitative proteomics using SILAC. Nat Rev Mol Cell Biol 7:952–958
7. Geiger T, Wisniewski JR, Cox J et al (2011) Use of stable isotope labeling by amino acids in cell culture as a spike-in standard in quantitative proteomics. Nat Protoc 6:147–157
8. Geiger T, Cox J, Ostasiewicz P et al (2010) Super-SILAC mix for quantitative proteomics of human tumor tissue. Nat Methods 7:383–385
9. Cox J, Mann M (2008) MaxQuant enables high peptide identification rates, individualized p.p.b.-range mass accuracies and proteome-wide protein quantification. Nat Biotechnol 26:1367–1372
10. Cox J, Mann M (2009) Computational principles of determining and improving mass precision and accuracy for proteome measurements in an Orbitrap. J Am Soc Mass Spectrom 20: 1477–1485
11. Cox J, Neuhauser N, Michalski A et al (2011) Andromeda: a peptide search engine integrated into the MaxQuant environment. J Proteome Res 10:1794–1805
12. Cox J, Mann M (2012) 1D and 2D annotation enrichment: a statistical method integrating quantitative proteomics with complementary high-throughput data. BMC Bioinform 13(Suppl 16):S12
13. Luber CA, Cox J, Lauterbach H et al (2010) Quantitative proteomics reveals subset-specific viral recognition in dendritic cells. Immunity 32:279–289
14. Cox J, Michalski A, Mann M (2011) Software lock mass by two-dimensional minimization of peptide mass errors. J Am Soc Mass Spectrom 22:1373–1380

Index

Bettina Warscheid (ed.), *Stable Isotope Labeling by Amino Acids in Cell Culture (SILAC): Methods and Protocols*, Methods in Molecular Biology, vol. 1188, DOI 10.1007/978-1-4939-1142-4, © Springer Science+Business Media New York 2014

MIX
Papier | Fördert gute Waldnutzung
FSC® C083411

Zeitfracht Medien GmbH
Ferdinand-Jühlke-Straße 7
99095 Erfurt, Deutschland
produktsicherheit@kolibri360.de